Fundamentals
of
Physical Geography

Fundamentals
of
Physical Geography

Arthur H. Doerr
University of West Florida

with cartographic assistance of
Jerome F. Coling
University of West Florida

 Wm. C. Brown Publishers

Book Team

Editor *Jeffrey L. Hahn*
Developmental Editor *Lynne M. Meyers*
Production Editor *Barbara Rowe Day/Harry Halloran*
Designer *Carol S. Joslin*
Art Editor *Janice M. Roerig*
Photo Editor *Mary Roussel*
Visuals Processor *Andé Meyer*

WCB

Wm. C. Brown Publishers

President *G. Franklin Lewis*
Vice President, Publisher *George Wm. Bergquist*
Vice President, Publisher *Thomas E. Doran*
Vice President, Operations and Production *Beverly Kolz*
National Sales Manager *Virginia S. Moffat*
Advertising Manager *Ann M. Knepper*
Marketing Manager *David F. Horwitz*
Executive Editor *Edward G. Jaffe*
Production Editorial Manager *Colleen A. Yonda*
Production Editorial Manager *Julie A. Kennedy*
Publishing Services Manager *Karen J. Slaght*
Manager of Visuals and Design *Faye M. Schilling*

Cover photo by Peter L. Kresan. View of Kenai Mountains and Kachemak Bay, Homer Beach, Alaska.

Jerome F. Coling
Figures 7.24, 17.26, 17.27, and 18.34

Susan Mahalick
Figures 1.13, 1.17, 2.3, 3.18, 4.26, 4.27, 5.6, 5.11, 6.5, 8.13, 10.29, 12.15, 14.1, 17.10, and 19.14.

Dennis Tasa
Figures 4.30, 5.2, 8.12, 12.11, and 12.32 D and E

Library of Congress Catalog Card Number: 89–61733

ISBN 0–697–07905–8

Printed in the United States of America by Wm. C. Brown Publishers, 2460 Kerper Boulevard, Dubuque, IA 52001

10 9 8 7 6 5 4 3 2

to Marc

Contents

Preface

*The chess-board is the
world; the pieces are the phenomena of the universe; the
rules of the game are what we call the Laws of Nature.*
T. H. Huxley, *A Liberal Education*

*G*eography, and especially physical geography, was one of the earliest of the sciences to be studied and developed. The Greeks were in the vanguard of those who formulated general principles concerning the nature of the earth and its place in the cosmos. Nature study and the naturalist emanated from early efforts to discover the secrets of the earth and the position of people on it. Because so much of the earth's surface was unknown territory, travels and voyages of discovery were essential to fill in blank spaces on the map and to add new knowledge concerning the nature of the earth. This *terra incognita* (unknown land) was gradually stripped away in early times, but the pace of discovery quickened during the Age of Exploration, and virtually every inch of the surface of the terrestrial globe has now been explored, at least in a perfunctory fashion.

As scientific inquiry developed and became more sophisticated, geography spawned other sciences like geology, meteorology, and climatology. As these individual sciences prospered, geography, notably physical geography, became impoverished in intellectual force and fashion. It was derided by some as a "synthetic science" and, in the last several years, the physical aspects of the field have yielded center stage to new techniques and approaches in the cultural realm. Physical geography is certainly a "science of synthesis" in which meaning is sought by integrating environmental parts into whole spatial patterns of complex interrelationships. The linkages and connections of various elements of the physical environment and associations with cultural landscape expressions of people give the holistic character to the discipline.

Briefly, physical geography deals with land, water, air, and life and their mutual interrelationships. The face of the land and its constant changes, the restless stirrings of the earth's waters, the invisible and varying breath of the atmosphere, and the pulse of life—all of which create and respond to a physical environment—are the stuff of physical geography. The temporal and spatial patterns, which develop from the interconnectedness of environmental elements, are of interest to and value for development and planning of human pursuits.

Physical geography is an intimate and vital subject. The flashy pyrotechnics of volcanism and the dramatic quivering of earthquakes as well as the subtle changes in the physical landscape wrought by agencies of weathering, erosion, and deposition are everyday concomitants of living. A warm spring day, a howling blizzard, or a roaring hurricane reminds us that weather and climate are ever-present features of daily life. Floods impress on us that land clearing and urbanization accelerate runoff and accentuate the effects of a downpour or seasonal precipitation, which is greater than normal. A drought makes us realize how vulnerable we are to climatological conditions that stray from the normal. Forest vastnesses or desert emptiness intrude on our consciousness. Our senses perceive the sights, sounds, smells, and feel of our physical world during every waking second. We are at once responsive to environmental conditions and responsible for a modification of the physical environment, while simultaneously creating a cultural landscape.

Fundamentals of Physical Geography provides an understandable operational framework of our physical world. With such an understanding, it is my fond and expectant hope that readers of this work will be able to make individual and collective rational decisions about living and environmental uses. The nature and nurture of our good blue-green earth require understanding and right actions.

This book provides information about the earth and the processes that continue to shape it. The illustrative materials add form, dimension, and interpretation to the face of the land.

Numerous illustrations are provided in the body of the text to clarify or reinforce verbal statements, but the perceptive reader will find that a good atlas is an appropriate companion piece. Any topic discussed in the body of the text could be expanded significantly, but this book is designed to serve the needs of students and instructors in a one-term course in physical geography. It will serve equally well for students who will have only this limited exposure to geography and for those who will undertake further geographic study.

Public agencies have produced a number of exceptionally useful illustrative materials. Among the most important are National Aeronautics and Space Administration (NASA), National Oceanic and Atmospheric Administration (NOAA), National Science Foundation (NSF), United States Department of Agriculture (USDA), and United States Geological Survey (USGS). As appropriate, these abbreviations will be used in credits for illustrative materials.

Arthur H. Doerr

As an instructional aid to those who have elected to use this text in their courses, several ancillary materials are available in conjunction with *Fundamentals of Physical Geography.* A set of 32 overhead transparencies are reproduced from selected textbook illustrations. An Instructor's Manual includes a test item file with questions from each chapter. All of the questions are available on WCB TestPak, a computerized service that enables you to create customized exams. For each exam, you can choose up to 250 questions; you can print it yourself or have WCB print it. Printing it yourself requires access to an IBM personal computer that uses 5.25- or 3.5-inch diskettes, an Apple IIe or IIc, or a Macintosh. Diskettes are available through your local WCB sales representative or by phoning Educational Services at 319–588–1451. If you don't have a computer, you can use WCB's call-in/mail-in service. First, determine the chapter and question numbers, and any specific heading you want on the exam. Then, call Pat Powers at 800–351–7671 (in Iowa, 319–589–2953) or mail information to: Pat Powers, Wm. C. Brown Publishers, 2460 Kerper Blvd., Dubuque, IA, 52001. Within two working days, WCB will send a test master, a student answer sheet, and an answer key to you via first-class mail.

Acknowledgments

The author of this book has had four decades of experience in teaching courses in physical geography, and he owes a debt of gratitude to former mentors and colleagues who helped in the forging of concepts, who fostered the development of a breadth of understanding, and who facilitated the initiation of new syntheses. To William E. Powers (formerly of Northwestern University), who helped to develop the concept of order and process; to Annemarie Krause (formerly of Southern Illinois University), who made learning the material a joy; to John Kesseli (formerly of the University of California at Berkeley), who first introduced me to the subject; to Thomas F. Barton (deceased—formerly of Southern Illinois University and Indiana University), who exuded enthusiasm for the field; to Lee Guernsey (Indiana State University), who shared research and writing in past ventures and who has remained a friend over more than half of my life; to Harry Hoy, John Morris (deceased), and Ralph Olson (all formerly of the University of Oklahoma), for waging the fight for physical geography, which built the Department of Geography there; to Stephen Sutherland (University of Oklahoma), who developed the teaching of the subject into a fine art; and to former students at Indiana University, Northwestern University, Eastern Illinois University, Central Washington University, George Peabody College, Western Washington University, University of the Philippines, Wisconsin State University at Steven's Point, University of Oklahoma, and The University of West Florida, who asked the right questions and issued the appropriate challenges, I offer my thanks and appreciation. To other colleagues, within the discipline and outside it, who have stimulated my thought processes over the years, I'm most appreciative.

To former students, colleagues, or friends, who provided materials or photographs to improve the presentation, I'm deeply grateful. Especially noteworthy have been the efforts of Daniel H. Ehrlich of Linn-Benton Community College, who was particularly helpful in taking some excellent photographs of some sites at remote locations, at my request. The quality of his work has greatly enhanced mine. Lee Guernsey, formerly of Indiana State University, provided me with an array of photographs for possible use in the book. Joseph Castelli of East Stroudsburg State University generously gave me permission to use photographs that he had provided years ago. Harry E. Hoy, formerly of the University of Oklahoma, gave me several of his excellent physiographic diagrams. Aaron Williams of the University of South Alabama loaned me several remotely sensed images. Wallace Akin of Drake University provided several excellent photographs for use in the book. John Lounsbury of Arizona State University provided me with a number of fruitful sources of information. Don R. Hoy, formerly of the University of Georgia, provided information about Guadeloupe.

Several agencies, public and private, were generous with their loan of material. As appropriate, they're cited in the credits section of the book.

The readers at various stages of manuscript preparation made numerous excellent suggestions for emendations, additions, and deletions. The work is stronger because of their contributions. They deserve credit, but they share no blame for existing shortcomings. Their names follow: Robert Fredericks, Chabot College—Valley Campus; Roland L. Grant, Eastern Montana College; R. T. Hill, Concord College; John G. Hehr, University of Arkansas; Diann S. Kiesel, University of Wisconsin Center—Baraboo/Sauk County; Roger L. Richman, Moorhead State University; Norman R. Roberts, Clark College; and Peter J. Valora, Ricks College.

In addition, I want to thank Ed Jaffe, Executive Editor of Wm. C. Brown Publishers, especially, because he recognized what I was trying to do and provided abundant encouragement and tangible assistance. Jeffrey Hahn assumed the task of editor with good cheer and grace. Developmental editor Lynne Meyers elicited a better manuscript. People in the art department of Wm. C. Brown supervised artists' renderings of my crude sketches to bring life to my ideas. Carol Joslin developed cover and interior designs. Barbara Day, production editor, worked tirelessly to clarify meaning and to eliminate printing glitches. When she withdrew from the project to pursue other endeavors, Harry Halloran ably brought the book to publication. He, like all of his Wm. C. Brown Company colleagues, was courteous, efficient, and very helpful. Each and all of them have my enduring gratitude.

For help in typing and editing, I thank Evelyn Grosse and Lucia Howe. For help in drafting and illustration preparation, I'm in debt to Michael Wylie. For skilled cartographic work, I'm grateful to Jerome Coling. Marcia Bennett was extremely helpful in indexing and assisting with the correlation of almost five hundred illustrations and in a host of other chores. Ashley Devine helped with arranging and typing the index. For dedication above and beyond the call of duty in facing the unforgiving screen of the word processor through numerous iterations of the manuscript, I'm particularly appreciative to Connie Works. She is skillful, patient, and forgiving of the author's idiosyncrasies. She is, in short, an outstanding professional.

The loving support of my wife Dale has been a steadfast anchor during stormy times. Encouragement during the long months of effort from Bark and Statia, Carolyn and Jack, Eleanor and John, Betty and Ted, and Betty and Ralph is warmly acknowledged. Beulah Doerr and Mary Wiehe exhibit their own profiles in courage and buoy my spirits at crucial times. Orville Hudgens helped in ways that will always be appreciated.

I'm grateful to all who have contributed directly or indirectly to the completion of the manuscript, but I alone am responsible for errors of omission or commission.

Arthur H. Doerr

1

The Nature of Geography

Rice planting on the Luzon Plain with clouds forming and mountains in background.

To a person uninstructed in natural history, his country or seaside stroll is a walk through a gallery filled with wonderful works of art, nine-tenths of which have their faces turned to the wall.

T. H. Huxley, *On the Educational Value of the Natural History Sciences*

*T*he etymology of the word **geography** provides the essence of the field (i.e., from the Greek, *geo*—the earth, and *graphein*—to write; hence, to write about or describe the earth). The description involves analysis and synthesis as well as the development of theories about relationships between the physical environment and the cultural realm. Constant changes initiated by natural forces as well as by people cause the field to remain dynamic and challenging. The parameters of geographic inquiry expand and contract, although the eclectic nature of the field makes it quite broad in concept and execution.

Several other definitions of the field, which are, essentially, variations on a theme, provide a broader appreciation of the area of study. For example, geography may be defined as the study of the areal differentiation of the earth's surface; or, geography is the significance of differences from place to place. These definitions recognize that there are physical and cultural patterns on the surface of the earth, and, although the stress is on differences, it's clear that similarities are implied as well. The interrelationships among these different areas, in terms of their physical and biological attributes, are significant to people's occupance of the earth, as well as their understanding and appreciation of it.

Figure 1.1
The ruins of Takht-i-Jamshid (Persepolis) illustrate that civilization is a long and persistent process. This city was begun about 500 B.C. and flourished during the height of the Persian Empire.

Other definitions include the following: geography is the study of the relationship of people to their environment. A more applicable variation might be: geography is a study of the mutual interrelationships between people and their environment. Some have said (while anthropologists wince) that geography is human ecology. These definitions all include people and the environment, and how people have adapted to or made modifications to their physical world while creating a uniquely human setting.

Spatial elements and the distributional patterns on, and immediately adjacent to, the earth's surface are central to all definitions of geography. The ultimate objective of geography and geographic inquiry is to understand, adapt to, and adjust to the world in which we live.

Geography occupies a special place among the academic disciplines since it serves as an intellectual bridge between the sciences and the social sciences. The field derives its power and persuasiveness from the synthesis of essential elements of the physical environment and elements of material culture. The holistic character of the discipline provides a unique perspective of the world as the home of people. This is a book about physical geography and is, therefore, scientifically oriented, but it is important to have some appreciation for those aspects of the field that lie outside the scope of this text; hence, a brief description of the social aspects of the field is in order.

Geography as a Social Science

Geographers consider elements of material culture as well as the physical environment. People's activities are not controlled by the physical environment, but natural forces and phenomena significantly affect economic and cultural developments. As people have become sophisticated users of numerous technological devices, the species has increasingly sought to modify the environment as well as to adapt to it. The more than 5 billion people now occupying the earth are constantly modifying its physical and cultural fabric.

In certain regions, a seemingly benign environment may foster people's economic and cultural pursuits, whereas in other regions, a hostile physical environment may deter human activities. Environmental shortcomings or attributes may be real or perceived, and perceptions may change over time or vary from group to group. One culture's problem is another culture's opportunity, and a problem at one time may become an opportunity at another. Environmental perceptions are powerful, and have a significant effect on the way a particular group of people relate to their physical world.

For example, before air travel became commonplace, mountainous terrain, deserts, oceans, and jungles were significant barriers to human penetration. Before irrigation and winter recreation, snowfall in the mountains was a nuisance, now it may be a recreational boon in the mountains and meltwater may foster agriculture in adjacent areas. Remote places have yielded their secrets and their treasures more readily as modern transportation and communication have made them accessible. For example, regions in high latitudes remained little understood, and detailed knowledge awaited twentieth century exploration. Now, permanent scientific stations exist at the South Pole where explorers died less than seventy-five years ago in a quest to reach the pole. The North Pole has yielded to repeated over- and under-the-ice sorties, and aircraft regularly criss-cross polar skies flying short great-circle routes in an area that was an inaccessible goal a century ago. Amateurs may be quickly transported into polar remoteness in this decade where experienced explorers feared to tread less than a century ago.

In using more effective tools and technologies, people have produced a dazzling array of cultural attributes (e.g., cultivated land, transportation facilities, manufacturing facilities, urban environments, and vast networks for sharing information). In addition, however, people have altered the physical environment in ways hardly dreamed about just a century ago. For example, human exploitation has destroyed literally thousands of species of animal life; vast areas have been devegetated, falling to the plow

Figure 1.2
These Qashqai nomads depend on their herds for survival.

Figure 1.3
The urban population of Manila requires a reliable water supply.

Figure 1.4
Transportation is an essential ingredient in modern civilization.
Association of American Railroads.

or the saw; soil erosion has been accelerated; air and water pollution have become commonplace; and environmental degradation has extended to even the most remote corners of the globe.

As the population soars, and as requirements for food, fuel, and amenities increase, people expand the cultural landscape while modifying the physical environment. The processes of nature have been altered and accelerated by the activities of people. More often than not, this human interference has degraded the pre-existing environment, although it could be argued that certain practices, like fertilization, have improved the pedological (soil) environment, at least in some places and from the perspective of the agriculturalist.

Several examples of geography as a social science include: economic geography, urban geography, human geography, and transportation geography. **Economic geography** examines how people earn a living, including the areal distribution and linkages of the several elements of production, distribution, and consumption of goods and services. Primary, secondary, tertiary, and quaternary activities are considered as part of an increasingly sophisticated and interconnected aspect of human occupance of

a region. **Urban geography** examines the patterns of urban agglomerations, including the distribution of different facilities, such as those associated with housing, manufacturing, warehousing, and commerce. **Human geography** considers the human response to environmental conditions, and finds expression in the modes of earning a living and in the facilities constructed to provide shelter and amenities. **Transportation geography** is a study of linkages, flows, and patterns in the movement of people, goods, or ideas (figure 1.4).

Figure 1.5
Primitive agricultural techniques like those employed by these rice winnowers in the Philippines are still used by many people in the developing countries.

Figure 1.6
Modern industrial states use an array of manufacturing and facilitative activities to provide goods and services to their citizens.

Basically, the social science aspects of geography include an assessment of the patterning of human responses to environment. The scientific aspects of geography, on the other hand, assess the environment in terms of patterns and relationships based on analysis of data according to principles of universality. The development of theories from either deductive or inductive processes is characteristic of all aspects of the discipline. In fairness, it should be pointed out that the unpredictability of human response to environment makes theory development and testing more difficult than in the physical portions of the discipline. Further, long-term temporal factors may resist testing within a reasonable time frame. Nevertheless, an active group of economic and cultural geographers are at work to develop principles of universality in those aspects of the field.

Geography as a Science

The scientific aspects of geography include a study of the physical components of the environment at or near the earth's surface as well as a synthesis of the various interrelationships that exist. The planetary relationships that permit life; the ebb and flow of seasons; the inexorable march of time; the elevation, slope, and relief of the land; the amount and rate of erosion and deposition; crustal disturbances resulting in mountain building and land formation or destruction; the quantity and distribution of precipitation; the march of temperature from place to place and season to season; and the ecological succession and patterns of animal and plant life distribution are all parts of the scientific considerations of geography. These and other elements of the environment are inextricably linked together in producing a physical world, which is the home of man and of the millions of species of plants and animals

Figure 1.7
Mt. Mayon in the Philippines is an active volcano.

that share this planet with us. Although the linkages and connections are almost infinitely complex, scientific investigations are slowly unraveling the associations while developing theories to serve as platforms for prediction.

Several elements of the physical environment are appropriate fields of study and inquiry in their own right, but a number of them are essential to the scientific study of the environment. **Geomorphology** is the study of landforms, including their orogeny, evolutionary development, and patterns of distribution. **Pedology** is the study of soil development, characteristics, and distribution (figure 1.8). **Hydrology** considers the origins, patterns, and distribution of water at or near the earth's surface (figures 1.9 and 1.10). **Meteorology** is the study of weather, including prediction (figure 1.11). **Climatology** is an analysis of climatic characteristics, a study of the distribution of climatic

Figure 1.10
The waters of the land support life while they etch and sculpt the surface.

Figure 1.9
The waters of the sea occupy most of the earth's surface.

Figure 1.11
The cirrus clouds in this photograph are a fair weather aftermath of the recent snowfall.

patterns, and projections of long-term trends or tendencies. **Biogeography** is a study of the ecological developments and biological distributions that exist at or near the earth's surface (figure 1.12). Further, since the areal phenomena on the earth's surface are central to an appreciation for an understanding of the physical environment, a study of **cartography** (the art and science of map making) must be considered.

Actually, these represent only some of the elements of that intricate network of physical factors that make up the physical environment (figure 1.13). The complicated interconnected aspects of the natural environment are fragile and subject to unraveling when one aspect goes awry or when man's activities upset a delicate balance. It is especially important to understand existing interrelationships to avoid or minimize catastrophic disruptions. Indeed, those who read this book are enjoined to learn to appreciate our earth and to do their part to ensure the continued habitability of the planet.

Essentially, the examples of geographic inquiry that have been described represent topical or systematic aspects of the discipline. These investigations are designed to permit the development of generalizations and principles. Geographers also undertake regional studies that provide an integrative analysis of the facets of a specific area, which give it a particular character.

This book involves primarily topical investigations, although elements of regional inquiry are also included. Elements of analysis and synthesis are introduced to provide an appreciation for patterns that are a part of the earth's mosaic. The spatial character of these patterns is particularly important, and a sense of place is integral to a description and understanding of the earth.

Figure 1.12
These Rocky Mountain sheep resting on a high alpine meadow adjacent to a woodland are a part of the earth's biological fabric.
© *Peter Kresan.*

Figure 1.13
The various elements of geography are closely related.

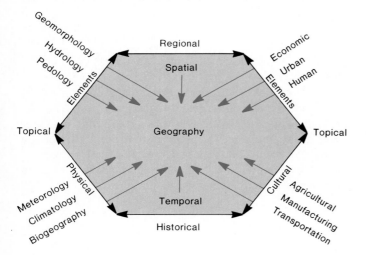

Nature of Geographic Inquiry

As various elements of the physical environment are examined, the basic principle of uniformitarianism is employed. Essentially, **uniformitarianism** asserts that all physical processes are operating in the same way and at the same rate as they always have. A Scotch geologist, James Hutton, proposed the theory in the face of a great deal of skepticism, especially by Christian clergymen who objected to the enormous time scale attaching to the evolution of earth and life forms. The continuing controversy between creationists and evolutionists is, essentially, a derivative from these notions of progressive change. Actually, uniformitarianism does not reject the notion of a divine plan. Process and change are certainly not antithetical to religious precepts and concepts. In a sense, uniformitarianism is a basic concept (with the exception that modern people may have the capacity to accelerate or upset several physical processes) that forms the basis for rational analysis of environmental patterns and inter-relationships.

It should also be recognized that uniformitarianism applies only to the last several million years of geologic history, since significant changes have occurred over the eons in the lithosphere, hydrosphere, atmosphere, and biosphere of the earth. For example, an earth without free oxygen weathered at a vastly different rate than one with the present atmospheric composition. The pre-oceanic world was vastly different from what we now know. An earth with significantly different biological assemblages weathered, eroded, and changed in different ways than have been experienced in the last several million years. It's clear that modern people, especially, because of numbers and technology may accelerate or slow natural processes in certain circumstances. Even primitive people caused environmental change, especially in the use of fire to remove vegetation. Nevertheless, uniformitarianism is the essential concept that undergirds modern physical landscape analysis. It seems clear that the theory is basically true for our world. It seems equally clear that a new corollary will have to be added to the concept as human activities affect the earth's landscape in broader and more pervasive ways.

Environmental Determinism Over time, geographers have flirted with the concept of **environmental determinism,** which suggested that human response was determined by the physical environment. It became obvious that this concept was too limiting. Different cultural groups living in the same or analogous settings developed quite different responses to environments. The same groups also adjusted to the same environment in different ways at different times.

Figure 1.14
Rich volcanic soils in the tropical Philippines have produced a plantation agriculture response.

Figure 1.15
A cool, humid environment close to a large city (Chicago) has made this dairy farm a profitable human response.

Possibilism Environmental determinism was succeeded by **possibilism,** which suggested that there were a large number of human responses possible within a given environmental framework (figure 1.15). Careful observers had noted different human responses to analogous situations. They had seen evolutionary changes in human adjustment over time. Modern people seem to be expanding the possibilities by the use of technology, and natural limits seem less significant than was once the case. Modern science and technology have permitted man to modify significant environmental limitations in such a way as to expand environmental parameters and to enhance opportunities for human occupance and use. There is every reason to anticipate that new technologies will broaden human choice even further.

Analysis in Geography

Basically and above all, analysis in physical geography demands a holistic approach (i.e., a consideration of all facets of the physical environment). It requires a search for interconnections, distributions, linkages, and patterns. Often these interrelationships may be very complex, and frequently they are obscure. Expanding investigations and sophisticated techniques are continuing to provide explanations for phenomena. Much has been accomplished. Much remains to be done. A constant objective is to simplify associations to allow the human intellect to understand our earthly home. In this, physical geography has the same objective as all science and scientists (i.e., to simplify explanations of complex phenomena). Sometimes, however, there are obvious connections that the keen observer can postulate, and that may be verified subsequently by close analysis of data to test a hypothesis.

Several examples will illustrate the point. A number of years ago, a colleague at another university led a group of students in a plant ecology class on a field trip to the semiarid area of eastern New Mexico. He asked the students to observe that the grass seemed to be thicker, greener, taller, and generally more lush immediately adjacent to the asphalt road than in nearby areas. When he asked for an explanation, students gave a number of responses. One student suggested that the nitrous oxide from automobile exhausts added a fertilizing element. Another suggested that the additional heat absorbed by the road was transferred by conduction to the grass. Another thought that grazing animals might be less prone to graze at the roadside. Yet another said that the disturbance of soil incident to road construction had released essential plant nutrients. After the professor pointed out that these were not nitrogen-deficient soils, that low temperatures in summer were not a limiting parameter for grass growth in this area, that numerous jackrabbits and certain other grazing animals had been killed on the road, and that the soils throughout the area had abundant quantities of minerals essential for the growth of grass; the most naive of the students (not a botany major like the rest) asked, "Could it be that when rain comes, the runoff from the highway provides more moisture for the plants adjacent to the road?" The light of discovery went on suddenly in the mind of every student in the class. Quickly, elaborations followed. The strip of asphalt contained no competing plants; dew sometimes formed on the road surface to be pushed to the thirsty soil at the side of the road by passing motorists; and the asphalt served as a capillary moisture loss barrier in the top few inches of soil.

Of course, detailed analysis might have demonstrated other causal factors, but the obvious connection, which the professor confirmed, had been suggested by the student who had the broadest vision of the problem . . . a geography major, it should be added. In a sense, the geography student had been able to see the grassland assemblage, while the others were intent upon individual blades

Figure 1.16
Cogon grass has succeeded in an area formerly covered by tropical rainforest.

of grass. Others had then been able to elaborate on the obvious relationship, once it had been pointed out. A comparable example of such analysis comes from the direct experience of the author of this book.

While leading a field course in the Philippines a number of years ago, I pointed out that the area covered by tropical forest seemed to be diminishing, and the area covered by cogon grass (*Imperata cylindrica*) was expanding (figure 1.16). Everyone could account for the reduction of the tropical forest.

Clearly, cutting the forest to exploit the wood and to expand agricultural land reduced the forest domain. The huts of squatters and subsistence farmers, girdled trees, and brush fires gave eloquent testimony to the assault on the rain forest; however, the cogon grass expansion was not so simply answered. The students struggled. Together, the students and I were able to see that farming ventures on newly cleared forest land were frequently unsuccessful because of the small mineral budget and fragile nature of the tropical soils. The delicate symbiosis between forest trees and soil, once broken, resulted in further soil deterioration. Trees were not effective colonizers of the impoverished soil, whereas cogon grass could gain a foothold. Once established, the grass could spread through rhizomes and could survive the frequent natural and manmade fires.

Seedling trees, if established at all, were destroyed by the frequent fires. The grass growth was actually enhanced by fires, since the removal of dead tops by burning facilitated the new green growth with the coming of the rainy season, and the ash contributed nutrients to the impoverished soils. There is a high likelihood that this is a permanent transformation. A grass savanna has been produced where once a magnificent tropical rainforest was dominant, and the forest has limited opportunity to reestablish itself in the newly created savanna. It seems clear that savannas have expanded at the expense of forests in historic, and perhaps prehistoric, times because of the interference of man, especially in the use of fire.

At another time, I observed a bit of medical detective work with clear geographic implications. A significant number of the children in Fars Ostan in Iran were dwarfs. It was noted, too, that a number of those children tended to eat clay if they were given the opportunity. Several questions were obvious: Was this a genetic defect? It didn't appear to be, since these were almost exclusively children of normal-size parents. Was there a toxic material in the clay? Plants or animals didn't seem to be harmed. Was there something missing in the soil? Subsequent investigation showed that there was a serious zinc deficiency in the soil. Poor children who did not get proper diets or vitamin-mineral supplements were afflicted in almost direct proportion to the level of zinc deficiency. If zinc supplements were introduced into the children's diet at an early enough age, dwarfism was thwarted.

Of course, this last situation reflects a medical problem, but it also has geographic implications. The nature of the soil and its distribution were ultimately key elements in finding a cause and, indirectly, a treatment for dwarfism in that part of the world. Mapping patterns of dwarf incidence provided initial clues as to where environmental conditions should be carefully studied. Other examples of geographic assistance in medical and epidemiological studies are plentiful (figure 1.17). Legionnaire's Disease, a particularly virulent form of pneumonia, secrets itself in air conditioning filters, usually in public buildings where the filters may be changed infrequently. The site of an outbreak can usually be quite quickly determined by tracing the victims' patterns of movement during the incubation period. The isolation of sources of contagion involves medical detective work that is spatially focused. The determination of a locus of infection enables public health officials to eliminate the cause of the contagion and to limit the number of those infected.

Kwashiorkor, a protein-deficiency disease, is endemic in poor areas of the humid tropics where starch constitutes most of the diet. The disease can be prevented or mitigated by the introduction of appropriate plant and/or animal protein foods. Mapping of disease incidence, when correlated with crop and dietary patterns, provided the early clues of the nature of the disease.

Figure 1.17
Maps of distribution of disease-carrying insects (such as the tsetse fly) assist in charting possible areas of infestation. The tsetse fly is the carrier of sleeping sickness.
Source: David R. Zimmerman, Mosaic, 1988.

Miles
0 200 400 600 800 1,000
0   400 800 1,200 1,600
Kilometers

Figure 1.18
The earth's rocky crust (lithosphere) supports vegetation—a part of the biosphere.

Figure 1.19
White River dam site in eastern Utah illustrating interrelationships of atmosphere, hydrosphere, lithosphere, and biosphere.
Nolan Preece/Biological Photo Service.

These examples all illustrate the mutual interrelationships between man and environment, and all suggest the need for a holistic and spatial approach to environmental analyses.

Although space may provide new frontiers for exploration and perhaps colonization, people must depend on the earth for the elements of survival in the foreseeable future. An understanding of and an appreciation for this lovely blue-green planet is essential for people to live effectively, now and in the future.

In subsequent chapters, various aspects of the several spheres of the earth will be analyzed, interpreted, and interrelated. These several spheres include the **lithosphere** (the earth's rocky crust), the **atmosphere** (the gaseous envelope surrounding the earth), the **hydrosphere** (the earth's waters), and the **biosphere** (the plant and animal associations that blanket the earth). Together, these spheres and their interconnections establish the earthly parameters for life and human occupance. Disturbances within and between them have short- and long-term ramifications for continuing habitability of the planet.

Figure 1.20
Elements of the environment are related in intimate and intricate ways. The only constant on the earth is change.

Study Questions

1. Which of the following are properly considered a part of physical geography? Of cultural geography?

   a. houses
   b. swamps
   c. airports
   d. railroads
   e. missile silos
   f. rocks
   g. bridges
   h. streams
   i. helioports
   j. hills
   k. parks
   l. hail
   m. snow
   n. rivers
   o. roads
   p. factories
   q. earthquakes
   r. tornadoes

2. In which area or field of geography would the following be emphasized?

   a. analyzing the influence of terrain on man.
   b. studying features resulting from human occupance of a region.
   c. describing and analyzing conditions in the corn belt of the United States.
   d. analyzing the characteristics of Chicago neighborhoods.
   e. studying the chain of occupance of the Great Plains.
   f. investigating malarial outbreaks in Third World countries.
   g. studying the distribution pattern and resource base of integrated steel manufacturing centers in the United States.
   h. analyzing the flora and fauna of a Florida swamp.
   i. studying the orogeny and existing terrain of the Rocky Mountain.
   j. analyzing the climate patterns and cycles in the American Great Plains.

3. From your experience, give several examples of how man has altered the *physical* environment.

4. What is a geographic region? Does such a region exist in the area where you live? What are its important characteristics?

5. Why is geography often considered to be a correlative subject—a science of synthesis?

6. Are typical boundaries in nature sharp and discrete, or are they zones of transition?

7. Is the dispersion of business activity from central cities to suburban locations a proper subject for geographic inquiry?

8. How may geography be considered to be human ecology?

9. What are the principal arguments against the concept of determinism?

10. What examples can you give of irreversible changes in the physical environment caused by human activities?

11. How does possibilism differ from environmental determinism?

12. Explain the basic tenets of uniformitarianism.

13. Explain how the concept of uniformitarianism may need to be modified in light of growing population and expanding technology.

14. What changes have you observed in the cultural environment in the area where you live? In the physical environment?

Selected References

Adams, F. D. 1954. *The birth and development of the geological sciences.* New York: Dover Publications.

Broek, J. O. M. 1966. *Compass of geography.* Columbus: Charles E. Merrill Publishing Company.

Doerr, A. H., and Guernsey, J. L. 1976. *Principles of physical geography.* 2d ed. Woodbury: Barron's Educational Series.

Fellows, D. K. 1985. *Our environment: An introduction to physical geography.* 3d ed. New York: John Wiley and Sons.

Gabler, R. E.; Sager, R. J.; Brazier, S. M.; and Wise, D. L. 1987. *Essentials of physical geography.* 3d ed. Philadelphia: Saunders College Publishing.

McKnight, T. L. 1984. *Physical geography: A landscape appreciation.* Englewood Cliffs: Prentice-Hall.

Oberlander, T. M., and Muller, R. A. 1987. *Essentials of physical geography today.* 2d ed. New York: Random House.

Strahler, A. N. 1969. *Physical geography.* 3d ed. New York: John Wiley and Sons.

2

The Earth and Its Planetary Relationships

All life on earth is made possible by the sun.
Utah Travel Council.

The kingly brilliance of Sirius pierced the eye with a steely glitter, the star called Capella was yellow, Aldeberan and Betelgeuse shone with a fiery red. To persons standing alone on a hill during a clear midnight such as this, the roll of the earth eastward is almost a palpable movement.

Thomas Hardy, *Far From the Madding Crowd*

*A*lthough our understanding of the universe and our place in it are subject to change as we gain more information about it, it seems reasonably clear that the universe began about 15 billion years ago when an incredibly dense mass of material exploded in the so-called big bang. The source of this material continues to be an enigma, just as the ultimate fate of the universe continues to be a matter of conjecture. Indeed, these questions occupy the attention of the best theoretical physicists and astronomers. The postulated explosion sent all of the material of the universe outward in all directions in a cosmos that is still expanding. The static that fills the electromagnetic spectrum to this day is residue of this cataclysm attending the birth of the universe. All of the galaxies, stars, planets, asteroids, and other bodies in the universe were formed, or are forming, from the gas and dust of this enormous explosion.

11

This process of stellar and satellite formation continues, while other suns and satellites die in the incredible explosions of supernovae or in the gravitational doom of a black hole. Whether this process will continue until all of the various solar fires have gone out and only blackened remnants will continue to move outward through the infinity of space, or whether gravity will eventually pull all the materials together again in preparation for another gigantic explosion remains an unfathomable mystery. Perhaps matter is being annihilated by antimatter at some unseen boundary of collision. As astronomers' eyes and ears are stretched further by more and more sophisticated equipment, old mysteries are solved and new ones are created. The intricacies and complexities attending the birth and evolution of the universe stagger and challenge the imagination.

Origin of the Solar System

Our own solar system was formed when a giant cloud of gas and dust began to clump together and rotate into a kind of flattened disk. The central nucleus condensed and heated, and ultimately a series of self-perpetuating thermonuclear reactions were initiated. The central nucleus had become a star—our sun. That event, which occurred about 5 billion years ago with the birth of the sun, will apparently be reversed in about 5 billion years from now when the last thermonuclear fires of the sun go out, after paroxysms have engulfed all the planets in the inner solar system in a bath of incandescent fire and radiation. If life still exists on earth four billion years from now, it is doomed to extinction several tens of millions of years before the ultimate demise of the sun. The sun will likely swell to enormous size bathing the inner planets, including the earth, in blistering heat and deadly radiation. The oceans will boil away and the atmosphere will evaporate. The astronomers' modern version of incineration of the earth is not greatly different from the biblical description of earth's destruction by fire. While the sun permits life in our solar system, the existence of planets at a reasonable distance from it is essential to its development and evolution. Planets did evolve in our solar system, and we are the thinking creatures who developed on planet earth.

Outlying eddies of gas and debris coalesced into planets and moons, and the balance of gravitational attraction and forces of rotation caused them to assume orbits around the sun. From the ignition of the thermonuclear fires, about 5 billion years ago, the initial conditions were set in place to permit the development and evolution of life on earth. As the planets formed about 4.6 billion years ago, and subsequently to about 3.8 billion years ago, they were subject to a merciless bombardment of asteroids and space debris. Most evidence of that early bombardment has long since been erased on the surface of the earth by tectonic movement and forces of weathering and erosion; however evidence of those impacts are omnipresent on the moon and most other planets. Manned and unmanned space flight continues to confirm this merciless bombardment, and the pocked and cratered surface of the moon and other planets bear silent testimony to the myriad impacts of meteorites. We continue to witness the fiery entry of meteorites into the earth's atmospheric envelope, but most of them are entirely incinerated before they reach the surface of the earth. A few specimens that survived the fiery trajectory through the earth's atmosphere are found in the dusty recesses of museums.

By a fortuitous set of circumstances, the earth developed at the correct orbital distance from the sun to permit life to develop. If it had been located closer to the sun, like Mercury or Venus, the atmosphere would have been boiled away, or the surface would have been subjected to killing heat or lethal doses of ultraviolet radiation. Mars, a short distance beyond earth, is not of a sufficient size to permit gravity to hold most atmosphere in a close envelope around the planet. Thin air and radical swings of temperature from day to night have apparently precluded life or, if it ever existed, the planet now appears to be biologically inert. Unmanned spacecraft have detected no life on Mars, although these craft have sent pictures reflecting geomorphological activity quite reminiscent of activity on earth. Perhaps manned exploration early in the next century will provide more answers about conditions on the so-called red planet. In 1988, the Soviets called upon the United States to undertake such a joint venture. Whether politics and resources will permit such an excursion remains to be seen.

The outer planets, Jupiter, Saturn, Uranus, Neptune, and Pluto, are much too cold for life as we know it to have evolved. This fragile blue-green ball is the only life-bearing planet that we are aware of in the entire universe. The laws of celestial mechanics, which dictate orbital position and speeds of revolution and rotation, along with appropriate mass and distance from the sun, are legacies of the earth apparently not shared by any other body in our solar system. Although there is constant conjecture about life on the postulated billions of planets elsewhere in the cosmos, there is no persuasive evidence to confirm or deny the presence of life elsewhere. Astronomers patiently aim their radio telescopes at various segments of the sky in the search for a signal that some other form of life is out there. If such a signal ever comes, it will raise the most profound scientific, psychological, sociological, and ethical questions.

Whether we are alone, or whether there are innumerable others—each a staggering concept to consider—it is important for us to understand our earthly home, for it's all we have and, barring unforeseen changes in technology, it's all we're ever likely to have. Large-scale settlements of space stations and significant colonization of other planets, science fiction notwithstanding, are extremely remote prospects for people in the near future. Although almost nothing is impossible, the difficulties of extraterrestrial colonization on a large scale are staggering, and it appears highly unlikely that such ventures will be undertaken for millennia, if ever.

Our solar system is located in an obscure corner of the Milky Way Galaxy, and our average-size sun is in no significant way distinguishable from literally billions of stars in the universe. The Milky Way extends about one hundred thousand light years in maximum diameter and about ten thousand light years in minor diameter. In overall shape, it is a little like a slightly warped phonograph disc. Galactic distances are almost incomprehensible when we consider that light travels approximately 186,000 miles per second (298,000 kilometers per second), but the Milky Way is only one of billions of galaxies in the universe.

The sun is, by standards of the earth, an enormous body possessing more than 99 percent of the material in the solar system. Its mass is about 330,000 times that of the earth, and its supply of hydrogen will cause it to burn for about 5 billion years more.

The earth is the platform from which we view the solar system, galaxy, and universe. It is a sphere that is slightly flattened at the poles because of the effects of rotation. The earth has an equatorial diameter of about 7,927 miles, or 12,757 kilometers. The polar diameter is somewhat less. The slightly flattened, spherical shape of the earth, technically an **oblate spheroid,** is now accepted by virtually everyone on the basis of perceived physical phenomena and imagery from space. A few eccentrics cling to the belief that the earth is flat and that all our proofs of sphericity are illusory. Adherents of the Flat Earth Society should stop reading here, because the discussion that follows hinges upon a spherical earth.

Motions of the Earth

Four principal motions of the earth are detectable: rotation, revolution, precession, and galactic rotation. Two of these motions, rotation and revolution, are very important

to people and their physical environment. The effects of precession on long-range climatological events and the earth's molten interior are under increasing study. The effects of galactic rotation, if any, are not understood.

More data suggest, however, that the slightly different tilt of the earth's axis relative to the plane of the ecliptic and the precessional oscillation of about 1° may have a significant influence on climate. Indeed, paleoclimatologists are investigating the intriguing possibility that past major climatological shifts can be tied to precession of the earth's axis, along with a cosmic wobble, and slight eccentricity in the earth's orbit. Further, there is some suggestion that this precession may influence the earth's molten interior, which, in turn, may change the earth's magnetic field and geothermal activity. In the chapter on glaciation, we will speculatively examine some of earth's movements as possible causes of aberrational climatic changes that were the precipitating factors that caused Pleistocene glaciation.

Rotation

The rising and setting of celestial objects occurs because the earth is rotating in a direction opposite to that which the celestial objects appear to travel. The earth experiences day and night because it is alternately lighted by the sun or is in its shadow. The eastward rotation of the earth causes the apparent westward march of the sun (figure 2.1). Since the earth rotates through 360° in approximately twenty-four hours, it rotates through about 15° in one hour, and through approximately 1° in four minutes. There is some evidence that the earth is gradually slowing in rate of rotation, but that reduction in speed is very small and, except for some very precise measurements of time, is irrelevant to human experience. There

Figure 2.1
The effect of the earth's rotation in twenty-four hours.
From Arthur Getis, Judith Getis, and Jerome Fellmann, Introduction to Geography, *2d ed. Copyright © 1988 Wm. C. Brown Publishers, Dubuque, Iowa. All Rights Reserved. Reprinted by permission.*

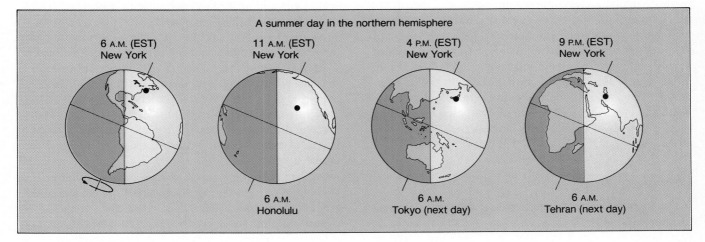

A summer day in the northern hemisphere

| 6 A.M. (EST) New York | 11 A.M. (EST) New York | 4 P.M. (EST) New York | 9 P.M. (EST) New York |

6 A.M. Honolulu — 6 A.M. Tokyo (next day) — 6 A.M. Tehran (next day)

are also minor wobbles in rotation, which may relate to protuberances or irregularities in the core. Again, except for very precise measurements, these wobbles are apparently of minimal significance to people, although they may have some long-term climatic influences, and conceivably may be influenced by or may influence activity within the core or mantle of the earth.

Revolution

The earth revolves around the sun in an elliptical orbit in about 365¼ days. The earth travels a distance of approximately 586,000,000 miles (somewhat more than 943,000,000 kilometers) in one revolution. It travels this distance at an average speed of approximately sixty-six thousand miles per hour, (about 106,000 kilometers per hour) although the speed varies somewhat depending on orbital position of the earth. Since the earth's orbital path is elliptical rather than circular and the laws of celestial mechanics dictate that the area between the center of the orbital path and the perimeter be constant for any period of time, the speed varies. Travelling along the path causes the earth to be closest to the sun in early January (**perihelion**) at a distance of about 91,500,000 miles (147,250,000 kilometers); while in July the sun is at its greatest distance away from earth (**aphelion**) at a distance of about 94,500,000 miles (152,079,000 kilometers).

The axis upon which the earth rotates is inclined at an angle of about 66½° to the plane of the ecliptic (orbital path) of the earth around the sun, or about 23½° from the vertical, and the axis in one place in the orbit is essentially parallel to all other orbital positions (figure 2.2). The effect of this **parallelism of the axis** causes the northern and southern hemispheres alternately to be pointed more directly towards the sun from season to season. This causes the sun's rays to strike most sections of the earth at a constantly changing angle. The varying altitudes of the sun result in variations in the length of day and the amount of solar energy received at the surface. Together, these variations in day length and sun angle affect the amount of insolation received from place to place and result in the march, or change, of seasons. During the summer season in the Northern Hemisphere, the sun's rays are more nearly overhead, while low-angle sun occurs during the winter season in middle and high latitudes. The length of daylight increases in latitudes away from the equator during summer, and it diminishes during winter. More direct rays of the sun and length of daylight period combine to heat the earth more effectively in the appropriate hemisphere during the summer. Less direct rays and shorter periods of daylight account for the colder period of winter.

Precession

If the earth's axis remained perfectly constant in direction, the seasons would occur at exactly the same time each year; however, the earth's axis makes a very slow circular motion opposite to the direction of rotation. This motion is called **precession,** and is responsible for the fact that seasons begin about twenty minutes earlier than the preceding year.

The precessional motion of the earth's axis is analogous to the slow conical motion the axis of a top makes when it is spun in a leaning position. For the earth, the motion is produced by the gravitational pull of adjacent celestial bodies and by the equatorial bulge of the earth. The precessional rotation outlines a large circle in the sky in about twenty-one thousand years resulting in a variation of the tilt of the axis of about 1°. In addition to the minor influence on the beginning of seasons, there is increasing speculation that precession may have an influence on long-term climatic and geophysical changes. Such causal relationships, if they exist at all, are still uncertain, but current speculation will be considered in the chapter dealing with glaciation.

Galactic Rotation

In addition to rotation, revolution, and precession, the earth exhibits another movement as part of the Milky Way galaxy. This is **galactic rotation,** which involves all the celestial bodies within the Milky Way. These bodies are speeding at about two hundred miles (322 kilometers) each second in the direction of a point in the sky near the star called Vega. There also appears to be a kind of generalized movement of all the galaxies into a kind of supergalaxy. The effects, if any, of galactic rotation or supergalactic movement on the earth are unknown, although a nagging suspicion suggests that some relationship to long-term climatic variations exists. It must be stressed that there is currently no evidence to associate galactic rotation and climatic change.

It requires approximately twenty-four hours to complete one rotation, about 365¼ days to complete one revolution around the sun, approximately twenty-one thousand years to complete a precessional cycle, and about 220,000,000 years for one galactic rotation. Obviously, only the first two of these motions are within the direct experiential understanding of most observers, although the seasonal lag influenced by precession may be noted by scientists or very careful observers.

Just as it is important to understand something of the earth's movements and the earth's position within the solar system and the galaxy, so it is important to have a sense of where things are located on the earth. Such locations may be described in a variety of ways, but certain conventions are used most commonly. Some aspects of distance and direction from given lines or points provide a frame of reference for accurate locations to be determined.

Without such a frame of reference, it would be impossible to navigate great distances with confidence, and great confusion about locations would exist. It should be understood that the conventional grid system of location is not the only possible system. It is, however, an excellent

Figure 2.2
The pattern of the earth's revolution around the sun.
From Arthur Getis, Judith Getis, and Jerome Fellmann, Introduction
to Geography, *2d ed. Copyright © 1988 Wm. C. Brown Publishers,*
Dubuque, Iowa. All Rights Reserved. Reprinted by permission.

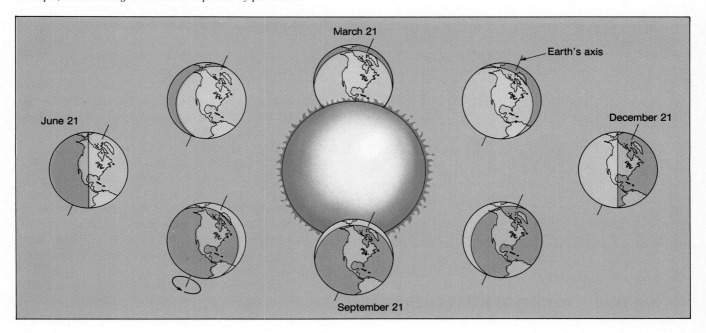

schema for determining positions on the earth's surface.
International convention in the use of the latitude and longitude grid system makes it practicable to determine precise locations over the surface of the entire earth.

Location

A set of imaginary reference lines has been established on
the surface of the earth to enable us to locate places with
some precision. The poles and the equator provide initial
reference points that can be used in establishing a location
grid network (figure 2.3).

The **equator** is an imaginary line that extends east–
west around the earth midway between the two poles, and
it divides the earth into two equal halves, or **hemispheres.**
At right angles to the equator, another imaginary line,
called the **prime meridian,** has been established running
through the Royal Observatory at Greenwich, England.
Obviously, an infinite number of meridians could have been
used to establish a principal meridian, but international
convention has settled on the Greenwich Meridian, although past nationalistic concerns caused a number of nations to attempt to establish a prime meridian through the
capitals of their countries. A number of different principal
meridians were used well into the nineteenth century. The
Soviets continue to use a Moscow Principal Meridian for
certain internal maps. Location may be determined by the
combination of distance and direction from the prime meridian and the equator.

Figure 2.3
Patterns of meridians and parallels on a sphere.
John P. Snyder, Map Projections Used by the U.S. Geological Survey,
Geological Survey Bulletin 1532, Second edition, United States
Government Printing Office, Washington, D.C., 1982.

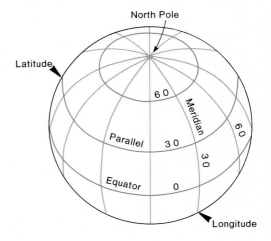

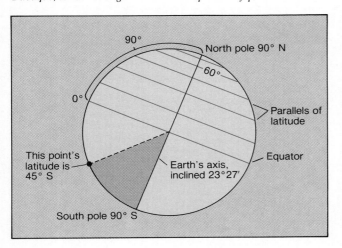

Latitude and Longitude

The conventional system used is based on the division of a circle into 360 degrees. Each degree may be divided into 60 equal parts called minutes, and each minute into 60 equal parts called seconds. One degree, one minute, and one second are used in this manner as angular distances and are written respectively as 1°, 1′, and 1″. Other locative grids may be employed by different agencies, but most location systems employ measurements east–west and north–south of fixed reference lines. In the case of locations at very high latitudes, polar coordinates may be used with either the North or South Pole being used as a principal reference, and locations described with reference to that set of coordinates.

Latitude is the angular distance north and south of the equator (figure 2.4). The latitude of the equator is 0°; the North Pole is a 90° N. latitude; and the South Pole is a 90° S. latitude. **Parallels** are lines parallel to the equator that are used to measure angular distances (latitude), north and south of the equatorial reference line.

Longitude is the angular distance east or west of the prime meridian. **Meridians,** used to measure longitude, are north-south lines converging at the poles. Since longitude is measured east and west of the prime meridian halfway around the earth, the maximum longitude is 180° E. longitude or 180° W. longitude (the same meridian).

Latitude and longitude measurements are not only of value in the precise location of places on the earth's surface, but they are also helpful in estimating and visualizing distances between places on the earth. A degree of latitude is about 69 statute miles, 111 kilometers, or 60 nautical miles at the equator (1/360 of the polar circumference of 24,860 miles, or 40,015 kilometers). A nautical mile is equal to 6,080 feet or about 1,853 meters. The length of a degree of latitude is approximately the same everywhere, although the polar flattening of the earth produces slight variations. On the other hand, a degree of longitude is approximately 69 statute miles (111 kilometers) at the equator, whereas it diminishes to zero at the poles where the meridians converge.

Time

The rotation of the earth on its axis in approximately twenty-four hours provides a basis for determining time. Since the sun is an imprecise timekeeper, however, a mean solar day of exactly twenty-four hours in length has been adopted. Since **apparent** solar time varies from **mean** solar time, it is necessary to employ a **time equation** to make the essential corrections. The elliptical nature of the earth's orbit causes the sun to be fast at some periods of time and slow at others. This equation of time can be represented by a graph known as an **analemma.** The analemma also provides information on the declination (the latitude where the sun's rays are vertical at noon) of the sun for each day of the year (figure 2.5).

Since distant stars are not affected by the same shortcomings that apply to solar time, the relative position of the stars may be used to determine **sidereal** time. Sidereal time is significantly more accurate than solar time, and even greater precision is determined by the decay of certain radioactive materials in an atomic clock. Less precise clocks and timepieces are suitable for the vast majority of human endeavors.

Time is of interest to man in the daily ordering of events, and the Greeks had two words for time: **kairos,** meaning a special moment in time; and **chronos,** meaning the passage of time. Time differences between two places provide a measure of the longitudinal differences between two locations. If one observes the meridional passage of the sun at noon, and compares local time with the time shown by a **chronometer** (a very accurate watch) for Greenwich, it is possible to determine the longitudinal distances between the two places. An accurate chronometer was quite late in coming, and, as a result, early maps that may depict latitude quite accurately are notorious for their inaccurate fixing of longitude. Latitude could be determined with some degree of precision by determining the altitude of **Polaris** (the North Star) or by measuring the elevation of the sun at noon with a quadrant or later on with the more precise sextant or octant. Indeed, the first acceptable chronometer was produced only in 1735, and a really accurate chronometer didn't exist until about 1840. The best of these expensive spring-powered chronometers, however, are inferior to even cheap battery-powered quartz watches commonly available today. If the chronometer is set on Greenwich time (GMT or Zulu), the time differences between observed local time and Greenwich time will yield the longitude of the observer.

Figure 2.5
The analemma.

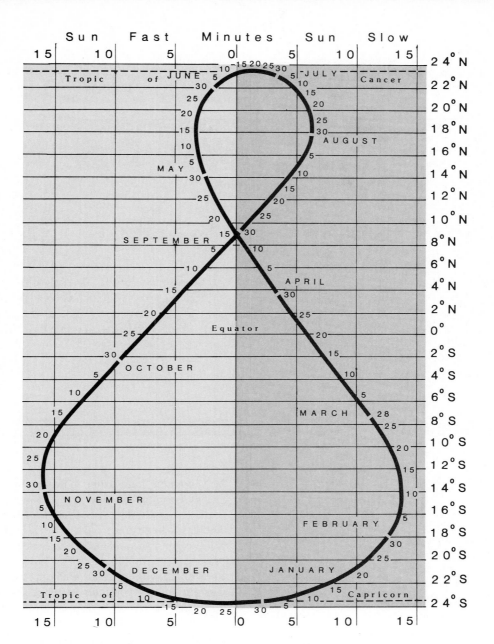

For example, if a chronometer set on Greenwich time registers 4:20 P.M. (1620 hours) when it is local noon (at the observer's position), the time difference is 4 hours and 20 minutes. Four hours and 20 minutes translates to 65° (4⅓ × 15°). The time is earlier at the observer's position, hence the location is 65° W. longitude.

If, on the other hand, the chronometer (Zulu) reads 10:20 A.M. (1020 hours) when it is local noon, the time difference is 1 hour and 40 minutes. The longitudinal distance between the observer and Greenwich is 25° (1⅔ × 15°). Since it is later at the observer's position, the location is 25° E. longitude.

From a practical standpoint, it is necessary in navigation to have very accurate information on time and solar declination for every day in the year. Such information can be obtained from publications like *The Air Almanac, The Nautical Almanac,* or *The American Ephemeris and Nautical Almanac.*

Figure 2.6
This 2,500-year-old bas-relief appears timeless, but it, too, will pass away.

Figure 2.7
World time zone map. Each time zone is approximately 15°
wide, but variations are made for political considerations.
The International Date Line attempts to avoid populated
places.

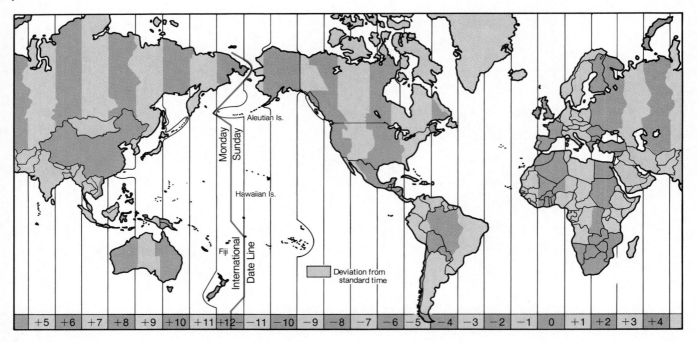

Standard Time Zones

Considerable confusion would exist if every place on earth used its local solar time. Any travel east and west would necessitate a constant change in time settings on watches. To minimize this difficulty, the earth has been divided into a series of standard time belts approximately 15° wide, so that areas 7½° on either side of a central meridian accept the time for that belt (figure 2.7). In practice, there are many irregularities in the belts because of the location of populous areas at or near the margin of a belt. In theory, in the United States, eastern standard time is based on the mean solar time for 75° W. meridian, central standard time on the 90° W., mountain standard time on 105° W., and pacific standard time on 120° W. This means that when the standard time is noon (1200) in London, it is 7:00 A.M. (0700) in New York, 6:00 A.M. (0600) in Chicago, 5:00 A.M. (0500) in Denver, and 4:00 A.M. (0400) in San Francisco.

Daylight Saving Time

Matters of time have been further complicated because of the introduction of daylight saving time. The initial purpose of advancing the clocks in the summer was to conserve fuel and energy in World War II. It has proven to be a popular innovation, however, since it still reduces energy consumption and provides more after-work hours of daylight, which may be used for recreational purposes. The clock is advanced in the spring and set back in the fall (i.e., in April and October); hence, the popular slogan, "spring forward—fall back." In 1986, a decision was made to extend daylight saving time approximately one month. Certain groups continue to resist daylight saving time citing problems like sending children to school in the dark and the imposition of much farm work in the dark. It's still true, of course, that certain areas do not recognize daylight saving time. Further, there is some consideration by Congress to continue daylight saving time until after election day, usually the first Tuesday in November, to facilitate a common time for the closing of polls to mitigate the effect of early television reports skewing the election results in the Western states.

International Date Line

Circumnavigation of the globe made it apparent that there was a need for a special place to establish the arbitrary beginning of a new day. A line, known as the **International Date Line,** was established by common agreement. It approximates the 180° meridian, although it has offsets in it to avoid populated places. As one crosses the International Date Line going westward, the calendar is set forward one day; as one moves eastward across the International Date Line, the calendar is set back one day. The antithesis is true with time, of course. As one goes west the time is earlier, and as one goes east the hour is later. These temporal relationships are important in a number of practical matters, since business and political relationships require some appreciation of time of day in other parts of the world.

Latitude and Sun Angle

Parallelism of the earth's axis at an angle of about 66½° to the plane of the ecliptic as it revolves about the sun affects the angle at which the sun's rays strike the earth. This has a major influence on climate, and that aspect will be discussed in later sections of the book. In addition, however, the relative position of the sun's rays can be used to determine latitude. As the earth revolves around the sun, the noon sun is vertical at the equator on or about September 21–23 and March 21 each year. These dates—when the sun is vertical at the equator at noon, and days and nights are equal everywhere—are known as **equinoxes.** On or about June 21–22, the noon sun is vertical at 23½° N. latitude (a parallel known as the **Tropic of Cancer**), and on or about December 21–22, the noon sun is vertical at the **Tropic of Capricorn** (23½° S. latitude). These dates are known as **solstices.** The analemma provides approximate dates when the sun is vertical at any other latitude. Precise information as to the position of vertical noon sun's rays may be determined from *The Air Almanac* or *The Nautical Almanac*. Of course, crude approximations can be made by the apparent migration of the sun through 47° (23½° N. to 23½° S.) in half a year (i.e., 182.5 days). If orbital speed were constant—which *it isn't*—the sun would make an apparent average migration of about ¼° per day or about 1° in four days (47° ÷ 182.5).

With this information, it is possible to determine latitude. On an equinox date, the sun's rays are vertical at noon at the equator (i.e., at an elevation of 90°). If an observer were located at 5° N. or 5° S., the elevation of the noon sun would be 85°. The observer at 5° N. would see the sun in his or her southern horizon, whereas the observer at 5° S. would see the sun in the northern horizon. The angle made between the horizon and line of sight to the sun is known as the **elevation angle,** or sun's altitude, while the angle between the line of sight to the sun and a point directly over the observer's head is known as the **zenith angle** (figure 2.8). The sum of the two is always equal to 90°. The zenith angle subtends an arc that is equal to the latitudinal distance between the vertical rays of the sun and the observer's position.

Figure 2.8
Diagram illustrating the relationship between elevation angle, zenith angle, and latitude. In the diagram, ∟A is the elevation angle, H-H' is the horizon line, and ∟Z is the zenith angle.

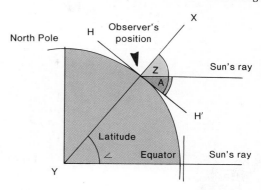

An example will illustrate the point. If an observer measures the elevation of the noon sun and finds it to be 40° in his or her northern horizon on December 22, what is the latitude of the observer? Moving methodically from the known to the unknown, latitude can be determined: the sun's vertical rays at noon are at the Tropic of Capricorn (23½° S.) on December 22; the observer saw the sun in the northern horizon, so the observer's position is south of the location of the sun's vertical rays; the sum of the zenith angle and the elevation angle is always equal to 90°; and the zenith angle is equal to the latitudinal distance between vertical rays of the sun and the observer's position. In this example, the observer is south of 23½°. If the elevation angle is 40°, then the zenith angle is 50° (90° − 40°); and the observer is 50° south of 23½° S., or at 73½° S. latitude.

Obviously, the opposite kind of information may be obtained. If the latitude is known, it is possible to calculate the elevation of the noon sun angle. For example, if one is situated at 30° latitude, it is easy to determine that the sun will be at a noon elevation of 60° on an equinox date. Consider, again, the known facts: the noon sun's rays are vertical at the equator on an equinox date, the latitudinal distance between the observer's location and the equator is 30°, the elevation angle plus the zenith angle is equal to 90°, and the latitudinal difference between the latitude of the vertical noon sun and the observer's position is equal to the zenith angle. In this case, the zenith angle is 30°; hence, the elevation of noon sun is 60° or (90° − 30°).

Using the sun and *Polaris* (the North Star), navigators have been able to determine reasonably accurate latitudinal positions for hundreds of years. Since Polaris is approximately situated over the North Pole, the elevation of Polaris above the horizon is equal to latitude.

Figure 2.9

Singapore is located at 1°20' N., 103°57' E. Determine the latitude and longitude of Kuala Lumpur.

From Arthur Getis, Judith Getis, and Jerome Fellmann, Introduction to Geography, 2d ed. Copyright © 1988 Wm. C. Brown Publishers, Dubuque, Iowa. All Rights Reserved. Reprinted by permission.

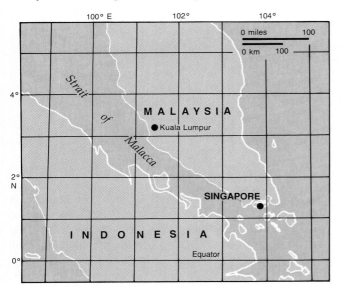

Obviously, Polaris is limited in utility, since it cannot be seen south of the equator at all. In addition, it is not a very bright star and is difficult to observe at low latitudes. Further, it does not lie precisely above the geographic pole, and, as a result, observations do not yield precise latitudes.

Any careful inspector of historical maps will note that longitudinal positions were frequently grossly distorted until relatively recent times. The development of precise chronometers quickly improved the accuracy and reliability of maps by providing accurate determination of longitude.

The twentieth century has brought dramatic improvements in navigation through the use of radio, radar, loran, inertial guidance systems, and satellite tracking. Clearly, the accurate determination of position is essential for the production of maps, for the movement at great speed over long distances, and for the aiming and positioning of offensive or defensive weapons of war such as ballistic missiles.

Modern surveying has been improved by the availability of satellite tracking data, which provides a level of accuracy heretofore unavailable. Such precise locations are very important in establishing property or political boundaries, thereby preventing disputes between political entities or individuals over the proper position of boundaries.

An array of globes, maps, and charts depend upon precise locations (figure 2.9). These and numerous other geographic tools will be described in the following chapter. The correct use of those tools and the accurate interpretation of data derived from them are crucial to geographic study and analysis.

Study Questions

1. Explain why the shape of our galaxy, the Milky Way, appears distorted to an earthbound observer.
2. Briefly explain why the Northern Hemisphere is warmer in summer than in winter in spite of the fact that the sun is closer to the earth in January.
3. Explain why the earth is apparently the sole home of life in our solar system.
4. Why does the earth bulge slightly at the equator?
5. What is the principal effect of each of the earth's motions?
6. What is the approximate latitude of the area in which you live? What climatic ramifications derive from such a location? What is the approximate length of daylight at your latitude in midsummer? In midwinter?
7. At the end of 1987, a leap second was added to the year to comport with temporal realities. Why was this necessary?
8. What are the advantages of daylight savings time? The disadvantages?
9. Why were early navigators able to determine reasonably accurate latitudinal positions many years before accurate longitudinal positions could be ascertained?
10. What is the location at the point directly opposite (antipode) to your own on the earth's surface?
11. If a person walks 10 miles north, then 10 miles east, and finally 10 miles south whereupon he or she has reached the spot where the journey began, what is the person's location?
12. If an observer sees Polaris at an elevation of 60°, what is his or her approximate latitudinal position?
13. Why would Polaris not be a useful navigation star if you lived and flew in Argentina?
14. What is the central meridian for the standard time zone in which you live?

Selected References

Brown, L. A. 1979. *The story of maps*. Boston: Dover Publications.

Cloud, P. 1978. *Cosmos, earth, and man*. New Haven: Yale University Press.

Doerr, A. H., and Guernsey, J. L. 1975. *Principles of geography*. 2d ed. Woodbury: Barron's Educational Series.

Fellows, D. K. 1985. *Our environment: An introduction to physical geography*. 3d ed. New York: John Wiley and Sons.

Gabler, R. E.; Sager, R. J.; Brazier, S. M.; and Wise, D. L. 1987. *Essentials of physical geography*. 3d ed. Philadelphia: Saunders College Publishing.

Hawkes, J. 1962. *Man and the sun*. New York: Random House.

McKnight, T. L. 1984. *Physical geography: A landscape appreciation*. Englewood Cliffs: Prentice-Hall.

Strahler, A. N., and Strahler, A. H. 1987. *Modern physical geography*. 3d ed. New York: John Wiley and Sons.

3

Geographic Tools, Methods, and Techniques

Western hemisphere satellite composite showing cloud cover. Note the well-established hurricane in the Gulf of Mexico. NOAA.

When I'm playful I use the meridians of longitude and parallels of latitude for a seine, and drag the Atlantic Ocean for whales.
Mark Twain, *Life on the Mississippi*

*A*major facet of geographic inquiry is to observe, report on, and interpret order in the landscape. To accomplish this objective, the geographer depends on a variety of tools and uses a number of methods and techniques. To understand the nuances of geographic inquiry, to appreciate the principles developed, and to recognize the significance of environmental interrelationships, it is necessary to become acquainted with some of the most common geographic tools and methods of analysis. Some of the tools are so commonplace as to be scorned by the scientifically sophisticated, but even the most ordinary device can yield significant data. All are valuable in augmenting the inherent powers of observation and interpretation of the competent observer. The geographer of the late twentieth century is fortunate to be able to use equipment and materials undreamed of a hundred years ago.

It should be recognized, however, that sophisticated tools and enormous arrays of data cannot substitute for the analytical and synthesizing power of the careful scientist. Tools and procedures extend the powers of observation and interpretation of the geographer, and they make interpretations and explanations more intelligible. Each of us should be grateful for the contributions of generations of scientists who preceded us. That scientific genius Sir Isaac Newton sounded the proper note of humility and gratitude when he said, "If I have seen further than others, it's because I stood on the shoulders of giants."

In developing concepts, it is often necessary to proceed from the general to the particular, although in other instances a great many particular phenomena or events may allow the development of generalization. The level of sophistication tends to increase as closer and closer observations are made. Greater detail demands greater precision, and the intricate can usually be adequately measured only with very precise and sophisticated instruments. Fortunately for us, the inventive mind of man and more powerful technologies are adding new devices.

Globes

In spite of the creation of new levels of sophisticated equipment, certain old and common devices continue to contribute to an understanding of phenomena. The physician still uses a hammer to test reflexes, and the most beautiful designs are still produced by human glassblowers. So it is with geography. The globe developed more than five hundred years ago, and improved to be sure, still is unparalleled in providing a realistic perception of the nature of the earth's surface. This situation illustrates that there is usually value in the old and the new. The trick is, as John Kennedy said, "to hold fast to the best of the past and move fast to the best of the future."

The globe is a close approximation of the shape of the earth and is, therefore, of great value in developing a perceptual framework for the various aspects of the environment. The globe depicts distance, direction, sizes, and shapes of various features accurately. Its diameter and area are proportional to that of the earth. There are a variety of globes that are suitable for depicting a host of physical and cultural features. They vary in quality and level of sophistication from models used in elementary schools to large-scale precise representations in certain scientific agencies.

A globe has increased utility with an appropriate locative frame of reference. Meridians and parallels provide such an appropriate location grid. Certain features may be stressed depending on the nature of the inquiry. Commonly, globes depict both physical and cultural features. Slated globes make it practical to add and modify various features being studied at a particular time. Such globes are of particular value in classroom settings, since they allow the teacher to make appropriate drawings for class perusal.

The usefulness of globes is reduced somewhat because size restrictions result in such a small scale that little detail can be shown. Further, one can observe only one hemisphere at a time, and most globes would be cumbersome to carry about. In fact, restricted transportability may be the most limiting aspect of a globe's use. They vary in utility from the small-sized decorative type in a den or office to large scale forms used to highlight a particular earth feature. Some globes are made with painstaking care and focus on certain salient features. Extraterrestrial globes have been made of the moon and several planets. Such globes are useful in simulation training for astronauts and in providing a better perception of extraterrestrial environments for the average person.

To overcome the problem of portability and to expand scale significantly, it is necessary to make two-dimensional representations of portions of the earth's surface. Maps and charts accomplish this objective. A discussion of their salient characteristics follows.

Maps

A **map** is a representation of all or a part of the earth's surface drawn to scale on a two-dimensional surface. To fully accomplish all of the objectives for which it was intended, a map must possess certain essentials: title, legend, direction, scale, latitude and longitude (or other locative grid), and date. Occasionally one or more map essentials may be eliminated if the map is being used with a text, or if common usage makes a particular essential universally known. It is, however, valuable to have all essentials included on the map, particularly if the map stands alone as a purveyor of information. Absence of the various map essentials may lead to confusing or deceptive interpretations of map data.

Title

The title of a map, like a book title, is highly significant. Since maps depict a variety of features, it is essential that the user be aware of the nature of a particular map. Data of all kinds may be depicted on maps. In fact, few aspects of the human experience are not suited to illustration on maps. Examples of commonly mapped features include landforms, soils, vegetation, streams, highways, population distribution, crop production, mineral occurrences, agricultural regions, and so on. The prospective user of the map must have no doubt as to the nature of the data shown. Without a title, maps of production, especially, are useless.

Legend

Legends should be clearly printed on maps to give an appropriate indication of symbols employed to represent features (figure 3.1). In a sense, a legend provides the tools to unlock the interpretation of the map; hence, the term *key* is sometimes used as a synonym for legend. Common usage may make it possible to eliminate certain symbols from the legend, or key. Prudence dictates, however, that the cartographer should err on the side of completeness.

Topographic Map Symbols

BOUNDARIES

National
State or territorial
County or equivalent
Civil township or equivalent
Incorporated-city or equivalent
Park, reservation, or monument
Small park

LAND SURVEY SYSTEMS

U.S. Public Land Survey System:
  Township or range line
    Location doubtful
  Section line
    Location doubtful
  Found section corner; found closing corner
  Witness corner; meander corner

Other land surveys:
  Township or range line
  Section line
Land grant or mining claim; monument
Fence line

ROADS AND RELATED FEATURES

Primary highway
Secondary highway
Light duty road
Unimproved road
Trail
Dual highway
Dual highway with median strip
Road under construction
Underpass; overpass
Bridge
Drawbridge
Tunnel

BUILDINGS AND RELATED FEATURES

Dwelling or place of employment: small; large
School; church
Barn, warehouse, etc.: small; large
House omission tint
Racetrack
Airport
Landing strip
Well (other than water); windmill
Water tank: small; large
Other tank: small; large
Covered reservoir
Gaging station
Landmark object
Campground; picnic area
Cemetery: small; large

RAILROADS AND RELATED FEATURES

Standard gauge single track; station
Standard gauge multiple track
Abandoned
Under construction
Narrow gauge single track
Narrow gauge multiple track
Railroad in street
Juxtaposition
Roundhouse and turntable

TRANSMISSION LINES AND PIPELINES

Power transmission line: pole; tower
Telephone or telegraph line
Aboveground oil or gas pipeline
Underground oil or gas pipeline

CONTOURS

Topographic:
  Intermediate
  Index
  Supplementary
  Depression
  Cut; fill

Bathymetric:
  Intermediate
  Index
  Primary
  Index Primary
  Supplementary

MINES AND CAVES

Quarry or open pit mine
Gravel, sand, clay, or borrow pit
Mine tunnel or cave entrance
Prospect; mine shaft
Mine dump
Tailings

SURFACE FEATURES

Levee
Sand or mud area, dunes, or shifting sand
Intricate surface area
Gravel beach or glacial moraine
Tailings pond

VEGETATION

Woods
Scrub
Orchard
Vineyard
Mangrove

COASTAL FEATURES

Foreshore flat
Rock or coral reef
Rock bare or awash
Group of rocks bare or awash
Exposed wreck
Depth curve; sounding
Breakwater, pier, jetty, or wharf
Seawall

BATHYMETRIC FEATURES

Area exposed at mean low tide; sounding datum
Channel
Offshore oil or gas: well; platform
Sunken rock

RIVERS, LAKES, AND CANALS

Intermittent stream
Intermittent river
Disappearing stream
Perennial stream
Perennial river
Small falls; small rapids
Large falls; large rapids

Masonry dam

Dam with lock

Dam carrying road

Intermittent lake or pond
Dry lake
Narrow wash
Wide wash
Canal, flume, or aqueduct with lock
Elevated aqueduct, flume, or conduit
Aqueduct tunnel
Water well; spring or seep

GLACIERS AND PERMANENT SNOWFIELDS

Contours and limits
Form lines

SUBMERGED AREAS AND BOGS

Marsh or swamp
Submerged marsh or swamp
Wooded marsh or swamp
Submerged wooded marsh or swamp
Rice field
Land subject to inundation

The legend, or key, provides the essential symbolic clues that make it possible for the user to interpret the map correctly. Without an appropriate legend, a prospective user has a useless series of symbols and lines that have minimal meaning, unless the symbols are so widely used as to be universally understood. Although there is a certain cross-cultural universality in map symbols, certain cultural idiosyncrasies can yield erroneous interpretations unless a particular symbol is specifically defined. For example, the crescent and star symbol for a mosque in an Islamic country might be mistaken in a Christian country. Similarly, a Star of David symbol for a cemetery might be perfectly understandable in Israel, but be misunderstood in Papua, New Guinea. Very complex maps or maps treating esoteric subjects must have detailed legends to ensure that they are correctly interpreted.

Scale

The scale may be defined as the ratio between map distance and earth distance. Scale may be shown as a **representative fraction (fractional scale)**; for example, 1/63,360 (i.e., one unit on the map equals 63,360 units on the earth's surface). Linear or graphic scales use a line or bar of specific length to represent distances on the earth or a portion of it. A linear scale equal to the representative fraction shown above is:

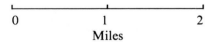

The same scale may be expressed verbally as one inch equals one mile or one inch equals 1.6093 kilometers.

A large-scale map has a larger representative fraction and shows greater map detail. A small-scale map can show a larger area on the same size paper, but in much less detail. Maps that are used to show large areas, such as the world, are, of necessity, small scale. Conversely, maps that may be used to highlight small areas in considerable detail are large scale. Relationships between scales may be illustrated as follows. A map with a scale of 1:250,000 is ¼ the scale of a map with a scale of 1:62,500, (i.e., 1/250,000 ÷ 1/62,500 = ¼).

Direction

A **compass rose,** showing true direction and magnetic direction, and frequently grid direction as well, is a desirable feature on every map. Nautical and aeronautical charts almost always have a compass rose. The compass rose can be eliminated on a map where parallels and meridians are properly oriented to show true direction. Certain maps may have **isogonic lines** (lines of equal magnetic declination) to make the user aware of true and magnetic north. In such maps a date is particularly important, since the migration of the magnetic poles changes the compass variation over time. An isogonic line of 0° magnetic declination (a line along which magnetic and true north coincide), is known as the **agonic line.** The most common orientation of maps is with north at the top, south at the bottom, east at the right, and west to the left. Certainly, that usage is not universal, however, and care must be exercised in orienting a map correctly. Indeed, Australians take delight in showing Yanks (Americans) maps of the world with south at the top. For those of us reared with north conventionally at the top, such an orientation seems strange indeed.

Latitude and Longitude

Without a locational grid system provided by intersecting parallels and meridians, or some other intersecting grid lines such as those used on certain military maps, a map is very difficult to use. Proper orientation and location is essential to map use and analysis, and proper orientation is difficult without a locational grid; hence, virtually every map has some grid network.

Date

Of all the major map essentials, the one most often omitted is the date. Every map should show the date surveyed or the date of the data used on the map as well as the date when the map was constructed. Those produced by the United States Geological Survey (USGS) are particularly good, since they clearly indicate not only survey dates, but dates of revisions. Many other mapmakers are not nearly so careful in providing dates, however. For example, a map showing population distribution may be quite misleading unless the prospective user is aware of the date of the census used. Many maps that could have great historical or comparative value are useless because it is impossible to ascertain the time of their development and publication. Modern day mapmakers are generally better about providing dates than many of their predecessors.

Other Useful Map Information

Although not as significant as the map essentials, certain other information about map construction may be important to the prospective user. For example, the source of data used or the technique of surveying employed may be valuable to the user in assessing accuracy of sources. Again, the USGS and most other government agencies are very good about citing data sources. This may enable the person using the map to choose alternative information sources or to use other means of acquiring information essential to the investigator's purposes. Any valid information that does not cause unnecessary clutter or undue expense in producing and publishing should be included. It is not normally possible to include too much information about data sources and timeliness.

Whether information is derived from satellite or other remote sensing sources, it is important to provide a note detailing the level of accuracy. The user should be apprised of the sources of data and the reliability of the depicted materials. Producers of maps should err on the side of inclusion rather than exclusion.

Map Projections

Although a globe is the only true representation of the earth's surface, it is usually inconvenient to use globes because of the difficulty in handling, transporting, and storing them. Although rotation will permit the user to observe the opposite hemisphere, complete patterns of world distribution cannot be observed simultaneously. **Cartographers** (mapmakers) have been faced with the insoluble dilemma of accurately representing the entire area of a three-dimensional globe on a two-dimensional surface. Obviously, distortions occur when any attempt is made to accomplish this objective. In an attempt to minimize inaccuracies in depicting a three-dimensional earth on a two-dimensional surface, cartographers have developed map projections. A **map projection** may be defined as a systematic arrangement of grid lines transferred from the globe to a flat surface. These developed coordinates then make it possible to depict physical and political patterns with considerable accuracy, and it is possible to manipulate coordinate systems in such a way as to highlight specific features or data.

Certain map projections are known as geometric projections since, in theory, they are developed when light is passed through a globe grid onto a tangent or secant plane, cone, cylinder, or other geometric figure (figure 3.2). Such projections would normally be part of a **planar, conic,** or **cylindrical** family. Most map projections are based on mathematical transfers, however, and are not dependent on theoretical light projection on a geometric surface.

Of the projections that are developed geometrically, the source of light can cause significant modification in the patterns projected on a geometric surface. A light source at the center of the earth produces a **gnomonic** projection; one on the opposite side of the globe to the point of tangency is known as a **stereographic** projection; whereas, one developed with the light source at infinity is called **orthographic.**

Desirable Characteristics for Map Projections

Cartographers try to make projections **conformal** (shapes of areas on the projection are true representations of comparable features on the earth's surface), **equivalent** or **equal area** (areas contained on a map have the same proportions

Figure 3.2
Theoretical light sources and projection on a planar surface for gnomonic, stereographic, and orthographic projections.

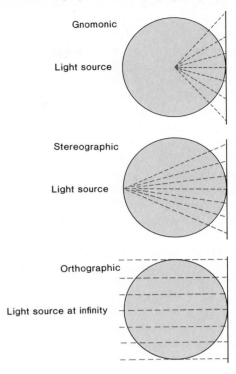

throughout as they truly have on the earth's surface), **equidistance** (distances from a specified point, or along a particular line, are correct), **azimuthal** or **zenithal** (directions from a certain point are correct). It is, as previously indicated, impossible to have all of these characteristics simultaneously in a single map projection. No projection can be both equal area and conformal at the same time; hence, certain projections are chosen to accomplish specific objectives. Some examples of map projections and their characteristics will be given in the following sections. Only a few of the most commonly used projections are described.

The Mercator Projection

The Mercator projection, developed from a cylinder placed tangent to the earth's equator, is one of the world's most widely used and misused projections. Its principal attribute is simplicity. On the Mercator projection, meridians and parallels are straight lines that cross at right angles (figure 3.3). Since the meridians do not converge as they do on the earth, areas are vastly distorted at high latitudes. For example, Greenland, which is about one-sixth the size of South America, is depicted on the Mercator projection as being as large as South America.

The Mercator is a *conformal* projection, and it has one other very desirable property. All straight lines (**rhumb lines**) cut all meridians and parallels at a constant angle. A navigator's course between two locations can be easily plotted. Indeed, the projection was developed initially to support the needs of ocean navigators. It has been used subsequently by aerial navigators and continues to be a useful chart in scores of ways.

The weaknesses of the Mercator projection make it ill-suited for world distribution maps, because the poles cannot be shown, the scale varies with latitude, and areas are grossly distorted at high latitudes. Nevertheless, the Mercator is frequently misused and may be in part responsible for the imperfect perceptions of world relationships that are held by many. For example, although the Soviet Union and Canada are very large land areas, they appear disproportionately large on a Mercator projection. The distortions of the Mercator have given whole generations of students a distorted idea of size relationships, especially in high latitude domains. There have been recent sporadic attempts to eliminate the Mercator for classroom instruction because of the gross misrepresentation of size relationships. It is unlikely that this will occur, but users must recognize inherent shortcomings of the Mercator, as well as all other projections. Indeed, for size relationships, especially, novices must constantly be referred to the globe to ensure that they have the proper perspective on size and distance relationships.

The Transverse Mercator

This projection is developed as if a cylinder were placed tangent to any two opposite meridians or to a great circle drawn oblique to opposite meridians (figure 3.4). Note that except for the meridians of tangency, the equator, and each meridian 90° away from the central meridian, none of the meridians or parallels are straight lines. This type of projection is especially useful along heavily travelled air routes since, like other Mercator projections, it is conformal, and scale distortion near the tangent portions is minimal. Other cylindrical projections are developed as if the cylinder was placed secant to the globe. Still others are mathematically modified to reduce area distortions.

The Gnomonic Projection

The gnomonic projection is developed from a plane placed tangent to the earth at any desired point (figure 3.5). The most important attribute of the gnomonic projection is that all straight lines represent arcs of great circles. **Great circles** are the surface intersections of the globe as if a plane

Figure 3.3
The Mercator projection is conformal and all rhumb lines (straight lines) cut meridians at a constant angle.
John P. Snyder, Map Projections Used by the U.S. Geological Survey, *Geological Survey Bulletin 1532, Second edition, United States Government Printing Office, Washington, D.C., 1982.*

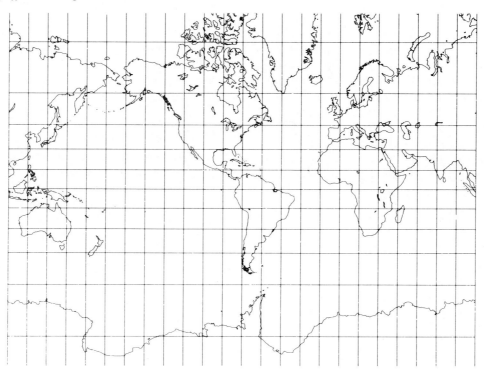

Figure 3.4
The transverse Mercator.
John P. Snyder, Map Projections Used by the U.S. Geological Survey,
*Geological Survey Bulletin 1532, Second edition, United States
Government Printing Office, Washington, D.C., 1982.*

had cut it into two halves, or hemispheres. Great circles are the shortest distance between two points on the earth's surface, and are desirable routes for aircraft to save time and energy.

Typically, gnomonic projections are used for plotting great circle courses, polar maps, and certain astronomical maps. Gnomonic projections have the serious disadvantage of distance, area, and shape distortions that are magnified away from the point of tangency.

Not infrequently, the gnomonic projection is used in conjunction with a Mercator projection in aerial navigation. A great circle course is plotted on a gnomonic projection, and then a series of known points along the course are transferred to a Mercator projection to permit the plotting of a series of rhumb lines that approximate a great circle without making it necessary to change compass headings constantly.

Modern inertial navigation and computers make it possible to fly an almost perfect great circle route, since heading adjustments are made constantly by automatic devices. Clearly, distance savings are important conservators of time and fuel.

Figure 3.5
The gnomonic projection in polar, oblique, and equatorial aspect.
John P. Snyder, Map Projections Used by the U.S. Geological Survey,
*Geological Survey Bulletin 1532, Second edition, United States
Government Printing Office, Washington, D.C., 1982.*

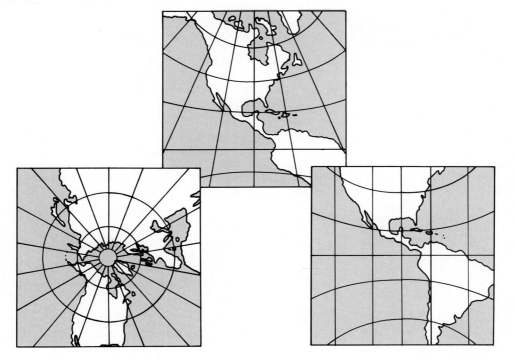

The Conic Projections

Maps of smaller areas, such as countries and states, are frequently developed from a cone or series of cones placed tangent or secant to the earth's surface (figure 3.6). These projections are easily developed, usually have a relatively accurate scale, and most such projections are conformal. Conic or polyconic projections are not usually suited to coverage of a large area, since inaccuracies increase from the line of tangency (figure 3.7). The polyconic projection, formerly the most used United States map, has been largely superseded by the **Lambert Conformal Conic** (figure 3.8).

   The Lambert conformal conic projection has been derived mathematically as if the cone was secant to the earth at two parallels. For the United States, the choice of the 33rd and 45th parallels results in minimum distortion of much of the area of the conterminous 48 states.

Since the projection is conformal and a straight line approximates a great circle, the projection is used for World Aeronautical Charts.

Other Projections

A very desirable map projection for showing world distributions is some form of the **Mollweide projection.** It is developed mathematically by making the equator twice the length of the central meridian. Except for the central meridian, which is a straight line, the meridians are ellipses with different foci, whereas the parallels are straight lines. As an equal area projection, it possesses inherent advantages for depicting distributions. On the other hand, sections in polar regions and at the periphery of the map are badly distorted. The Mollweide may be "interrupted" by removing wedges from the map in areas of minimal

Figure 3.6
The simple conic projection with one standard parallel.
From Arthur Getis, Judith Getis, and Jerome Fellmann, Introduction to Geography, *2d ed. Copyright © 1988 Wm. C. Brown Publishers, Dubuque, Iowa. All Rights Reserved. Reprinted by permission.*

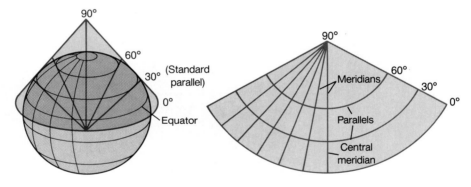

Figure 3.7
The polyconic projection.
John P. Snyder, Map Projections Used by the U.S. Geological Survey, Geological Survey Bulletin 1532, Second edition, United States Government Printing Office, Washington, D.C., 1982.

Figure 3.8
The Lambert conformal conic projection.
John P. Snyder, Map Projections Used by the U.S. Geological Survey, Geological Survey Bulletin 1532, Second edition, United States Government Printing Office, Washington, D.C., 1982.

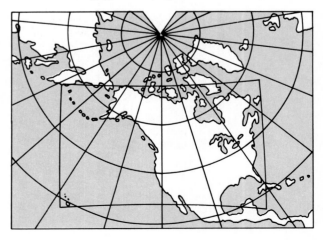

concern. These interruptions allow the projection to be squeezed together to give greater clarity to the areas being stressed.

The **sinusoidal projection** is similar to the Mollweide in that it is equal area and the parallels are straight lines. The meridians are developed from sine curves, however, rather than from ellipses. The sinusoidal has essentially the same shortcomings and attributes as the Mollweide. It is also amenable to interruptions of areas with little interest in a particular projection (figure 3.9).

The **homolosine projection** represents a "marriage" of the Mollweide (sometimes termed a homolographic projection) and the sinusoidal projection. The sinusoidal projection is used for the area between 40° N. and 40° S., whereas homolographic characteristics are used for the remainder of the projection. The homolosine projection is frequently interrupted along certain meridians to focus attention on the continents or oceans depending upon the nature of the features being stressed. It is most frequently interrupted only in the Atlantic Ocean in the Northern Hemisphere and in the Atlantic, Pacific, and Indian Oceans in the Southern Hemisphere. This allows an adequate focus on terrestrial areas being considered. Of course, it is possible to interrupt continental areas, if the focus is on oceanic regions.

Figure 3.9
The interrupted sinusoidal projection.
John P. Snyder, Map Projections Used by the U.S. Geological Survey, *Geological Survey Bulletin 1532, Second edition, United States Government Printing Office, Washington, D.C., 1982.*

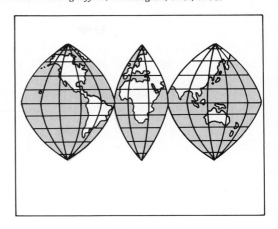

Figure 3.10
The orthographic projection in polar, equatorial, and oblique aspect.
John P. Snyder, Map Projections Used by the U.S. Geological Survey, *Geological Survey Bulletin 1532, Second edition, United States Government Printing Office, Washington, D.C., 1982.*

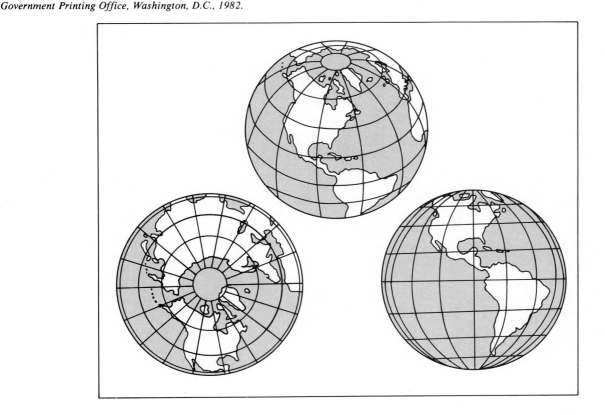

The **Van der Grinten projection** is a useful projection for showing distributions, since all of the world is shown on a single sheet without interruptions (figure 3.11). It is easily read and is a compromise between area-distorted and shape-distorted projections. In spite of its several useful attributes, it has been largely superseded by other projections.

The hundreds of projections that have been developed can logically be aggregated into several groups of projections (i.e., cylindrical, elliptical or circular, azimuthal, and conic) (figure 3.12). Each family possesses significant attributes, and all, of course, have shortcomings as well.

Cylindrical projections, like the Mercator, are usually conformal, characterized by a rectangular grid, and exhibit major distortions away from the point of tangency. **Elliptical** or **circular projections,** like the Mollweide, are bounded by ellipses, sine curves, or circles; they are typically equal area, and usually show all of the earth on a single sheet. The *conic* group, like the Lambert Conformal or polyconic, may be modified in a variety of ways to produce desirable characteristics, and the projections are frequently employed for areas with a large east–west extent like the United States, U.S.S.R., or Canada. **Azimuthal projections,** like the gnomonic, are planar, and directions are accurate at the point of tangency.

From the hundreds of projections that have been developed, one can usually be found that is especially suitable to a specific use (figure 3.13). Others are being tailored regularly to meet esoteric purposes. It is worth reemphasizing that no map can be completely distortion free. Only a globe can meet that test. But because new compromises are constantly being sought, it is fair to say that map projections will continue to multiply.

Again, the user must be careful to recognize the shortcomings as well as the attributes of a particular projection. Shapes, forms, and patterns have a way of assuming an element of authenticity, unless the user of projections exercises great care to ensure that distortions do not assume such an aura of truth. Misconceptions about size, shape, direction, and place once acquired are often difficult to overcome. It behooves us, therefore, to prevent such misconceptions from occurring.

Figure 3.11
The Van der Grinten projection.
John P. Snyder, Map Projections Used by the U.S. Geological Survey, *Geological Survey Bulletin 1532, Second edition, United States Government Printing Office, Washington, D.C., 1982.*

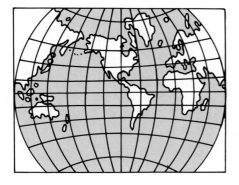

Figure 3.12
Conceptual relationships of projections on a plane, cone, and cylinder.

John P. Snyder, Map Projections Used by the U.S. Geological Survey, *Geological Survey Bulletin 1532, Second edition, United States Government Printing Office, Washington, D.C., 1982.*

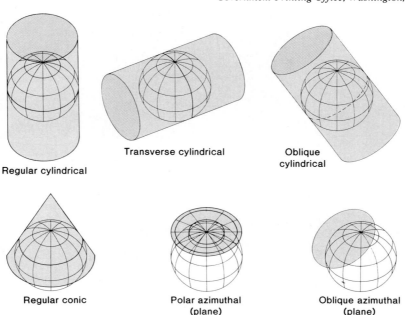

Regular cylindrical

Transverse cylindrical

Oblique cylindrical

Regular conic

Polar azimuthal (plane)

Oblique azimuthal (plane)

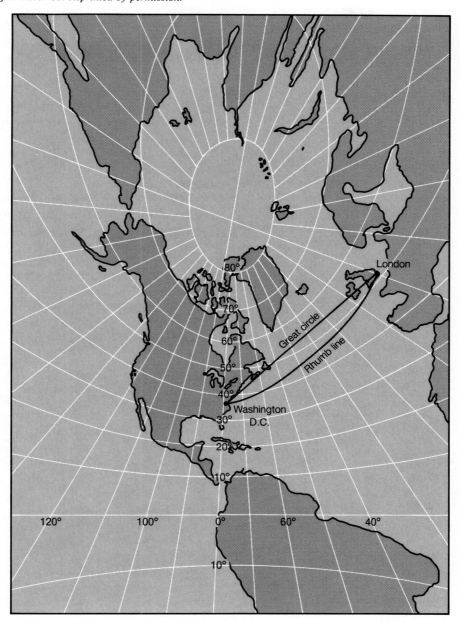

Figure 3.14
*Hachuring used on an early Coast and Geodetic Survey Map
of Anacapa Island.*
Morris M. Thompson, Maps for America, *United States Geological
Survey, U.S. Government Printing Office, Washington, D.C., 1979.*

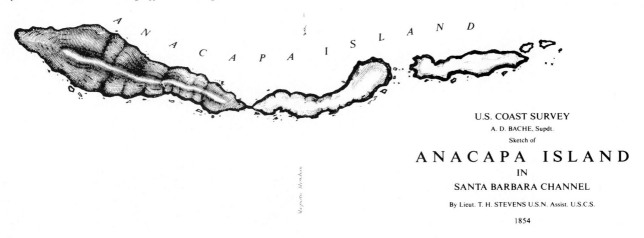

U.S. COAST SURVEY

A. D. BACHE, Supdt.

Sketch of

ANACAPA ISLAND

IN

SANTA BARBARA CHANNEL

By Lieut. T. H. STEVENS U.S.N. Assist. U.S.C.S.

1854

Figure 3.15
A digital elevation model of Yosemite developed by computer.
Morris M. Thompson, Maps for America, *United States Geological
Survey, U.S. Government Printing Office, Washington, D.C., 1979.*

Figure 3.16
This hypothetical island illustrates the relationship between elevation and contours.
From Arthur Getis, Judith Getis, and Jerome Fellmann, Introduction to Geography, 2d ed. Copyright © 1988 Wm. C. Brown Publishers, Dubuque, Iowa. All Rights Reserved. Reprinted by permission.

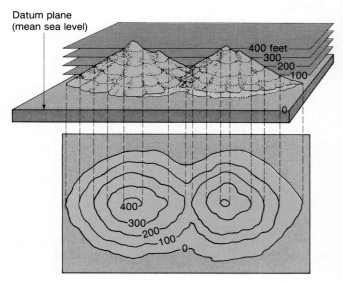

Figure 3.17
A bench mark.
USGS.

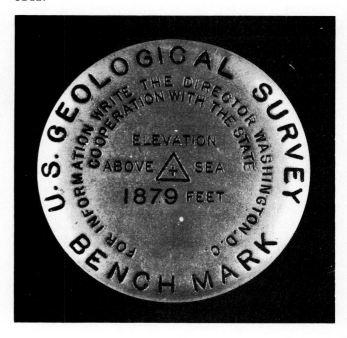

Relief Maps

Since maps are drawn on two dimensional surfaces, it is something of a challenge to depict relief effectively. Several techniques have been employed to illustrate terrain differences, and new approaches are constantly being considered.

Hachure Maps

The hachure map is one of the simplest types of relief maps (figure 3.14). Lines, called hachures, drawn up and down slopes, give the impression of a third dimension. Varying lengths and widths of lines, when skillfully drawn, illustrate changes in individual slopes, but do not indicate elevations. The construction of hachure maps is a laborious and time-consuming task requiring skilled technicians and much hand labor. High labor costs in the United States generally rule out the manufacture of hachure maps in this country. The Europeans, especially the French, have long been known for the quality of hachure maps they have produced. New techniques for showing relief, and a stress on machine production, have resulted in a decline in the number and quality of hachure maps being produced.

There is every reason to assume, however, that increasingly sophisticated holographic work by computers may make it possible soon to produce excellent quality hachure maps mechanically. Indeed, several quite competent computer programs have been developed to produce maps or diagrams that effectively depict relief (figure 3.15).

Contour Maps

A contour map uses lines drawn through points of equal elevation (**contour lines**) to give the impression of relief. The use of a constant **contour interval** (the vertical distance between any two consecutive contour lines) on a given map enables the user to ascertain specific elevation, form, shape, and relief of the land (figure 3.16). Where specific elevations have been determined by triangulation or more sophisticated means, a spot elevation, called a **bench mark** (B.M. and the specific elevation) may be shown (figure 3.17). **Depression contours** have small ticks or hachures on the inner edge of the contour line to indicate that a depression is present (figure 3.18).

Figure 3.18

This diagram illustrates the relationship between contours and depression contours. If this is the diagram of an island and point B is at sea level, with a contour interval of 50 feet, point A has an elevation of 50 feet. Point C has an elevation > 250 feet < 300 feet. D has an elevation of 100 feet. E has an elevation > 100 feet < 150 feet. F has an elevation < 50 feet > 0 feet.

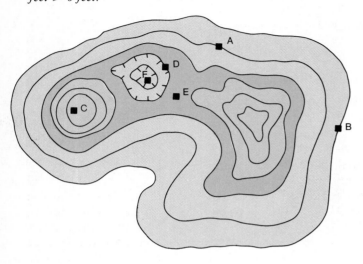

In the United States, the Geological Survey has produced topographic maps of much of the country at varying scales for many years (figure 3.19). In early years, a quadrangle of 15′ of latitude and longitude was produced at a scale of 1:62,500, or approximately one inch to one mile. Early maps of half that size, 1:125,000, or twice that size, 1:31,250, were convenient multiples or divisions. A 7½′ minute series has been more widely used in recent years, and at a scale of 1:24,000 has the advantage of providing much greater detail. Other scales have been used for specific purposes. A constant program of surveying and cartographic work continues to cover areas not previously surveyed, to expand large scale coverage in areas previously surveyed, and to provide revisions to obsolete maps. Fortunately, it has been possible, using remote sensing techniques, to produce maps less laboriously and with greater speed than was the case three or four decades ago. The days of the foot surveyor with plane table, alidade, and surveying rod have largely been supplanted by machine interpretation of remotely sensed images, except for very precise and very large scale maps.

Five colors are typically used on a USGS topographic map: brown, black, red, blue, and green. Brown depicts elevation; black and red show cultural features like roads, houses, churches, and so on; blue is used for water features; and green shows vegetation. Occasionally purple is used to show limits of urban development, and it highlights map updates made from aerial photographs.

USGS contour maps have an abundance of information for the layman as well as the scientist. Skill in using and interpreting such maps is essential for accurate field observations. It is important for the student of geography to gain familiarity with and skill in using such contour or topographic maps.

Shaded Maps

Relief may be shown by careful use of light and shadow (figure 3.20). Slopes of hills are highlighted as they might appear in early morning or late evening light, whereas the other side is left in "shadow." This technique produces a pattern not unlike that seen when flying in an airplane when flying over rough terrain. Frequently, shaded overlays add contrast to contour maps to enhance the perception of relief.

Hypsometric Maps

A specialized type of contour map that uses color tints between sucessive contour lines is known as a **hypsometric map.** The contour interval is usually in thousands of feet or hundreds of meters, and such maps are used for small-scale maps, such as those employed as wall maps in a classroom. On such a map, low elevations are usually shown in shades of green, intermediate elevations in yellow, high elevations in brown, very high elevations in red or purplish red, and areas with permanent ice or snow are shown in white. Water is shown in shades from light to dark blue with lighter blue representing shallow water and darker blue representing greater depths.

Other Relief Devices

Relief may be illustrated by a variety of other methods including the relief model, profile, block diagram, and physiographic diagram. The **relief model,** made from plaster or plastic, is a very effective device, since it actually shows the third dimension, which can be seen and felt. Such models have the handicap of reduced portability and, inevitably, there is significant distortion of the vertical scale compared to the horizontal scale, but they are useful in explaining various geomorphological features and processes. The distortion of vertical scale has to be carefully considered and understood, or the inexperienced scholar can develop incorrect assumptions concerning the nature of terrain. It is usually important to state the vertical exaggeration. A vertical exaggeration of 10:1, for example, means that relief and elevation as depicted are ten times higher and steeper on the model than the real situation on the earth.

Profiles show the earth's surface as if a giant cleaver had cut through a segment of the earth's crust, and one edge was tilted up for scrutiny. **Block diagrams** are drawn

Figure 3.19
Templet CR-2 used by USGS for most commonly employed quadrangle scales. Note that the templet also includes a compass rose.
Morris M. Thompson, Maps for America, *United States Geological Survey, U.S. Government Printing Office, Washington, D.C., 1979.*

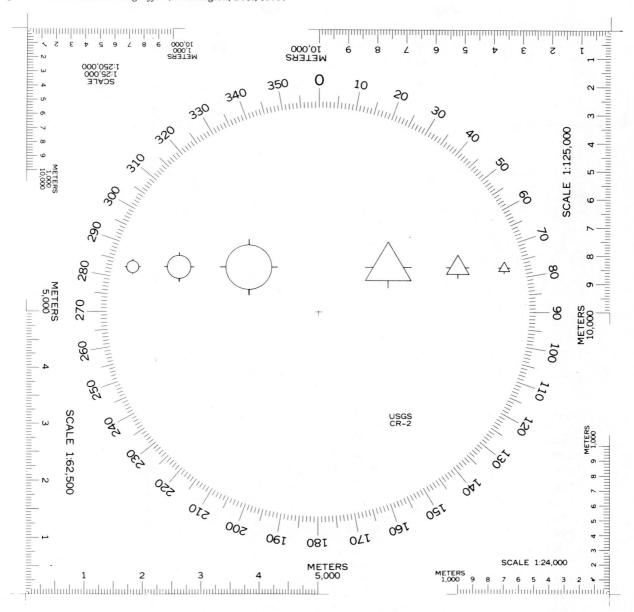

in such a way that the edges illustrate a land profile and usually subsurface stratigraphy; whereas, the top of the block represents a panoramic view. The **physiographic diagram** uses symbols, some clever hachuring, and occasional shading to give the impression of an oblique airplane view of surface terrain (figure 3.21).

All of these devices give a "feel" physically or psychologically for the third dimension. The nature of the terrain is an important element in appreciating the patterns of landscape, and the devices used to enhance

perceptions of relief improve understandings of surface landform features. Any technique that can enhance a perception of and an understanding about terrain features is a valuable adjunct to understanding the physical landscape. The surface beneath our feet is an omnipresent fact of life, and maps or devices that help us to depict the surface are very useful in helping us understand and appreciate the earth. There are many other devices and tools that are useful in geographic inquiry. A description of some of them follows.

Figure 3.20
Portion of a shaded relief map printed over contour lines.
Lake McBride, Kansas, 7 1/2 minute series, USGS.

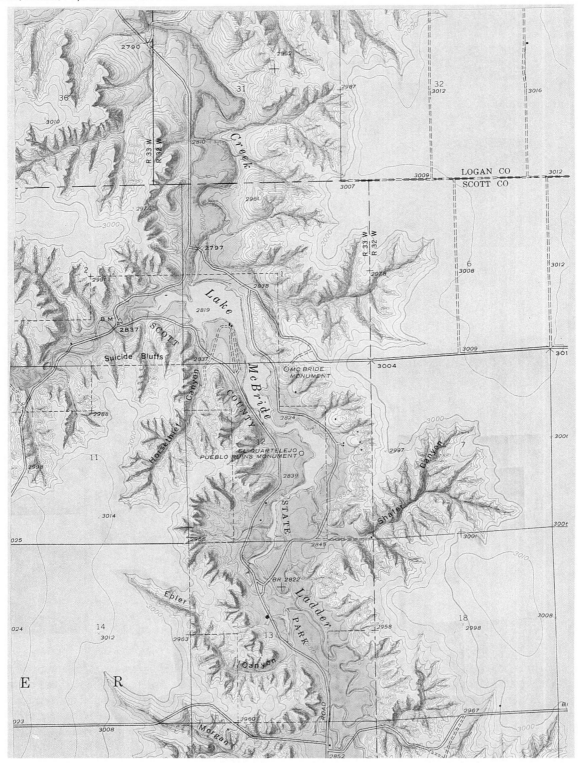

Figure 3.21
Physiographic diagram of Sri Lanka (Ceylon).
Harry Hoy.

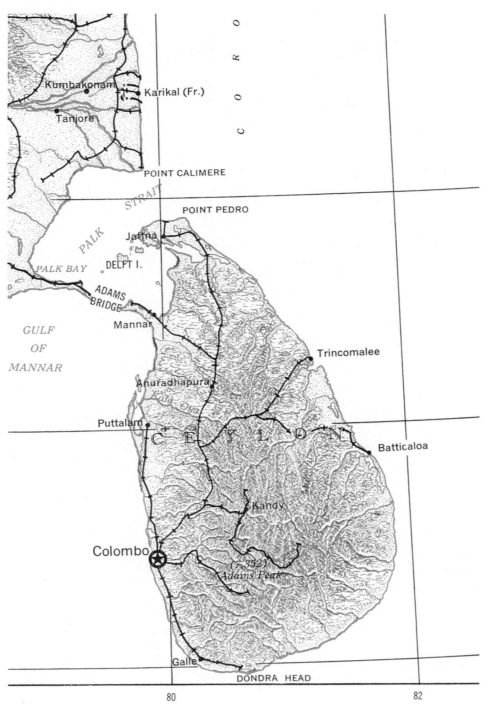

Figure 3.22
Population distribution in western Washington state for 1950. These volume spheres have reduced statistical data to graphic form.

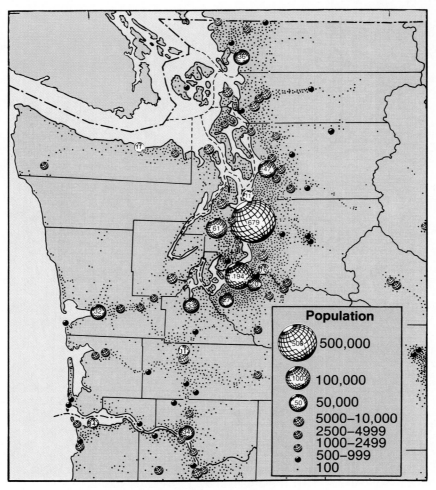

Other Geographic Tools

Numerous other types of geographic tools are used including charts of various kinds, cartograms, graphs, aerial photographs, remote sensing imagery, computers, and an array of survey data of various kinds.

Climatic Graphs or Charts

One of the most common types of graph is the climatic graph that typically shows annual precipitation and temperatures by monthly means as a series of bars and a curve on a rectangular graph. It is also possible to use a circular graph where the precipitation bars radiate out from a central hub, and the temperature curve is continuous in an ovoid form.

Transect Charts

Transect charts are designed to show surface profiles and a large array of other kinds of information. To illustrate, in addition to depicting terrain features such as mountains and valleys, such charts might also show other important aspects of the physical and cultural environment in each segment of the chart. Vegetation patterns, crops, and soil characteristics, as well as roads, houses, and other cultural features could be shown at the surface of the profile.

Cartograms

Cartograms are specialized maps that reduce statistical data to graphic form. For example, area circles or volume spheres might be used on a map to show varying intensity of manufacturing, extraction of minerals, production of crops, and any other human activity (figure 3.22). Cartograms might also show areas drawn to represent the size or dimension of the element considered (e.g., states drawn to the size indicating their population).

Aerial Photographs

Aerial photographs are widely used for map construction and as a direct map base for plotting data (figure 3.23). Surveying and mapping based on aerial photographs is known as **photogrammetry.** Aerial photographs are especially well adapted for revealing patterns not observable from the ground. The quality of such photographs has improved steadily over the years as there have been significant advances in cameras, film, and processing.

The scale of an aerial photograph is determined by the focal length of the camera and the altitude of the plane. Scale equals focal length divided by altitude. For example, a camera with a focal length of six inches flying at an altitude of 10,000 feet would have a scale of 1:20,000 (½ ÷ 10,000).

A stereoscopic effect may be obtained by use of a binocular instrument and overlapping parts (usually overlapping by 60 percent) of vertical aerial photographs. The three-dimensional effect obtained permits accurate interpretation of landscape features. In addition, when a **stereoplotter** is used, it is possible to produce contours directly from the photographs. In actual practice, most contour maps are now produced by using a small number of accurately surveyed ground control points and stereographic photographic coverage along with a stereoplotter.

Oblique photographs may increase shadowing and enhance relief interpretation. Obliques are especially useful for acquiring some data where overflights are not permitted. High flying planes near the borders of prospectively hostile nations are able to obtain substantial data with oblique photographs.

Other Remote Sensing Techniques and Tools

In addition to aerial photographs in black and white, color film and color infrared film have been employed to expand contrasts and to reveal certain aspects of the physical or cultural environment. Other wavelengths in the electromagnetic spectrum have been used by aircraft and satellites to probe various aspects of terrain or earth's features. These images, often computer-enhanced, are used to map and interpret a broad array of physical and cultural features at the earth's surface. With appropriate ground checks, it is possible to school computers in the recognition of certain electromagnetic wave-band signatures to produce maps of various phenomena. Better quality images and more sophisticated computer programs hold out the prospect for even greater accuracy and work reduction in the production of quality maps.

Radar is used to penetrate the veil of atmospheric moisture and is quite valuable in providing images for terrain analysis (figure 3.24). Previously unsuspected physical or cultural patterns are frequently revealed by radar. Side-looking radar is valuable in producing images that assist in terrain analysis. Radar is also very useful in storm analysis and weather prediction. **Sonar** is used to probe the ocean depths and provide a more realistic picture of the nature of the ocean floor.

Figure 3.23
Aerial photographs reveal aspects of both the physical and cultural environment.
USDA.

Figure 3.24
Radar imagery has added a new dimension to the collection of geographic data. This image was taken from 40,000 feet through total cloud cover in turbulent air.
Courtesy of Aero Service Division, Western Atlas International, Inc.

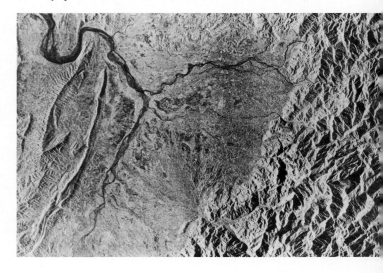

Figure 3.25
Color infrared photography has added a new dimension to geographic interpretation.
NASA.

Figure 3.26
Satellite photograph of weather conditions over a significant portion of the Northern Hemisphere.
NOAA.

Satellite imagery is an invaluable tool in analysis of landscape, in probing for minerals and other economic resources, and in providing very useful information in weather prediction and analysis (figures 3.25 and 3.26). Indeed, the failure of the United States to launch a number of satellites subsequent to the Challenger tragedy in early 1986 handicapped our study of the environment of the earth and the solar system, while we have been somewhat less secure militarily because of the reduction of surveillance satellites. The successful acquisition of geosynchronous orbit of another weather satellite in March 1987, greatly enhanced our ability to predict weather on a global basis.

Virtually everyone is familiar with radar imagery of weather phenomena on the TV news. Better radar and more effective interpretation add materially to our understanding of weather phenomena in both scientific and lay circles. Accurate determinations of areas of turbulence using radar reduces dangers and discomfort in flying during inclement weather.

Geographic tools are numerous and are constantly expanding in number and sophistication. The brief discussion presented in the preceding sections is not intended to be a comprehensive list. To the contrary, it represents the briefest kind of analysis consistent with the philosophy of this book—that is, to provide essential information to be supplemented by additional reading and study.

Whether the instrument is as simple as a block diagram or a rock pick, or as complex as a stress-strain gauge or a chromatographic analysis of atmospheric pollutants, it is important to collect data from a variety of sources to provide a meaningful synthesis of the earth's environments. The analysis and synthesis of data, however acquired, is the critical element in understanding earth environments; but abundant data acquisition and sophisticated devices are no substitute for the analysis of astute observers and mature scientists. Such scientists approach tasks of analysis and synthesis in rational, orderly ways. A brief description of some of those approaches and techniques follows.

Geographic Techniques

The geographer is first and foremost a keen observer of the physical and cultural habitat in which he or she lives. The field is his or her laboratory. By using direct observations, interviews, maps, remotely sensed materials, and appropriate data sources, the geographer observes, makes reasonable deductions, and accurately reports them after they have been thoroughly tested. These observations and deductions reveal many different kinds or degrees of areal relationships between physical and cultural phenomena. Electronic computers assist in the storage, manipulation, and interpretation of data.

Since geographers are vitally concerned with areal patterns, much of their time is spent accurately mapping and assessing distributions of phenomena. These maps are often still derived from data gathered in the field, or they may be compiled from a staggering array of statistical information, remote sensing imagery, or reports from other scientists. The evolution and analysis of patterns, connections, linkages, flows, and relationships permits the geographer to develop theories concerning the nature and ordering of events, to provide the basis for rational planning of most facets of human existence, and to develop a sense of synthesis that brings order to a sometimes seemingly disordered landscape. Even apparently chaotic situations may, however, reveal discrete patterns when subjected to sophisticated mathematical analysis.

An essentially new field of chaos investigations holds the prospect of producing order out of seeming disorder. If the promise of chaos investigations is fulfilled, interesting new truths about such matters as turbulent flow in water or atmosphere, or biosphere–lithosphere interconnections may be revealed.

Ultimately, the geographer uses the techniques of analysis to produce synthesis. Geographers assemble seemingly disconnected bits of information gathered from the field or library, and organize these pieces into a meaningful mosaic of patterns. The computer has simplified the task enormously by permitting the assembly, statistical testing, and array of enormous quantities of data. The geographer's task, made simpler by modern technology and equipment, is to present and interpret an integrated view of significant interrelationships between land, air, water, biota, and humankind's cultural environment in space over the earth's surface. Perhaps the geographer's greatest contribution is the development of a holistic view of environmental associations.

Sophisticated mathematical and statistical modeling and testing have added reliability to geographic theory development, and an array of computer programs has simplified the task of dealing with massive accumulations of data. Both inductive and deductive reasoning is used in theory building and theory testing. In using these types of reasoning, geographers are like all scientists. Since human beings are involved in the geographic landscape, however, predictions are more difficult, and reproducibility of a given phenomenon is often impractical or impossible because human behavior is unpredictable. Even in the most sophisticated realms of physical geography, the events and features do not always conform with the available data. In many realms, theories have resisted testing because of an enormous time frame or because of the difficulty of introducing essential controls.

With meteorological phenomena, constant repetitions of essentially analogous situations provide ready and frequent tests of developed theory. Vegetation patterns provide less frequent, but still obvious opportunities for theory testing. On the other hand, soil genesis and geomorphological developments occur over a longer time frame, making the testing of prevailing theory more inferential. Simulations using computer models and laboratory analogues are helping to transcend limitations imposed by time and enormous scales.

Tools, techniques, procedures, and theories all combine to help the seasoned observer to understand the rationality or order in the landscape. Certain secrets of nature will continue to resist proper analysis well into the future. People will discover new scientific associations and connections, in part to satisfy an innate curiosity shared by members of the species *Homo sapiens* and in part to help bend environments to the human will.

Study Questions

1. List the map essentials. Which one is the most often deleted?
2. What is the area of a map that is 12 × 12 inches and has a scale of 1:63,360? Would a map with the same dimensions and a scale of 1:250,000 cover a larger or smaller area?
3. Why is the Mercator of particular value in navigation? What are the major shortcomings of that particular projection?
4. Explain why the bottom of a map isn't always south.
5. What kind of map would have top, bottom, and sides all being south?
6. What map projection would you choose to accomplish the following objectives?
   a. plotting a great circle as a straight line.
   b. depicting the distribution of world population.
   c. a map of the United States.
   d. an equal area map.
   e. a map having true compass directions.
7. What is a hypsometric map? What are such maps often used for?
8. Why is a great circle the shortest distance between two points on the earth's surface?
9. What uses do geographers make of side-looking radar?
10. Why are hachure maps used less and less frequently?
11. Why are earth satellites so important in modern weather prediction?
12. Recount the ways you have used maps of any sort in the past six months.

Selected References

Doerr, A. H., and Guernsey, J. L. 1976. *Principles of physical geography.* 2d ed. Woodbury: Barron's Educational Series.

Espenshade, E. B., Jr., Ed. 1986. *Goode's world atlas.* 17th ed. Chicago: Rand McNally and Company.

Gabler, R. E.; Sager, R. J.; Brazier, S. M.; and Wise, D. L. 1987. *Essentials of physical geography.* 3d ed. Philadelphia: Saunders College Publishing.

Greenhood, D. 1964. *Mapping.* Chicago: University of Chicago Press.

Raisz, E. 1962. *Principles of cartography.* New York: McGraw-Hill Book Company.

Robinson, A. H.; Sale, R. D.; and Morrison, J. L. 1978. *Elements of cartography.* 4th ed. New York: John Wiley and Sons.

Strahler, A. N., and Strahler, A. H. 1987. *Modern physical geography.* New York: John Wiley and Sons.

4

Elements of the Weather

Hurricane Frederic nears landfall along the
Alabama Gulf Coast—1979.
NOAA.

*There is a sumptuous
variety about the New
England weather that
compels the stranger's
admiration—and regret.
The weather is always
doing something there;
always attending strictly
to business; always
getting up new designs
and trying them on
people to see how they
will go. But it gets
through more business in
Spring than in any other
season. I have counted
one hundred and thirty-
six different kinds of
weather inside of twenty-
four hours.*

Mark Twain, *Speech to New England
Society of New York*

*V*irtually all human activities are affected by the **weather,** which may be described as the condition of the atmosphere at a given time. A study of weather conditions is a basic necessity for the understanding of the geography of an area. The omnipresence of weather, the rapidity of atmospheric change in most areas, and the pervasive influences on the physical environment ensure people's interest in and concern for weather elements. Destructive storms, high winds, heat waves, cold snaps, sunshine and cloudiness, rain or the absence of rain, snow, hail, thunder, and lightning are but a few of the observable weather phenomena that occur in the atmospheric envelope that surrounds the earth. A geographer must also consider other weather elements because of their influences on people and because of their value in weather prediction. Such elements as pressure, precipitation, temperature, wind, humidity, and clouds prove to be invaluable in predicting weather sequences and occurrences. More sophisticated atmospheric models, a closer net of weather stations, availability of satellite information, and automated procedures improve general forecast accuracy, but it is still fair to say that weather prediction is an art based on a science. Like diagnosis and prognosis in medicine, results are not always predictable either in the course of an illness or in the weather forecast. Certainly, everyone is familiar with forecasts that have gone badly awry, and most forecasts are now hedged with probability predictions. The chances of rain occurring, for example, are expressed in percentage probabilities (e.g., there's a 60 percent chance of rain today). Only rarely does the meteorologist of the middle latitudes put his or her reputation on the line and declare that there's a 100 percent chance of rain or any other weather phenomenon.

The mixture of atmospheric gases surrounding our planet swirl together in a turbulent flow, generated by solar radiation and the dynamics of angular momentum resulting from the rotation of the earth, to create an ever-changing array of weather events. Few elements of the physical environment change so rapidly and affect people and their works so dramatically. Doubtless, weather is the universal topic of conversation in every quarter of the globe. The obvious impact of weather elements on human activities may mask more subtle effects on chronic diseases, like arthritis, or a variety of respiratory diseases.

The folk wisdom about weather in many parts of the United States and the world hinges on change and unpredictability. The expression, "If you don't like the weather, wait a minute," is common in many sections of the country. This simply reflects the fact that weather created by a dynamic atmosphere is, in fact, a constantly changing element in the human experience, especially in the middle latitudes.

In addition to its production of weather, the earth's gaseous envelope supplies the oxygen that we breathe and serves as a reservoir for the products of animal respiration that are used by plants. Plants, which use the CO_2 of animal respiration to carry on photosynthesis, provide oxygen as a by-product of that process. The symbiotic relationship of plants and animals tends to keep oxygen and carbon dioxide in reasonable balance, although the increased burning of the hydrocarbon fuels is adding to the CO_2 burden in the atmosphere. The shifting percentages of certain atmospheric elements, notably water vapor, influence weather developments. An analysis of atmospheric constituents and zonal characteristics follows.

The Atmosphere

The **atmosphere** is a mixture of several gases. There are about ten chemical elements that remain permanently in gaseous form in the atmosphere under all natural conditions. Nitrogen makes up about 78 percent and oxygen accounts for about 21 percent of these gases. Several other gases comprise the remaining one percent of dry air: argon, carbon dioxide, hydrogen, krypton, neon, and xenon. The amount of water vapor in the air varies from near zero to almost 5 percent by volume. This small amount of water vapor, and its variations in quantity and distribution, is of extraordinary importance in weather conditions and changes.

The atmosphere also holds in suspension great quantities of dust, pollen, mold spores, smoke, salt spray, and a variety of other substances that are always present in varying amounts. Quantities of particulate matter are expanding steadily as billions of people add to the natural atmospheric burden through an array of activities: clearing and cultivating land, power-generation, transportation, and industrial activities. It is estimated that in excess of 800,000 tons (725,760,000 kilograms) of dust falls to the earth each day. Of that amount, more than 200,000 tons (181,440,000 kilograms) are generated by human activity. That particulate burden has generally diminished long-range visibility over most sections of the globe and has added to the respiratory difficulties of those with chronic lung disease.

The atmosphere has no definite upper limits, but gradually thins until it becomes essentially imperceptible. In space, near-vacuum conditions exist, although an absolute vacuum is apparently nonexistent in nature. Until relatively recently, it was assumed that the air above the first few miles gradually grew thinner and colder at a constant rate. It was also assumed that the upper air had little or no influence on weather conditions near the earth's surface. Increased observation of atmospheric conditions at various levels using more sophisticated equipment, satellite observations, missile probes, and manned orbiting observation vehicles has increased our understanding of the earth's atmosphere. It is characterized by several individual layers, or zones, with quite definite physical characteristics.

Troposphere

The layer of the air next to the earth, which extends upward 6 to 10 miles (9.7 to 16.1 kilometers), is known as the **troposphere.** On the whole, it makes up about 75 percent of all of the weight of the atmosphere. It is the warmest part of the atmospheric envelope because most solar radiation is absorbed by the earth's surface, which then heats the atmosphere adjacent to it. A steady decrease of temperature with increasing altitude is a distinguishing characteristic of the troposphere. This decrease in temperature with increasing altitude is known as the **standard lapse rate** or **normal lapse rate** and amounts to approximately 3.5° F/1,000 feet (6.5° C/kilometer).

Within the troposphere, winds and air currents distribute heat and moisture from place to place. The active weather that most affects people occurs within this zone. It is this cauldron of ever-changing mixtures, pressures, and fluxes that causes the day-to-day weather changes that influence human activities and, at the same time, affect all biological and geomorphological activity.

Strong winds, known as **jet streams,** discovered during World War II, are located near the top of the troposphere. It seems clear that they are involved in heat transfer from higher to lower latitudes, and it has been observed that variations in their position are responsible for the movement of storm tracks and the weather associated with them. Jet streams may also be responsible for the invasion of very cold or very warm air over the middle latitudes leading to cold waves and heat waves. The oscillatory wave-like motions of the jet streams are regularly observed, but the cause for the shifts is uncertain, and the long-term prediction of those shifts is a very imperfect art.

Certain patterns of the jet stream are indicative of prevailing weather conditions, and a shift in position provides essential clues in weather prediction. In North America, the Omega Ω form, for example, which shows a marked northward bulge of the jet stream, is characteristic of heat wave conditions in the northern United States and southern Canada in the summertime. Conversely, a marked dip, or bend, to the south in the winter, a kind of inverted Omega form, is a precursor to a cold wave. These northward or southward bulges of the normal east-west trending jet streams are precursors to invasions of warm or cold air masses.

At the upper reaches of the troposphere is a transition zone to the stratosphere called the **tropopause.** Within the tropopause, which extends from about 10 to 20 miles (16 to 32 kilometers) above the earth's surface, the temperature remains fairly constant at about −70° F (−56.6° C).

Stratosphere

The **stratosphere** extends from the upper reaches of the tropopause to approximately 35 miles (56 kilometers). The lower portions of the stratosphere contain a layer of ozone that filters out a high percentage of the ultraviolet rays of the sun. Without this filtering effect of the ozone, the sun would burn our skin, blind our eyes, and eventually result in the destruction of the human species. Concern has been expressed in the late 1980s that a large hole in the ozone layer over Antarctica may be enlarging. If this is so, and if there is no mechanism to reestablish the layer, life in the area may be at considerable hazard. There is some disquieting evidence, too, that the ozone layer is thinning elsewhere especially above the Arctic. If this thinning continues, it is reasonable to assume that the rate for skin cancer will increase because harmful solar rays will be less effectively screened. Within the stratosphere, temperature and atmospheric conditions are relatively uniform.

From approximately 35 to 50 miles (56 to 80 kilometers) above the earth's surface is a zone call the **stratopause.** Temperatures there are about the same as those at the earth's surface, but the conductivity of temperature is reduced dramatically because the air is so thin.

Mesophere and Thermosphere

From about 50 to 200 miles (80 to 322 kilometers) above the earth are two zones called the mesophere and the thermosphere separated by the **mesopause.** In the **mesosphere,** there is a reduction of temperature with increased elevation; in the mesopause, temperatures are relatively constant; and in the **thermosphere,** the temperatures increase to almost 2,000° F (almost 1,100° C) at noon, but the air is so thin that conductivity is minimal.

These zones were formerly known as the **ionosphere** because of the ionization of molecules within the layer. The **northern lights (aurora borealis)** originate within this highly charged portion of the atmosphere. The Southern Hemisphere experiences a comparable display known as the **aurora australis.**

The effects of these far upper reaches of the atmosphere on the earth's weather, if any, are largely unknown. Additional research continues, and these zones of the atmosphere are regularly probed by both manned and unmanned satellites. As space travel becomes more prevalent, it behooves us to be knowledgeable about the upper margins of the atmosphere at the near reaches of space. The hazards these zones may hold and the possibilities they afford need to be thoroughly understood as people venture increasingly into proximate realms of space and prepare for more distant voyages in the years ahead.

The various zones of the atmosphere possess distinctive characteristics. Nevertheless, the weather conditions that affect us are largely restricted to the troposphere. The various elements of weather that affect us most will be considered in the following sections.

Weather Elements

Several weather elements combine to produce a measure of weather conditions at a given time and place. These principal elements include temperature, humidity, precipitation, pressure, winds, air masses, fronts, clouds, and storms.

Temperature

Of all the weather elements, probably the one most readily observed by people is **temperature.** The general distribution of temperature is based mainly on the amount of solar energy received. Solar energy variations relate primarily to the angle at which the sun's rays strike the earth's surface, the length of day (figure 4.1), and the amount of reflectivity from clouds and other atmospheric constituents, termed the **albedo.**

The sun radiates energy into space at a fairly constant rate. Although it is a slightly variable star, the amount of variation in radiant energy emitted is considered to be negligible in terms of influence on the heating of the earth. The **solar constant** amounts to 1.94 calories per square centimeter per minute—a unit of measure known as a **langley.** Slightly less heat is received per surface unit when the earth is at the greatest distance from the sun. The angle at which the sun's rays strike the earth is, however, of far greater significance in the heating of the earth's surface than the distance of the earth from the sun, since the concentration or dispersion of rays concentrates or diffuses solar heating resulting in effective or ineffective heating of the surface (figure 4.2).

The highest average temperatures are found in low latitudes, since the sun's rays are most nearly vertical in those latitudes for most of the year. Conversely, temperatures diminish as one approaches the poles, since the sun's rays are farther from vertical in those higher latitudes. Seasons reflect the incidence of solar radiation received (i.e., periods of high sun are warmer, whereas times of low sun are cooler). The length of the summer daylight in middle and high latitudes also exceeds the darkness, so more heat is received typically during the day than is reradiated to space in periods of darkness. This results in a gradual warming in the summer season. The opposite situation occurs in the winter in similar latitudes when a low-angle sun heats the earth inefficiently, and lengthened periods of darkness result in progressive cooling.

There is some lag in heating and cooling the earth during the seasons, so July, which comes after the summer solstice in the Northern Hemisphere, is usually the hottest month, whereas January, which follows the winter solstice, typically is the coldest month. The **temperature lag** is a common characteristic of middle- and high-latitude climates. There is a similar lag of cold and warm seasons in the Southern Hemisphere.

A comparable temperature lag occurs on a daily basis so the warmest temperature is experienced about three o'clock in the afternoon. Often the coolest time of the night is just before dawn.

Figure 4.1
The effect of the earth's tilt on length of day. Away from the equator, the days lengthen until the summer solstice and become shorter from the summer solstice until the succeeding winter solstice.

From Arthur Getis, Judith Getis, and Jerome Fellmann, Introduction to Geography, *2d ed. Copyright © 1988 Wm. C. Brown Publishers, Dubuque, Iowa. All Rights Reserved. Reprinted by permission.*

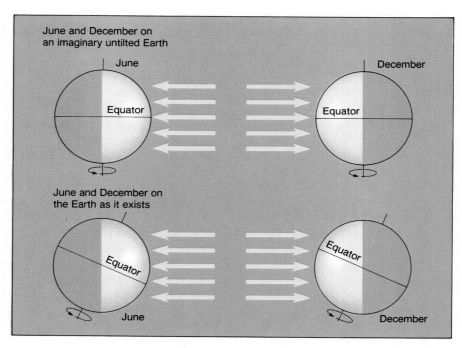

Atmospheric Heating

Radiant energy from the sun is affected not only by the angle at which the sun's rays strike the earth and the length of day, but also by the condition and character of the earth's atmosphere. As the sun's rays pass through the atmosphere, about 27 percent of the energy is reflected back into space by clouds and the earth's surface; approximately 5 percent is scattered by minute particles in the air and is returned to space as diffuse radiation; about 20 percent reaches the earth as diffuse radiation after having been scattered; almost 30 percent reaches the earth's surface as direct radiation; and 18 percent is absorbed by the ozone layer, dust, and clouds (figure 4.3). These figures represent approximate averages for the whole earth and may vary considerably from place to place and time to time.

Table 4.1
Influence of Sun Angle on Insolation

Sun's altitude	0	5	10	20	30	50	70	90
Relative thickness of atmosphere in units	35.5	10.2	5.6	2.9	2.0	1.3	1.1	1.0
Proportion of solar radiation reaching earth's surface	0.00	0.05	0.20	0.43	0.56	0.69	0.74	0.75

Figure 4.2
The varying effectiveness of the sun's rays as heating agents under equinox conditions. Sun's rays close to the equator are concentrated, whereas they become more diffuse at higher latitudes.
From Arthur Getis, Judith Getis, and Jerome Fellmann, Introduction to Geography, *2d ed. Copyright © 1988 Wm. C. Brown Publishers, Dubuque, Iowa. All Rights Reserved. Reprinted by permission.*

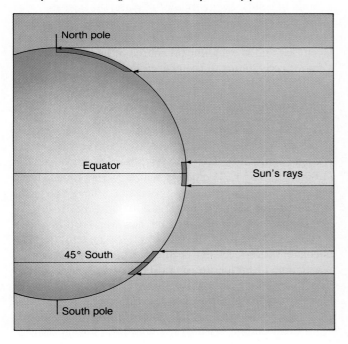

Figure 4.3
The average effect of various reflecting and absorbing agencies upon receipt of insolation at the earth's surface. The fifty percent, or thereabouts, of energy received at the surface is eventually released to the atmosphere and reradiated into space.
From Arthur Getis, Judith Getis, and Jerome Fellmann, Introduction to Geography, *2d ed. Copyright © 1988 Wm. C. Brown Publishers, Dubuque, Iowa. All Rights Reserved. Reprinted by permission.*

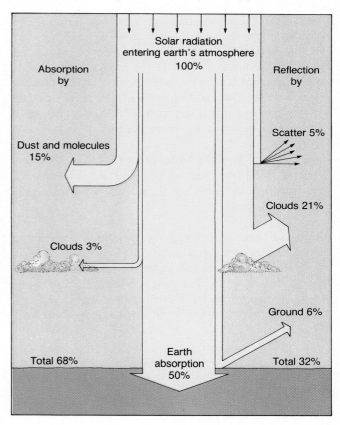

Obviously the energy budget is in a constant state of balance, which means that about 47 to 50 percent of solar insolation is returned to the atmosphere by a number of processes. These processes include radiation, convection, conduction, advection, and change of state of atmospheric moisture. Much of the warming of the earth's atmosphere is accomplished by heat reradiating from earth back to space.

Radiation   The radiant energy from the sun in the form of short-wave energy is reradiated as long-wave energy by a warmed earth to heat the lower portion of the atmosphere. Bodies that are warmer than their surroundings radiate heat energy to the cooler areas around them. Radiators in a home heating system radiate heat to a room because they are made warmer than their surroundings by heat energy derived from burning some type of fuel. The warmed earth, in a similar way, warms the surrounding atmosphere. Heat flows from warmer bodies to cooler surroundings through **radiation.**

Convection   Air near the surface of the earth that has been heated tends to rise, since it is less dense than cooler air that surrounds it. As it rises, cooler air moves in from each side to replace the air that has risen. The vertical movement of the warm air transfers heat vertically in a process known as **convection.** It is essentially analogous to the circulation established within a boiling kettle of water (i.e., warmed water bubbles up (boils) because of the addition of heat to the kettle).

Conduction   Heat flows from warmer to cooler parts of a body. Molecules that come in contact with each other have this flow of energy from warmer to cooler areas. Just as heat flows from a poker left in the fire to the handle, so does heat move between two bodies in contact. **Conduction** warms only the lower portion of the atmosphere in contact with a warm earth. It is responsible for the penetration of heat to some depth in the soil or the earth's rocky crust.

Advection   The lateral transfer of heat is called **advection.** Winds blowing from warmer or cooler areas modify the temperature of the earth's surface or air with which they come in contact.

Change of State of Atmospheric Moisture   A portion of the solar energy that strikes the earth's surface is used in evaporating water from the oceans, seas, rivers, lakes, and other bodies of water. The ranges of temperature in the atmosphere permit water to exist as a gas, liquid, or solid. As water changes from the liquid to the gaseous form, heat

Figure 4.4
Temperature lapse rate under normal conditions.
From Arthur Getis, Judith Getis, and Jerome Fellmann, Introduction to Geography, *2d ed. Copyright © 1988 Wm. C. Brown Publishers, Dubuque, Iowa. All Rights Reserved. Reprinted by permission.*

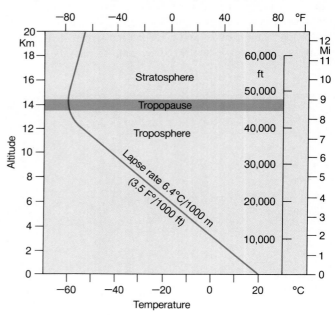

energy is taken up in the **latent heat of vaporization.** Subsequently, as this moist air is lifted and cooled, condensation occurs (i.e., the gaseous water vapor is changed to a liquid or solid). In this **change of state,** the heat that was taken up in the evaporation process is released. This **latent heat of condensation** modifies the temperature of the adjacent air and is a significant source of heat transfer and a major contributor to the energy in storms.

Temperature Distribution in the Atmosphere

As previously indicated, temperature in the lower reaches of the atmosphere normally decreases with increasing elevation (figure 4.4). This normal lapse rate, about 3.5° F /1,000 feet (6.5° C/kilometer), will vary from time to time and place to place. It does represent an average temperature gradient within a parcel of air that is not rising.

A parcel of air that is lifted, however, cools at a rate of about 5.5° F per 1,000 feet (1° C/100 meters), so long as condensation does not occur. The **adiabatic lapse rate** is a change of temperature that occurs as a parcel of air is rising or sinking without significant gain or loss of energy. This **dry adiabatic lapse rate** is greater than the normal lapse rate in average circumstances, since the air is expanding and doing work as it rises, thus causing cooling at a more rapid rate. Air that is forced downward is compressed and heats at the same **dry adiabatic lapse rate.**

Figure 4.5
General pattern of isotherms for January.
From Arthur Getis, Judith Getis, and Jerome Fellmann, Introduction
to Geography, *2d ed. Copyright © 1988 Wm. C. Brown Publishers,
Dubuque, Iowa. All Rights Reserved. Reprinted by permission.*

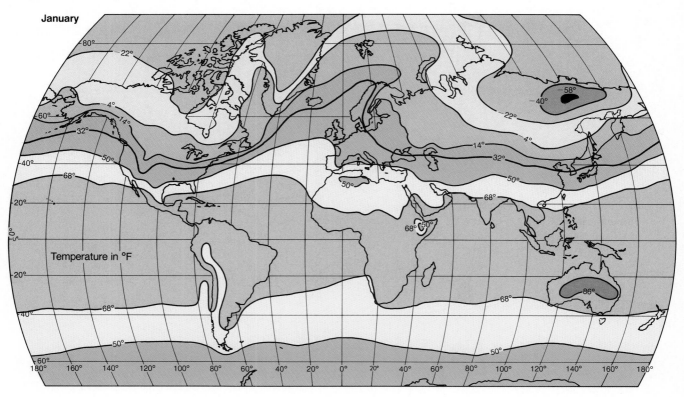

The **wet adiabatic lapse rate** is less than the dry, since, although cooling continues to occur, the addition of the **latent heat of condensation** slows the overall cooling rate. The wet rate varies, but typically approximates the standard lapse rate.

The effects of insolation and the distribution of heat over the surface of the earth result in a pattern that tends to create lower temperatures in high latitudes than in low latitudes and tends to show greater extremes over land areas than over oceans or seas. Temperature distributions over the earth's surface are shown by **isotherms** (lines connecting points with the same temperature) whose values have been reduced to sea level, so the effects of altitude on temperature have been eliminated. The general trend of isotherms for average temperatures is similar to that for parallels of latitude. Higher temperatures are found at lower latitudes, and lower temperatures are characteristic of higher latitudes. Latitude has the greatest single effect upon temperatures, although a number of other factors modify the temperature distribution patterns.

In January, the highest temperatures occur over the Southern Hemisphere continents. Conversely, the lowest temperatures occur over the Northern Hemisphere continents. In January, isotherms bend equatorward over the Northern Hemisphere continents and poleward over the Northern Hemisphere oceans, reflecting the moderating effect of water (figure 4.5).

During July, the higher temperatures are over the land areas of the Northern Hemisphere. Those isotherms bend northward over Northern Hemisphere continents and southward over adjacent oceans and seas, again demonstrating that greater extremes in temperature are experienced over land (figure 4.6).

These seasonal bendings of the isotherms over land and adjacent water bodies reflect the fact that land heats and cools more rapidly than water. Land masses experience a greater range in temperature from season to season than do large water bodies in comparable latitudes.

Figure 4.6
General patterns of isotherms for July.
From Arthur Getis, Judith Getis, and Jerome Fellmann, Introduction
to Geography, *2d ed. Copyright © 1988 Wm. C. Brown Publishers,*
Dubuque, Iowa. All Rights Reserved. Reprinted by permission.

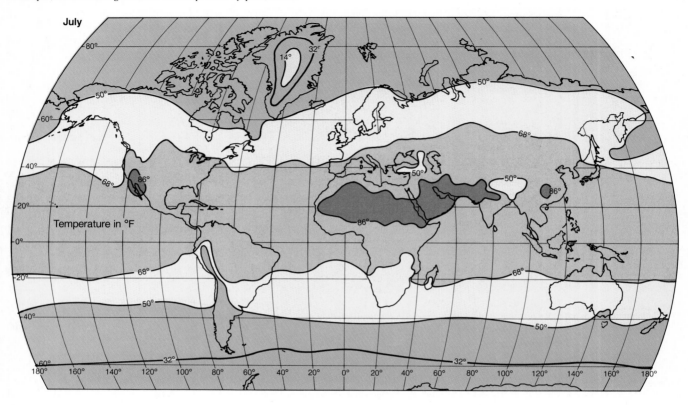

Measurement of Temperature

The temperature of the atmosphere is measured by an array of instruments called **thermometers.** One of the most commonly used thermometers is the **mercurial thermometer,** which is a glass tube with a bulb that contains a quantity of mercury. As temperatures increase, the column of mercury expands within the graduated tube, and temperature is determined by how high the column rises or how low it falls. Alcohol is also commonly found as the liquid within the tube and must be used if very low temperatures are anticipated, since mercury freezes at very cold temperatures. Commonly used thermometers establish reference points by using the freezing and boiling temperature of water at standard sea level pressure.

A **maximum thermometer** is generally a mercury thermometer with a very narrow constriction in the tube immediately above the bulb. As temperatures increase, the mercury column rises to its maximum height; but if cooling then occurs, the contraction causes a break in the column at the narrow constriction leaving the mercury column at the highest level attained. A common mercury thermometer is used to determine body temperatures. The user shakes the thermometer after use to reestablish the connection in the mercury column.

A **minimum thermometer** measures temperature by a small metallic marker that rides the surface of the liquid down to the lowest level attained in a given period. When temperature increases, the rising column of liquid cannot carry the index marker upward against the force of gravity. The index marker is set by a small magnet to permit minimum readings during another time cycle.

Temperature may also be measured by a device that uses the differential rates of expansion of two dissimilar metals. The expansion and contraction causes a curling effect when the two metals are connected. If the metallic strips are attached to a pointer that has a clockwork mechanism that turns a drum where temperatures may be recorded over a period of time, the device is known as a **thermograph.** These devices are especially useful in demonstrating the march (change) of temperature from one time of day until another and from one day to the next.

A number of scales are used to measure temperature. The most commonly used in meteorological and climatological studies are the **Fahrenheit scale,** in which the freezing temperature of water is 32° and the boiling temperature of water is 212° at sea level; and the **Celsius** (also known as **centigrade scale**), where the freezing temperature is 0° and the boiling temperature for water at sea level is 100°. Both temperatures are used throughout this book, since the Fahrenheit scale is commonly used in the United States and is most familiar to beginning students. Most of the non-English speaking world and the scientific community use the Celsius scale. Conversion from one temperature scale to another is important and may be accomplished by using the following formulas: $C = 5/9(F-32)$ and $F = 9/5C+32$. Students should become adept at making these conversions, and attempts should be made to recognize approximate realms of comfort or discomfort on the Celsius as well as Fahrenheit scales.

The Kelvin scale is used when temperatures approach absolute zero (i.e., $-406°$ F or $-273°$ C). Although such a scale has broad applications in low temperature physics, it is not used in normal meteorological study. Because electrical conductivity is greatly enhanced by very low temperatures, considerable practical effort is being expended to achieve and maintain temperatures close to absolute zero. Recently, however, superconductivity at substantially higher temperatures than was formerly possible has been achieved by using new rare-earth alloys. The ultimate benefits in the conservation of energy are certainly worth the quest.

Moisture

Moisture exists everywhere in the atmosphere as invisible water vapor or in the visible form as droplets of water or minuscule ice crystals. In the visible form, we see the water as fog or as clouds of differing types. The amount of water vapor in the air varies greatly from place to place and at different times. When air contains the maximum amount of water vapor possible at a given temperature, it is said to be **saturated.** Under certain circumstances, especially in air with little turbulence, air may become **supersaturated** (i.e., moisture continues to exist in the gaseous form after saturation temperatures have been reached). This is a transient phenomenon, but it does occur.

The moisture content of the air is referred to as **humidity.** The weight of water vapor in a given volume of air is **absolute humidity.** It might be expressed as grains/cubic foot, or as grams/cubic meter. The ratio between absolute humidity and saturation at a given temperature is called **relative humidity** and is expressed as a percentage of saturation. Indeed, relative humidity is most understood by the laymen and is reported regularly by the TV weather forecaster. **Specific humidity** is the weight of water vapor in a given weight of air. Meteorologists commonly use **vapor pressure,** which is the percentage of atmospheric pressure contributed by ambient water vapor (the amount of water vapor present in the air).

Warm air has a greater capacity for holding moisture than does cold air; hence, without the addition of moisture from some source, it is essential to cool air to cause condensation and precipitation. Cooling to condensation levels is most often accomplished by lifting a parcel

of air. If temperatures are lowered further when saturation occurs, moisture will change from the gaseous to liquid or solid form, producing fog or clouds. Under the right conditions, precipitation may then fall.

Devices used for the measurement of humidity include the **hygrometer,** which uses human hair as the activating element. As a bundle of human hair (blonde or light brown female Nordic hair works best) absorbs moisture, it tends to expand and lengthen. The expansion is measured through a series of gear linkages to a pointer, which shows relative humidity on a calibrated scale. Certain types of chemically impregnated paper may also be used as the activating mechanism in hygrometers. If such a device is connected to a drum being rotated by a clockwork mechanism, it is termed a **hygrograph.** As in the case of the thermograph for temperature, the hygrograph trace provides useful information about the humidity conditions throughout a protracted time period. A thermograph and a hygrograph may be coupled in the same device to produce a simultaneous trace of both temperature and humidity. Such an instrument is known as a **hygrothermograph.**

The **psychrometer** uses two identical thermometers, except that one has a tiny cloth sock placed over the bulb. The cloth sock is wetted, and evaporation causes the temperature to decline in the wet bulb thermometer, since evaporation of the moisture from the sock takes heat from the thermometer. Dry air causes rapid evaporation and a significant drop in wet bulb temperatures, whereas humid air means slow evaporation and lesser differences between the wet bulb and the dry bulb thermometers. A calibrated chart makes it possible to determine relative humidity by comparing the wet and dry bulb temperatures.

Condensation The **dew point** is the temperature of the air when it is saturated (i.e., the relative humidity is 100 percent). Further cooling of the air will normally cause condensation. The smaller the difference between prevailing air temperature and the dew point, the higher the relative humidity. **Condensation** occurs when water changes to a liquid or solid state. Technically, the change to a solid state is called **sublimation,** but the term "condensation" is usually used to refer to the change of state from gas to liquid or solid. Condensation occurs when air is cooled below the dew point in the presence of small particles in the air such as dust, pollen, salt, or similar nuclei called **hygroscopic nuclei,** or **condensation nuclei.** Air may be supercooled (i.e., chilled well below the dew point) before condensation actually occurs and, in some instances, water may exist in the same parcel of air in the gaseous, liquid, and solid forms. Often, too, in very still air, supercooling without condensation may be the norm. Turbulence may facilitate the condensation process, and, in almost every instance, it facilitates precipitation where condensation has already occurred. Turbulence increases the coalescence of water droplets by contact to larger sizes, which succumb to the force of gravity and fall as precipitation.

The source of atmospheric moisture is the great reservoirs of water on the earth's surface such as oceans, seas, rivers, lakes, and other bodies of water. Vegetation also gives off enormous quantities of moisture in a process called **transpiration. Evapotranspiration** is the term applied to evaporation and transpiration taken together. Moisture may also be released from sources deep within the earth's crust during volcanic activity.

Clearly, evaporation is most rapid when warm, dry air passes over a source of moisture. Cooler air, which holds less moisture, and very humid air, which also has a limited capacity for additional moisture, are less effective in evaporating moisture from existing sources. Warm or hot and dry air is most efficient in evaporating moisture. Wind also speeds the rate of evaporation, since it brings new air into contact with the moisture source, thus increasing the opportunity for further evaporation.

Moisture condenses in a variety of different circumstances, and the forms include dew, frost, fog, and clouds. **Dew** condenses on surfaces at night when cooling of the earth causes the air just above it to reach the dew point at above freezing temperatures. Moisture condenses on the surfaces that have been cooled below the dew point. Clear skies and relatively calm atmospheric conditions are usual prerequisites for dew formation. **Frost** forms under similar conditions when the dew point is below freezing. Fog is of two principal types, radiation fog and advection fog, although other special circumstances may also yield fog. **Radiation fog** occurs when, on a clear night, rapid radiation of heat from the earth's surface results in the lower levels of the atmosphere being cooled by contact with the cooler earth below the dew point. If enough hygroscopic nuclei exist and the air is relatively calm, the tiny droplets will condense into a fog. A **fog** is, essentially, *a cloud at very low elevations.* As the sun warms the earth in the morning, as well as the air immediately above it, the fog "burns off" (i.e., the moisture is vaporized and becomes invisible). Usually, such radiation fogs have cleared by late morning. They tend to be transitory events.

An **advection fog** occurs when a warm, moist air mass is cooled below the dew point as it moves in over a cold ground or sea surface. Such fogs are especially common when warm, moist gulf air moves in over the chilled earth of a midwestern American winter. Such fogs are also especially common at contact points between cold and warm ocean currents. The warm, moist air above the warm current is cooled to the dew point as it mixes with the cooler air over adjacent cold ocean water. Advection fogs tend to be more persistent and slower to disperse than radiation fogs.

Areas like the United Kingdom, which are bathed by the warm waters of the North Atlantic Drift, experience frequent and persistent fogs as warm, moist air from over the water moves across cooler land. Several west coast desert areas, especially those in the Southern Hemisphere are bordered by cold ocean currents. As onshore winds blow across these currents, they may be cooled in the basal,

or lower, portion by the cold currents, producing conditions of fog. Frequently, these fogs are very persistent. In some instances, such as at the margins of the Kalahari Desert in southwestern Africa, enough moisture may be condensed in contact with desert sands to support an intricate biological community.

Mist fogs occur when the moist air above lakes, swamps, or ponds is cooled as the result of radiation cooling during the night. These fogs may hang over the water body during late evening or early morning hours, but they are quickly dissipated by the heat of the sun.

Upslope fogs may develop as slowly rising air in hilly or mountainous areas is cooled adiabatically to and below the dew point. Such fogs may occupy lower mountain slopes while upper reaches of the same mountainous area are in bright sunshine.

Fogs are transitory events, which tend to dissipate as the air warms or as the wind velocity increases markedly. Where they do exist, fogs may be hazardous to navigation and the cause of vehicular accidents. Obviously, fogs are infrequent occurrences in desert interiors.

Clouds

Clouds, which are composed of tiny droplets of water or minuscule ice crystals are of major significance in weather analysis and prediction. They not only provide clues as to future weather, but, under the right circumstances, can yield precipitation, shade the earth, and affect the amount of insolation received. For study and analysis, clouds are divided into four families: high clouds, middle clouds, low clouds, and clouds with vertical development. **High clouds** are found at an elevation above 20,000 feet (6,000 meters), usually extending up to and slightly beyond 35,000 feet (10,600 meters); **middle clouds** extend between about 7,000 to 20,000 feet (about 2,000 to 6,000 meters); and **low clouds** are found at elevations below 7,000 feet (2,000 meters). Actually, there are scores of cloud types, but only a few type examples are included within the scope of this book. The types included are readily recognizable, and they may be used, in conjunction with other weather elements, in making semieducated predictions of the weather. Principal cloud designations and their characteristics are categorized by the U.S. Weather Bureau. These categories and their descriptions follow.

Cirrus **Cirrus** clouds are thin, wispy, featherlike clouds composed mainly of ice crystals (figure 4.7). They tend to diffuse sunlight or moonlight shining through and, occasionally, the ice crystals will act like miniature prisms producing a halo effect, especially as moonlight shines through. Under certain conditions, cirrus arranged in parallel bands mark the boundaries or margins of high altitude jet streams. Cirrus with frayed ends are sometimes described as mare's tails. When found alone, cirrus clouds tend to be harbingers of fair weather. When they thicken or are found in association with other clouds, they often portend significant weather change.

Figure 4.7
Cirrus clouds are high, thin, wispy, ice crystal clouds.
NOAA.

Figure 4.9
Cirrostratus clouds. Note the halo effect resulting from ice crystal refraction of light.
NOAA.

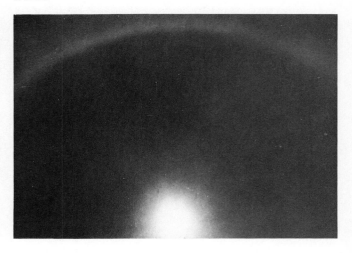

Figure 4.8
Cirrocumulus clouds.
NOAA.

Figure 4.10
Altocumulus clouds.
NOAA.

cirrostratus are especially characterized by the halo effect of moonlight shining through. These clouds often mean weather change. The halo effect is responsible for the folk wisdom, "Ring around the moon, rain coming soon."

The middle clouds range from about 7,000 to 20,000 feet (2,000 to 6,000 meters) in altitude. The principal types and their descriptive characteristics follow:

Altocumulus These are white or gray patches or layers of clouds with the cloud elements having a rounded appearance (figure 4.10). They produce the so-called mackerel or buttermilk sky described especially by seafarers, although some may also refer to cirrocumulus as a buttermilk sky. **Altocumulus,** as part of the cloud sequence in a middle latitude cyclone, often signal impending weather change.

Cirrocumulus **Cirrocumulus** are thin clouds, the individual elements of which appear as small white flakes or patches of cotton, usually showing a glistening quality, suggesting the presence of ice crystals (figure 4.8). The flake-like portions are of smaller dimension and higher altitude than altocumulus, which are somewhat similar in appearance. They often precede significant weather changes, since they frequently develop as a part of a frontal sequence of clouds.

Cirrostratus These clouds are thin, whitish cloud layers appearing like a sheet or veil; they may be diffused, or sometimes stratified or fibrous (figure 4.9). Thin layers of

Figure 4.13
Stratocumulus clouds. NOAA.

Figure 4.12
Stratus clouds. NOAA.

Altostratus These clouds form a dense gray veil or layer of clouds with a fibrous appearance. Such clouds frequently cover the entire sky. **Altostratus** are often precedent to frontal passage and usually suggest impending significant weather change.

Nimbostratus **Nimbostratus** clouds may be in the middle- or low-cloud groups. They are shapeless, thick dark gray cloud layers accompanied by rain or snow (figure 4.11). Indeed, the prefix nimbo or suffix nimbus when attached to cloud types indicates the presence of precipitation.

The low clouds range from near the surface to approximately 7,000 feet (2,000 meters). The principal types and their descriptive characteristics follow.

Stratus **Stratus** clouds are low, uniform, gray, sheet-like clouds that obscure the sun or any celestial bodies (figure 4.12). Their sullen, lead-gray color often adds to the gloom of middle and high-latitude humid climate winters. In such areas, they may be persistent for days or even weeks at a time.

Stratocumulus These clouds typically have large rolls, usually soft and gray, with darker shading (figure 4.13). The base of **stratocumulus** is essentially flat, whereas the tops have a rounded cauliflower-like appearance. The thickening of such a cloud deck usually foretells precipitation or weather change.

Clouds with vertical development extend from bases as low as 1,500 feet (about 500 meters) or as high as 10,000 feet (about 3,000 meters), sometimes to great heights, perhaps in excess of 50,000 feet (15,000 meters) in the tropics. Even in middle latitudes these clouds frequently soar to 35,000 feet (more than 10,000 meters) in height. Their vertical development reflects conditions of rising air and they are characterized by varying degrees of turbulence.

Cumulus **Cumulus** are dense dome-shaped clouds with flat bases, and white rounded cauliflower-like upper surfaces (figure 4.14). They are the clouds of warm summer afternoons with the constantly changing shapes, which cause any observer to see imaginary forms in the cloud appearances. They may exist as a few isolated clouds, or they may occupy a large part of the sky. When little vertical development is present, these clouds tend to be characteristic of fair weather with the promise of more to come.

Towering Cumulus Cumulus clouds may show great vertical development building into interesting intermediate forms in the process (figure 4.15). A cloud with the essentially flat base that has grown to towering heights may be called **cumulus castellatus;** subsequent growth may produce **cumulus congestus;** and great vertical development may yield the thunderstorm cloud **cumulonimbus.** The whitish color may change to gray or dark

Figure 4.14
Cumulus clouds.
NOAA.

Figure 4.16
Cumulonimbus clouds with precipitation.
NOAA.

Figure 4.15
Advancing squall line with towering cumulus and cumulonimbus clouds.
NOAA.

gray as precipitation falls or shadows become prevalent in various segments of the clouds (figure 4.16). Very large cumulonimbus clouds with a great deal of turbulence or overturning at high elevations may produce pendulous, bulbous, drooping forms called **cumulomammatus.** Thunderstorm clouds (cumulonimbus) may extend to elevations of 50,000 feet (15,000 meters) or even higher. The tops of such enormous anvil-topped clouds may flare out in response to prevailing winds and assume the characteristics of cirrus clouds. Enormous energy is released in thunderstorms, and accompanying lightning or high winds may be very damaging.

There are a variety of other clouds that occur in specific regions or under particular circumstances. Some may be of value in weather prediction. Others should be avoided because of turbulence or icing conditions. They are not described here in detail because of space limitations and because they are rarely observed. **Lenticular cumulus,** for example, are cumulus clouds with a lens-like shape that are seen fairly often along the Front Range of the Rockies and a few times a year in mountains of the Pacific Northwest, but they are rarely observed in other areas of the United States.

The presence of clouds signals moisture, and thickening or perceptible vertical development are often signs of impending precipitation. As a rule, areas with a great deal of cloudiness experience rainfall more often and in greater amounts. In some circumstances, clouds, usually stratus, may persist for days in middle and high latitudes during the winter season without yielding much precipitation.

Precipitation When the temperature of the air falls below the dew point, adequate hygroscopic nuclei are present, and atmospheric mixing occurs, some of the water vapor contained in the air condenses, clouds form, and **precipitation** may result in the form of rain, snow, sleet, or hail. Obviously, if the relative humidity of a parcel of air is high, only a small amount of cooling is necessary for the dew point to be reached. Conversely, if the relative humidity is low, it is necessary to cool the air significantly for condensation to occur.

Figure 4.17
Temperature inversion resulting from subsiding air. Note that the air temperature increases at about 1,500 to 2,500 feet.
From Arthur Getis, Judith Getis, and Jerome Fellmann, Introduction to Geography, *2d ed. Copyright © 1988 Wm. C. Brown Publishers, Dubuque, Iowa. All Rights Reserved. Reprinted by permission.*

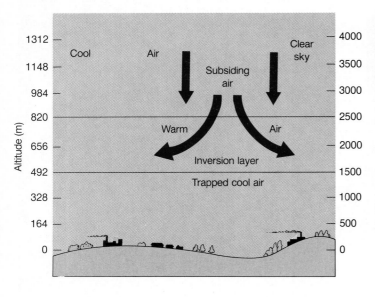

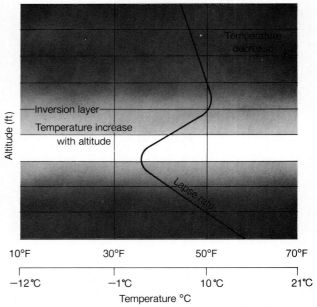

Conditions of the atmosphere determine whether air will be lifted and cooled sufficiently for condensation to occur. If air is stable, a rising parcel of air will tend to sink back to its original position without cooling to the dew point. **Stability** exists when the dry adiabatic lapse rate of about 5.5° F/1,000 feet (1° C/100 meters) is greater than the standard lapse rate. Under these conditions, a rising column of air quickly becomes cooler than its surroundings, and being cooler and more dense, such air tends to settle back towards the earth's surface. Such settling air warms as it descends and attains a level, often above the surface and occasionally trapping cooler air below it. When such a layer of warm air is temporarily trapped above colder air, a temperature **inversion** exists (figure 4.17). Inversions may trap air especially in cities surrounded by topographic barriers like Los Angeles or Portland (figure 4.18). This inversion layer may be responsible for trapping pollutants near the surface and creating unhealthy conditions for area citizens.

If, on the other hand, the lapse rate is 6° F/1,000 feet (> 1° C/100 meters), the air is **unstable** (i.e., a rising parcel of air continues to be warmer than its surroundings and may rise until the dew point has been reached and precipitation may occur). Some form of instability is necessary for precipitation to occur.

In the illustration just given, if air at a temperature of 70° F at sea level is lifted adiabatically, it has cooled 11° F (2 × 5.5° F) at 2,000 feet, or it has a temperature at that elevation of 59° F. Since the lapse rate in this example is 6° F/1,000 feet, air temperature at an elevation of 2,000 feet is 12° less (2 × 6° F), or 58° F. Since the rising parcel of air is still warmer than its surroundings, it will continue to rise. This is a condition of instability, and the air is said to be unstable. If the lapse rate is significantly greater than the adiabatic rate, air will rise rapidly creating turbulence and great mixing, of a kind common in cumulonimbus clouds. If it is only slightly greater, as in the example in this paragraph, the rate of rise of a parcel of air will be slow, and turbulence will be minimal.

Conversely, if the lapse rate had been 5° F/1,000 feet, the rising air would quickly become cooler than its surroundings and, as a result, would sink back to the earth. Such air is stable. It loses its opportunity to cool, and the possibility for precipitation to result disappears.

If rising air achieves the same temperature as its surroundings, it has neutral buoyancy, which means that it will neither ascend nor descend. Such air is, essentially, stagnant. If it continues to receive atmosphere pollutants, it may become unhealthy to breathe and downright dangerous for people with respiratory or cardiovascular illnesses.

As unstable air rises, the dew point also decreases. An increase in elevation in a rising parcel of air results in a cooling of about 5.5° F per 1,000 feet, but at the same time the dew point recedes at about 1° F per 1,000 feet;

hence, the air has come closer to the dew point by a net
of 4.5° F. After condensation occurs, saturation temperature recedes at the same rate as the wet adiabatic lapse
rate.

Rising air in unstable conditions caused by the effects of surface heating, especially during the summer half
of the year in the middle latitudes (although all through
the year in the tropics), results in **convectional precipitation** (figure 4.19). Convectional precipitation is quite
prevalent in the tropics, and it is pronounced in the southeastern part of the United States. Characteristically,
showers may come as heavy downpours and may be extremely localized. It is not unusual at all for areas only a
short distance apart to experience quite dissimilar weather
conditions. One area may receive a downpour of rain,
whereas a nearby area may enjoy bright sunshine. Areas
a few miles apart may contrast droughty (dry) conditions
with the perhumid (very wet). When such rain does fall,
it is often a veritable deluge. Frequently, such heavy rain
may drastically reduce visibility, and it is not at all uncommon for cars to stop at the side of the road for the
sake of safety until the thunderstorm has passed.

The author's home is about three miles from his
office. On many days during the summer season, one or
the other of these sites may have a violent thunderstorm
with copious quantities of precipitation while the other is
bathed in bright sunshine. It's a good idea to keep an umbrella handy during summer in the American Deep South.
Other mechanisms trigger precipitation in different environments. Hilly or mountainous terrain often intercepts
moisture bearing winds, and precipitation may result.

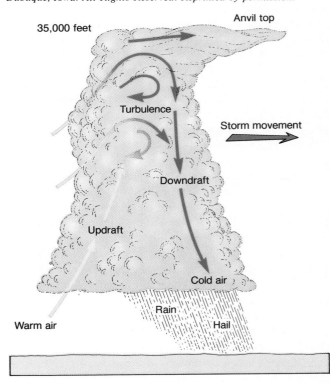

Figure 4.20
Orographic precipitation. The windward slope is wet, whereas the leeward slope lies in a rain shadow.
From Arthur Getis, Judith Getis, and Jerome Fellmann, Introduction to Geography, 2d ed. Copyright © 1988 Wm. C. Brown Publishers, Dubuque, Iowa. All Rights Reserved. Reprinted by permission.

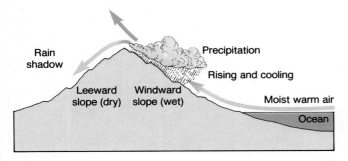

Figure 4.21
The wave form development of a cyclonic cell in the Northern Hemisphere. A wave (A) begins to form along the polar front; (B) the movement of cold air from the north and warm air from the south begins to produce fronts; (C) the warm air pocket is isolated; and (D) the storm dissipates as the polar front is reestablished.
From Arthur Getis, Judith Getis, and Jerome Fellmann, Introduction to Geography, 2d ed. Copyright © 1988 Wm. C. Brown Publishers, Dubuque, Iowa. All Rights Reserved. Reprinted by permission.

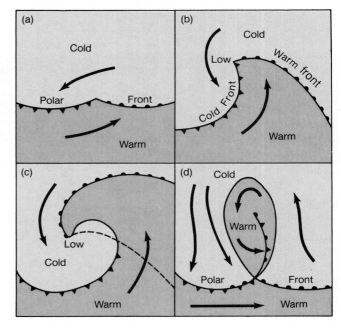

Orographic precipitation occurs when humid air is forced to rise over topographic barriers (figure 4.20). As the air ascends, it cools adiabatically. If the dew point is reached, condensation occurs and precipitation results on the windward slopes of the barrier. The leeward slopes are dry, because the descending air is being compressed and warmed. As the air is warmed, its capacity to hold moisture is increased, and the possibilities for precipitation are forestalled. Of course, mountain barriers are not continuous, and elevations are not constant, so typically there is some spillage of moisture and precipitation to the leeward side. Nevertheless, windward slopes are almost universally wetter than leeward slopes in such circumstances.

In orographic precipitation, the warm air movement may be related either to the general circulation of the atmosphere or to the circulation around a storm center. The important point is that a landform acts as a wedge, or barrier, over which warm, moist air is lifted and cooled to the temperature at which precipitation occurs. The contrasts between rainy windward and dry leeward slopes may be dramatic. The dry side of the mountain is said to be in a **rain shadow.** Such rain shadow areas may be semiarid or even arid, whereas the windward slopes may be humid.

Orographic precipitation seldom occurs as downpours, but the fixed position and orientation of mountain barriers relative to the prevailing winds may produce high precipitation on windward slopes and rainfall totals may be high because of a succession of rainy or snowy periods. In some instances, thunderstorms may develop or be accentuated by orographic lift, resulting in copious precipitation on windward slopes.

Cyclonic (frontal) precipitation is associated with the familiar storm type that tends to produce general rains over broad areas, particularly during the winter half of the year in middle latitudes (figure 4.21). Cyclonic precipitation occurs mainly in the middle latitudes where westerly winds predominate, but such storms also affect high and low latitudes. It occurs when cold, relatively dense air from higher latitudes meets warm, relatively light air

from low latitudes. These air masses tend to retain their individual characteristics over protracted periods of time, and they do not readily mix. If the colder and denser air wedges beneath the warmer air or if the warm air slides up over the surface of the colder air, the rising air may be cooled adiabatically to the dew point and precipitation may occur.

Over much of the United States, the dominant cause of precipitation is cyclonic or frontal activity; however the causes of precipitation may be augmented or combined in any region. A cyclonic cell moving through a mountainous region may be influenced by orographic lift. A rapidly moving front may generate heavy convectional thunderstorms at or near the contact between the cold and warm air masses. At other times and places, a slow moving system may be characterized by prolonged periods of cloud cover and drizzle. Areas with marine west coast environments, such as the northwest United States and northwest Europe, may have little sunshine and long periods of rain, especially in the winter season. Such rains are usually cyclonic in origin. In contrast, those areas that receive abundant convectional precipitation may experience hard downpours followed by brilliant sunshine within the space of a very short time.

Forms of Precipitation

The weather bureau has classified all forms of precipitation into about fifty specific types. The most common are rain, snow, sleet, glaze, and hail. **Rain** is simply the precipitation of liquid drops of water from clouds. Water droplets that condense around hygroscopic nuclei collide and mix, and eventually fall as rain droplets. Such droplets vary considerably in size from those of a fine mist falling as drizzle to large droplets issuing from a thunderstorm. **Snow** occurs when the dew point is below freezing and the moisture condenses into hexagonal ice crystals, which reach the surface in that form. It is said that no two are exactly alike, although the laws of probability suggest that the countless numbers of snowflakes that have fallen over time must have been replicated at some time and place. Snows may be dry and fluffy, or heavy and wet depending on temperature and moisture conditions at the time of the snowfall. **Sleet** is frozen rain droplets in the form of ice pellets. It occurs when liquid raindrops are frozen as they fall through a layer of the atmosphere that is below freezing. **Glaze** occurs when rain falls on below-freezing surfaces and the liquid is frozen into a sheet of ice. **Hail,** a product of thunderstorms, is composed of clear or opaque concentric rings of ice that are formed when droplets of water move successively in and out of freezing zones of the cloud by air turbulence. As frozen particles again intersect a zone of saturation, a concentric layer of ice is added. The number of concentric rings and the size of the hailstone are indicators of turbulence within a particular thunderstorm. A large hailstorm with several concentric rings of ice suggests violent updrafts and downdrafts within the cumulonimbus cloud and serves as a gauge of turbulence.

Under certain conditions, it is possible for two types of precipitation to fall simultaneously. For example, snow and rain may be mixed when the temperature is close to freezing. Sleet and rain may also be mixed during conditions close to freezing. Rain and hail may fall from a thunderstorm system at the same time because different degrees of turbulence within the same cloud may cause hail to form in one portion of a cloud, while water droplets fall directly to earth in others. Very large droplets of water falling from thunderstorms may represent hailstones that melted before they reached the surface. In addition to form, other significant characteristics of precipitation are the intensity or rate of fall, and the amount of precipitation that falls within a given time frame.

Measurement of Precipitation

Measurement of precipitation is accomplished by the use of a cylindrical container with a funnel-like top. The funnel collects moisture that falls and flows into a small cylindrical tube contained within the larger one. This inner tube has an area one-tenth that of the collector funnel, so the liquid will rise ten times as high within it as it would in the larger cylinder. This enables the observer to make accurate readings of even very light precipitation. A calibrated stick is used to measure the height of water within the tube. This device is known as a **rain gauge.** Recording rain gauges are designed to weigh the amount of water contained within a gauge and to translate that weight to a linear measure.

When snow is to be measured, the inner tube is removed, and the snow is melted to determine the water equivalent. Although snow varies significantly in water content from heavy, wet varieties to those that are light and fluffy, a rough equivalent for an average snow is ten units of snow equals one unit of liquid water.

Distribution of Precipitation

As a general rule, precipitation diminishes from low to high latitudes and from the margins of continents towards the interior. Typically, the windward slopes of mountain ranges adjacent to coasts are wetter than the leeward margins. There are, however, major exceptions to these generalizations. For example, the interior of Brazil in the Amazon Basin is generally wetter than the coastal margins. Several areas in middle to high latitudes, such as the coastal region of North America, are substantially wetter than regions at lower latitudes. A number of the world's driest deserts are found along coastal margins dominated by high pressure cells and bordered by cold ocean currents.

The distribution and intensity of precipitation from day to day, season to season, and year to year is important to people because the rhythm of life is affected significantly by rainfall patterns (figure 4.22). Light precipitation may result in crop failure and attendant famine, whereas an overabundance of moisture may result in flooding. In the analysis of precipitation data, the meteorologist or climatologist is concerned about total amounts, amounts received during different time intervals, the type and intensity of precipitation, and the variability. Generally, as the amount of precipitation diminishes, the variability increases; hence, very wet environments tend to receive reliable amounts of moisture from year to year, whereas dry environments frequently cannot rely on receiving average amounts.

Although rainfall amounts and intensity are primarily considered to be weather elements, rainfall is significantly connected to hydrology and geomorphology. More rainfall of high intensity typically yields greater runoff, which facilitates weathering and erosion, modifying landforms, and the nature of the hydrological patterns. Less intense falls of precipitation give greater opportunities for moisture to soak into the ground and to recharge groundwater supplies. Other elements of the weather are significant in a variety of ways. The pressure exerted by the atmosphere, and the winds generated by pressure difference are omnipresent facts of life.

Figure 4.22
Temperature and precipitation zones in (A) *winter and* (B)
summer.
From Arthur Getis, Judith Getis, and Jerome Fellmann, Introduction
to Geography, *2d ed. Copyright © 1988 Wm. C. Brown Publishers,
Dubuque, Iowa. All Rights Reserved. Reprinted by permission.*

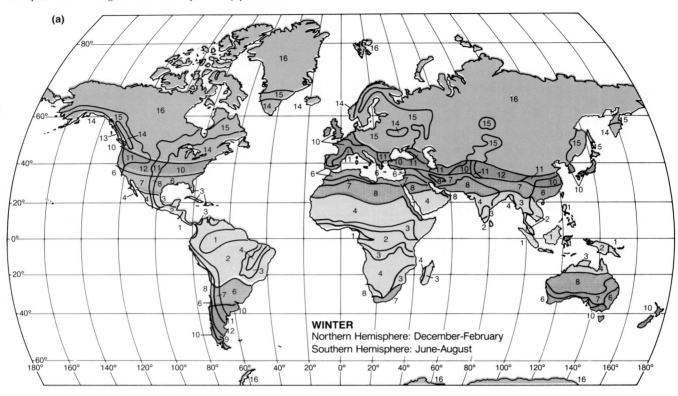

WINTER
Northern Hemisphere: December-February
Southern Hemisphere: June-August

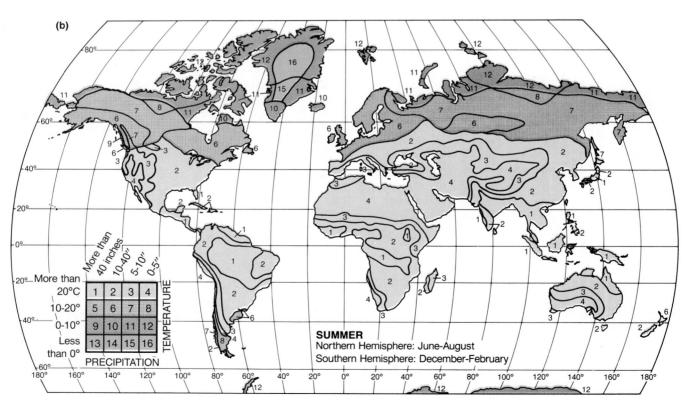

SUMMER
Northern Hemisphere: June-August
Southern Hemisphere: December-February

	More than 40 inches	10-40"	5-10"	0.5"
More than 20°C	1	2	3	4
10-20°	5	6	7	8
0-10°	9	10	11	12
Less than 0°	13	14	15	16

PRECIPITATION

TEMPERATURE

Pressure and Winds

Atmospheric **pressure** is exerted by the interaction of the molecules that make up the gases of the air. This pressure is exerted in all directions, and the greater density of the atmosphere at lower levels is responsible for the higher pressure that exists at the earth's surface. Pressure decreases with increasing elevation, and the pressure exerted at 18,000 feet (approximately 5,500 meters) is approximately half normal atmospheric pressure at the earth's surface at sea level.

There is considerable variation of pressure from one region to another over the earth's surface, and there are significant variations over time in the same place as well. Some of these broad regional variations are related to differences in temperature, whereas others are dynamically induced. Higher temperatures close to the equator are in significant part responsible for the creation of an interrupted belt of low pressure that spans the equator. In a zone about 30° north and south of the equator is an area of subsiding air that produces high atmospheric pressure. The areas adjacent to the poles have a zone of subsiding colder air that produces a zone of higher pressure. At about 60° north and south of the equator, the dynamics of earth rotation create an interrupted belt of lower pressure. Higher temperatures tend to produce low pressure, and lower temperatures tend to produce high pressure, although the effects of earth rotation produce a dynamic element that is very important in the evolution of pressure cells.

Weather conditions are closely related to variations and shifts in atmospheric pressure. Consequently, a knowledge of the world's distribution of atmospheric pressure and prevailing winds is fundamental to an understanding of broad climatological patterns as well as the daily meteorological consequences of pressure differences and wind characteristics.

Pressure is represented on maps by a series of lines called **isobars,** which connect points of equal pressure. Highs may be thought of as pressure ridges or domes, whereas lows may be conceived to be troughs or depressions. Isobars on a pressure diagram are closely analogous to contours on a topographic map. The closer together the isobars, the greater the pressure gradient and the more intense the pressure cell. Conversely, isobars spaced some distance apart reflect a low pressure gradient, which spawns light and variable winds.

Tropical low pressures are characterized by locations close to the equator throughout the year, but they extend south of the equator in January and northward in July. Pressures are high over the Northern Hemisphere continents in January, and well-defined low pressures occur over adjacent oceans in January. Such seasonal shifts of pressure are principally responsible for the great monsoon circulations with their alternating wet and dry seasons.

During July, the middle latitudes of both the Atlantic and Pacific Oceans are under the dominance of high pressure. Well-developed low pressure areas occur at the same time in southern Asia and North America. A more nearly perfect zonal alignment of wind and pressure belts occurs in the Southern Hemisphere because there are fewer land–water contrasts (i.e., the Northern Hemisphere has a much higher percentage of the earth's lands than the Southern Hemisphere).

Differences in pressure generate the horizontal movement of air that we call **winds** (figure 4.23). The vertical movement of air produces air currents. Winds are

Table 4.2
Alveolar Pressure (A Measure of Oxygen Entering Bloodstream)

Elevation (ft)	Alveolar pressure (%)
Sea level	100
5,000	83
10,000	67
15,000	54
20,000	42
25,000	34
30,000	25

Figure 4.23
Air blows from a high pressure area to a low pressure area. A modest pressure difference produces light winds, while a significant difference produces stronger winds.

From Arthur Getis, Judith Getis, and Jerome Fellmann, Introduction to Geography, *2d ed. Copyright © 1988 Wm. C. Brown Publishers, Dubuque, Iowa. All Rights Reserved. Reprinted by permission.*

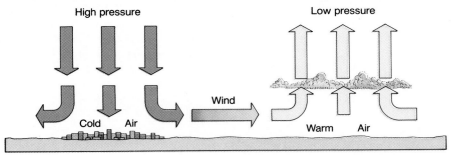

Figure 4.24
*The effects of Coriolis force on wind direction in the
Northern Hemisphere.*
From Arthur Getis, Judith Getis, and Jerome Fellmann, Introduction
to Geography, *2d ed. Copyright © 1988 Wm. C. Brown Publishers,
Dubuque, Iowa. All Rights Reserved. Reprinted by permission.*

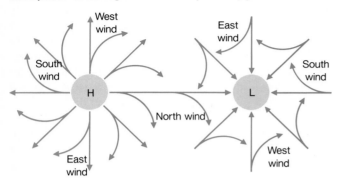

Figure 4.25
Circulation around pressure cells in each hemisphere.

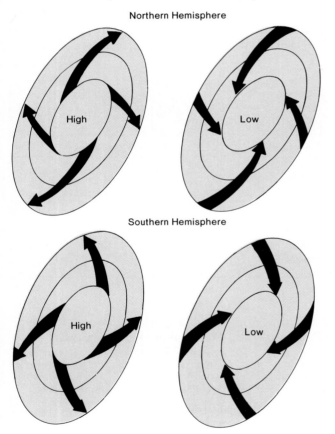

essential to human life. They distribute heat from low latitudes to higher latitudes; they are responsible for the invasion of colder air from high latitudes; they transport moisture from oceans and seas and drop it as precipitation; they sweep atmospheric pollution from industrialized areas; and they transport surface soil or sand. Without the actions of winds, much less of the planet would be habitable, and modern man would soon suffocate under the noxious blanket of atmospheric pollution that is generated by human activity. Winds are in a very real sense the breath of life.

Winds blow out from the center of a high pressure and in towards the center of a low. The **velocity** is determined by the steepness of the **pressure gradient** (i.e., the differences in pressure between two adjacent cells or the difference in pressure from the center to the periphery of a cell).

Because the earth is rotating from west to east, there is an apparent deflection of motion to earthly observers. This deflection of direction of motion on the earth's surface is called **Coriolis force.** Because of it, motion in the Northern Hemisphere is deflected to the right, and in the Southern Hemisphere it is deflected to the left. There is zero deflection at the equator and maximum deflection at the poles.

Because of air movement in towards the center of a low and out from the center of a high, frictional drag with the surface of the earth and coriolis force cause the circulation around a Northern Hemisphere **low** to be counterclockwise; around a Northern Hemisphere **high,** it is clockwise (figure 4.24). A reverse rotational pattern about pressure cells occurs in the Southern Hemisphere (i.e., a clockwise circulation around a low and a counterclockwise circulation around a high) (figure 4.25).

In the Northern Hemisphere, if the wind is at one's back, the lower pressure is at the left of the observer and the higher pressure is on the right. The converse is true in the Southern Hemisphere. These relationships are known as **Buys-Ballot's Law.**

General Planetary Pattern of Winds and Pressures

In a belt adjacent to the equator where pressure change is slight, winds tend to be light and variable. This belt is referred to as the **equatorial low** or **doldrums.** In a latitudinal band extending to almost 30° in each hemisphere, winds are northerly (from the north) in the Northern Hemisphere, and southerly in the Southern Hemisphere (figure 4.26). These winds have their origins in the subtropical belt of high pressure and blow towards the equatorial low. The effect of earth rotation causes a deflection of these winds producing the **northeast trades** and **southeast trades** respectively. Since these winds often have long

Figure 4.26
Idealized planetary wind and pressure patterns.

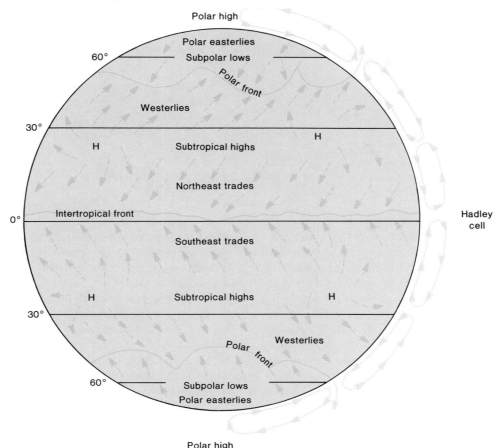

trajectories over water, and since they are moving towards lower latitudes and are being warmed in the process, they tend to carry abundant moisture. Unless lifted, the trades tend to dry up the areas they move over because they are heating as they move toward the equator. If topographic barriers are encountered, copious quantities of precipitation may be dropped on windward slopes. Indeed, the wettest place on earth is the windward slopes of mountainous areas in Hawaii, which intercept the trades that have had a long overwater trajectory, or fetch, in the Pacific. In contrast, the leeward side of the mountain has much drier conditions. Low-lying areas in trade wind zones, however, do not receive much moisture, since there are few mechanisms to produce the cooling essential for precipitation to occur. Several low-lying islands in the trade wind belt, especially in the Caribbean, are quite dry.

The **subtropical highs,** sometimes referred to as the **horse latitudes,** are a belt of descending air in high pressure centers. As this descending air moves to the surface,

some flows poleward, and some moves equatorward. These winds have a southerly component in the Northern Hemisphere and a northerly component in the Southern Hemisphere, but they tend to be dominantly westerly in both hemispheres because of the increasing Coriolis effect. These winds are somewhat less constant and persistent than the trade winds, but they are principally responsible for the movement of weather from west to east in the middle latitudes. They carry the name **westerlies** or **prevailing westerlies.** Within this zone, a succession of cyclonic cells or low pressure areas move with considerable frequency. Since much of the conterminous United States lies within the zone of the westerlies, most weather systems move from west to east in this country. The westerlies tend to undulate north and south as the predominant zonal flow is from west to east. These undulations or oscillations are known as **Rossby waves.** These oscillations are principally responsible for the shift of storm tracks from time to time.

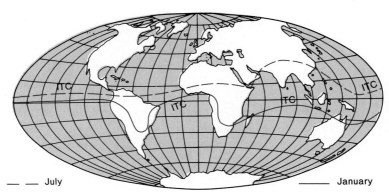

— — July   ——— January

The zone of **subpolar lows,** situated at about 60°, attracts the westerlies from the subtropical high and the easterlies from the polar high. Although the amount of moisture carried by the easterlies is limited, periods of vicious weather can exist because of high-pressure gradients and low-frictional drag.

Wind belts shift with the pressure belts in response to the angle at which the sun's rays strike the earth, as well as the length of day. These shifts cause a given area to experience more than one type of wind or pressure dominance in a year. For example, at about 7° to 15° north and south latitude, the rising air currents of the equatorial low prevail in the high sun season along an area of convergence of the trades called the **intertropical front** (ITF) or the **intertropical convergence zone** (ITZ or ITC). The intertropical convergence zone is an area of unsettled weather characterized by a great deal of convectional activity (figure 4.27).

This same area, about 7° to 15° north and south, is under the influence of the trades during the rest of the year. These latitudes tend to experience wet and dry seasons each year. Wet periods occur during the intertropical convergence zone dominance during the high-sun season, whereas dry conditions prevail during periods of trade wind dominance during the low-sun season.

Between 30° and 35° of latitude is a zone that tends to be dominated by the subtropical high during the summer half of the year and by the westerlies at other times. During high-sun periods under high-pressure dominance, such areas are dry, whereas westerly wind dominance in the low-sun season produces the wet season. Other winds and pressure belts show comparable shifts, expansions, and contractions in concert with the shift of the sun's rays and the length of daylight and darkness from season to season.

Areas that experience such seasonal shifts are often characterized by alternating wet and dry seasons. These asymmetries will be discussed in subsequent chapters dealing with climate.

The low-pressure belt adjacent to the equator and the area along the intertropical front are characterized by hot and humid weather with abundant precipitation because of the effects of convergence and convection. Winds in this area tend to be light and variable. The desiccating effect of the trade winds, except on elevated windward slopes, has been alluded to previously.

Weather in the subtropical highs tends to be dry, and several of the world's great deserts are located in these latitudes. This zone of descending air currents tends to have light winds and a high percentage of bright, sunny days. The subsiding air is warming by compression and tends to be a drying air. Further, the highs serve as buffers steering middle latitude storms away from the region. The effects of descending and warming air conspire to produce and sustain dry conditions, while the presence of cold currents offshore tends to accentuate the aridity.

Within the westerlies, there is a steady progression of weather that typically moves from west to east. The movements of lows through this region tend to bring precipitation and unsettled conditions, whereas highs often bring periods of bright, sunny weather. The frequency of movement of depressions through these areas is responsible for the variability of climatic characteristics so typical of westerly wind belts.

The subpolar lows are areas that often experience unsettled weather and a high percentage of cloudiness. Relatively cool air can contain limited quantities of moisture, however, so these areas tend to receive low amounts of precipitation except on exposed mountain slopes.

The polar easterlies are warming en route to lower latitudes, and they normally yield only limited amounts of precipitation. Similarly, the polar high, a region of subsiding air, tends to produce very limited amounts of moisture. Indeed, polar and subpolar environments are characterized by low precipitation amounts. They are, in essence, cold deserts.

This classical circulation model of the earth is predicated on the assumption that the rotating earth is a homogeneous body. Obviously, this is not the case, but the trends or tendencies are pronounced enough to make it a useful descriptive vehicle for understanding global wind and pressure patterns.

The circulation pattern may be considered, essentially, a giant heat engine focused on the equator. Greater amounts of insolation are received at low latitudes and, as the air is warmed, it rises. It cools as it rises and slides north and south, and at about 30° of latitude it tends to sink; a portion of the descending air moves towards the equator as the **tradewinds,** and a portion moves poleward as the westerlies. This circulation is sometimes referred to as a **Hadley cell.** At the poles, a dome of high pressure builds, and descending air slides equatorward. Compression, movement towards lower latitudes, and the dynamic effect of earth's rotation produces a band of lower pressure at about 60° of latitude.

The land and water bodies break what would be pressure belts into large semipermanent high- and low-pressure cells. The prevailing winds, although modified by moving highs and lows, and the interruptions of land and water bodies tend to follow the patterns described in the idealized model. Understanding of these basic patterns is essential to an appreciation for and understanding of the nature of climatic distributions and the procession of weather in different areas of the globe.

Measurement of Pressure and Winds

Atmospheric pressure is measured by a device known as a **barometer. A mercurial barometer,** which is a sophisticated version of a device invented hundreds of years ago, consists of a large glass tube about a yard long, sealed at one end and inverted at the open end in a container or receptacle of mercury. As air pressure exerts force on the container of mercury, the column of mercury within the tube will rise or fall, depending upon the amount of atmospheric pressure exerted. Standard atmospheric pressure at sea level will cause the mercury column to rise to a height of 29.92 inches, or about 760 millimeters. Another way to express standard sea level atmospheric pressure is 1013.2 millibars; in English measure, it is about 14.7 pounds per square inch.

For many purposes, it is inconvenient to use a mercurial barometer. It is large and unwieldy, and difficult to transport. A cheaper and more portable instrument is the **aneroid barometer.** This instrument uses a flexible, airtight metal box from which the air has been evacuated. When pressure increases, the metal box is compressed; when pressure diminishes, the box expands. A pointer, connected at one end through gear linkages to the box, indicates the atmospheric pressure. A device called the **barograph** uses the aneroid barometer and a clockwork mechanism that turns a rotating cylinder, which allows a trace of atmospheric pressure over a substantial period of time to be made. Pressure traces provide opportunities for observing day to day variations and for noting trends or tendencies over longer periods. Aneroid barometers scaled to read elevation are called **altimeters.** Since pressure decreases with increasing elevation, it's possible to record those drops in pressure in feet or meters, rather than in the use of measures of pressure. However, because of the natural changes of pressure, large commercial aircraft use other altimeter devices, such as radar altimeters, to assure greater accuracy than is possible with the pressure activated altimeter. Accurate pressure altimeter settings for the beginning point of a trip might be grossly inaccurate at destination, and, in periods of inclement weather requiring instrument flying, could result in false altitude readings, which could lead to serious mishaps. Since many small private planes have pressure altimeters, many accidents can be traced to failure to insert accurate altimeter settings at the beginning of a trip or because of significant pressure changes en route.

Wind direction is indicated by a freely rotating and exposed **wind vane,** which keeps shifting to point to the direction from which the wind is blowing. **Anemometers** measure wind velocity by ascertaining the horizontal pressure exerted by wind into a tube, or by a device with cups that rotates on an axis at varying speeds depending upon the wind velocity. Typically, anemometers and wind vanes are combined in a single instrument that records wind speed and direction instantaneously, or that provides a trace of wind speed and direction over a period of time.

Air Masses

Air masses, with very specific conditions and characteristics, frequently move over the earth's surface as distinct entities. They maintain their distinguishing characteristics of temperature and humidity for protracted periods. As they move from place to place, they modify preexisting weather conditions in a given place, and, as they clash with dissimilar air masses, they are major weather makers.

Figure 4.28
Source regions of major air masses affecting North America.
Source: After Hayes, U.S. Department of Commerce.

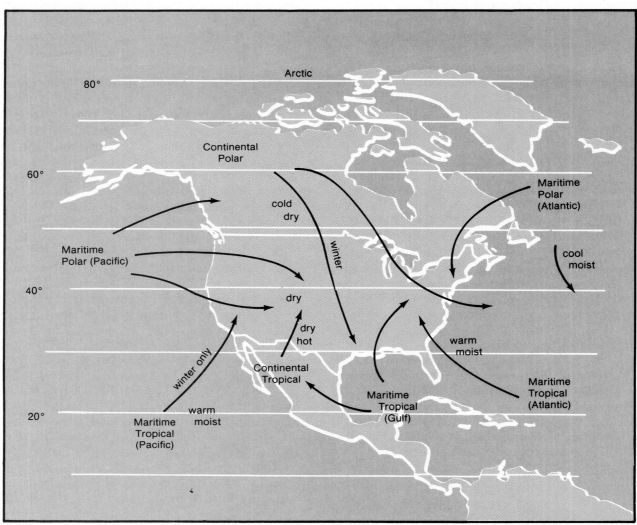

Indeed, in the middle latitudes, the invasion of and conflict between succeeding air masses is a principal factor in changes of the various elements of the weather.

Air masses are described according to their source regions and the physical characteristics that they exhibit (figure 4.28). Tropical air masses (T) are warm; polar (P) are normally cold; continental (c) are usually dry; and maritime (m) are frequently wet. They are also described in terms of their temperature condition relative to the land surface beneath them; hence, they might be described as w (warm) or k (cold). So that an air mass carrying the symbol of cPk would be continental polar cold. In contrast, mTw would refer to a maritime tropical warm air mass.

To develop their significant characteristics and effects, air masses must stagnate in a source region for some period of time, and the underlying surface must be reasonably regular. To influence climatic conditions in other regions, they must migrate from their source area.

Maritime tropical (mT) air has a source region over tropical or subtropical oceans. It tends to be characterized by high humidity and high temperatures, and it tends to bring convectional rain in summer and drizzly conditions in the winter. When such an air mass comes in contact with cP air, there tends to be an active zone of weather and precipitation at or near the contact zone because warm air is actively wedged upward by the cold as it slides over a zone of colder air. As it is lifted, it is cooled. Frequently the dew point is reached and precipitation results.

Polar continental air (cP) develops over large land masses at high latitudes; it tends to be dry and stable; and it brings cool conditions in summer and cold conditions in the winter. Along the leading edge of such a cold air mass, a very active weather zone can be expected if there is contact with milder air with higher humidity.

Polar maritime (mP) air develops over oceans in the higher latitudes; it tends to bring mild temperature and high humidity along with abundant cloudiness and dreary weather in winter. During the summer, such air masses tend to produce mild, fair weather.

Tropical continental (cT) air masses develop over continents in subtropical latitudes. The air is characterized by low relative humidity and clear skies.

All four types of air masses influence American weather. Polar continental air masses with a source in northern Canada and Alaska invade the Great Plains and the Midwest, often extending to the Gulf of Mexico and beyond. The air masses are pushed eastward and out to sea by the westerlies. There is some leakage of such air across the Rockies and into the Great Basin, but the topographic barriers cause it to be modified in transit.

Maritime polar air moves into the country from the North Pacific and North Atlantic. The air masses with a Pacific origin tend to have the greatest influence on American weather, since they are steered eastward by the westerlies. Air masses with an Atlantic source region typically affect smaller areas of the United States, because they tend to be pushed offshore by the same westerly winds. Occasionally, they are borne on the winds of a northeaster, especially to the New England and Middle Atlantic states, and they may bring abundant cloudiness, lowering temperatures, and generally nasty weather.

Maritime tropical air affecting the United States comes from three sources: the subtropical areas of the Pacific, the Atlantic, and the Gulf of Mexico. Typically, the Pacific air has influence limited to California and the Southwest because mountain barriers intervene. Atlantic air has less significant influence also, since the influence tends to be moved offshore by prevailing winds. Gulf air, on the other hand, is very influential in the area east of the Rockies. Clashes between the warm, humid gulf air and cooler, drier continental air from the north and west yield moisture and a variety of meteorological pyrotechnics. Indeed, these collisions between Canadian air and gulf air produce some of the most severe weather known, especially if the cold, dry air is advancing rapidly against a warm air mass pregnant with moisture.

Continental tropical air, which has its origins over the Mexican plateau, sends tongues of warm, dry air into the Great Basin country and the American Southwest. This air tends to accentuate the dry conditions of the region.

As indicated earlier, air masses tend to retain discrete characteristics for substantial periods of time. Dissimilar air masses that come in contact with each other are weather makers at and near their zones of contact.

During World War I, a Norwegian meteorologist by the name of Bjerknes noted that the contact zones between dissimilar air masses tended to be areas of weather development. He applied the term *front* to these zones of contact. His choice was probably influenced by the presence of the active fighting fronts of the allied and the central powers. The concept of fronts in terms of development, movement, and decay is an essential element of modern meteorology. No discussion of middle- and high-latitude weather is possible without a discussion of frontal systems and accompanying weather phenomena.

Fronts

The zone of contact between dissimilar air masses is called a front. In a **cold front,** cold air is actively displacing warm air, whereas in a **warm front** the warm air is displacing the cooler air. These frontal systems and others to be discussed subsequently are major weather makers in middle and high latitudes. Of primary importance in producing weather changes in the United States are the changes of pressure and the invasions of contrasting air masses.

A high usually develops as a tongue of cold air bulges into a mass of warmer and less dense air. These polar outbreaks are responsible for cold waves in the wintertime and refreshing, cooling periods in summer. The high, which is characterized by descending air currents and air moving out from the center, is typically responsible for fair weather; however, at the margins of the air mass, cold air may wedge under and lift warm humid air forcing it to rise, cool, and reach the dew point.

Within a low-pressure cell, air of dissimilar character is converging towards the center of a low. The circulation within a middle-latitude low typically brings colder and drier air from the north and west in contact with a tongue of warm, frequently moist air from the south. At the contact zone, or cold front, the colder, denser air slides under the warmer and moister air forcing it to rise. If the warm air is unstable and the cold air is of considerable thickness and moving rapidly, there is likely to be abundant precipitation preceding the passage of the front and for some distance behind it. After the front has passed, the stormy weather is replaced by clearing skies, a rising barometer, lower humidity, and cooler or colder temperatures (figure 4.29). If the moisture content in the warm air is less, the cold air is thinner, or the passage of the front slower, there tends to be a diminished intensity of precipitation and a slowed rate of clearing.

Figure 4.29
The cold front has passed city A and is approaching city B.
The warm front has passed city B and is moving away from
it.
From Arthur Getis, Judith Getis, and Jerome Fellmann, Introduction
to Geography, *2d ed. Copyright © 1988 Wm. C. Brown Publishers,*
Dubuque, Iowa. All Rights Reserved. Reprinted by permission.

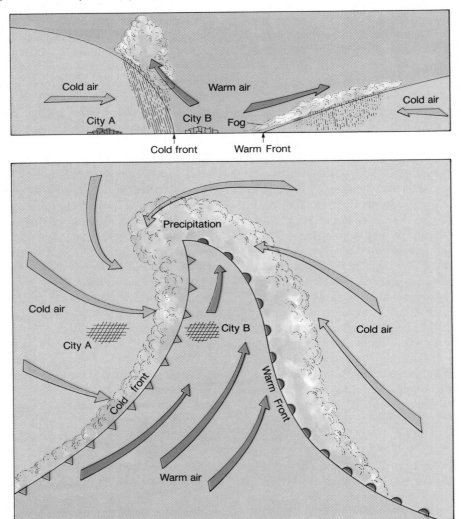

Within the cyclonic cell, as the warm air moves northward along the leading edge of the cell, the warm air tends to slide up over the cooler air at the location of the warm front. Precipitation will result if humidities are high enough and the lift is great enough. The passage of a warm front is marked by a rise in temperature, a fall in pressure, a shift in wind direction, and usually decreased cloudiness.

Cold fronts normally move faster than warm fronts, often overtaking them after the cell has been in existence for some time. The warm air is lifted as this overriding occurs, producing an **occluded front** (figure 4.30). Occlusions may also occur as a warm front in a succeeding system overtakes the cold front in a system that has moved through at an earlier time and has subsequently been slowed. Occlusions frequently produce drizzly, unsettled conditions, as virtually all the warm air at the surface has been lifted to higher elevations.

When fronts stall and show little movement, they are termed **stationary fronts.** Such areas may experience protracted periods of unsettled weather if warm, moist air continues to move in and is lifted over the colder air at or

Figure 4.30
A plan view and cross section of an occluded front.

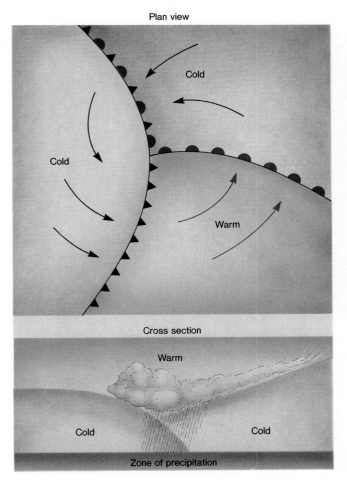

Plan view

Cross section

Figure 4.31
Satellite view of cloud cover along a cold front, a warm front, and an occluded front.
NOAA/NESDIS/NCDC/SDSD.

Figure 4.32
Idealized diagram showing position of jet stream relative to open and closed cellular clouds (i.e., clouds with significant vertical development).
Ralph K. Anderson, et al., Application of Meteorological Satellite Data in Analysis and Forecasting, *ESSA Technical Report NESC 51, NOAA, 1974.*

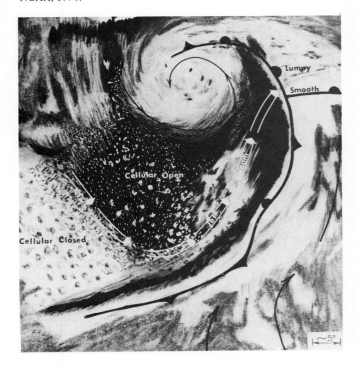

near the surface. They may be reinvigorated by a pulse of air within either of the air masses advancing again as a cold or warm front. Obviously, when this happens, they lose their stationary character and become, again, either a cold or warm front.

The normal elements of the weather vary from minute to minute, hour to hour, and day to day. Some regions experience very rapid changes; others, especially those under the dominance of a persistent high-pressure cell, tend to have many days with few significant weather changes.

Although cyclonic cells with accompanying frontal systems may produce some violent weather, they may also be benign invaders bringing periods of welcome rain or relief from searing heat or biting cold. Thunderstorms that may accompany them can destroy lives and property. They tend to punctuate prosaic weather changes with violent, often dangerous weather. Such storms will be discussed in the following chapter.

Elements of the Weather   69

Figure 4.33
Cirrus with straight line marking the boundary of the subtropical jet stream.

Study Questions

1. Name and briefly describe the various atmospheric layers.
2. What are the principal weather elements? What instrument(s) is used to measure each element?
3. Define the following terms:
   a. albedo
   b. langley
   c. normal lapse rate
   d. dry adiabatic lapse rate
   e. wet adiabatic lapse rate
   f. isotherm
   g. absolute humidity
   h. hygroscopic nuclei
   i. transpiration
   j. radiation fog
4. Explain how heat is transferred from one place to another.
5. List and characterize the various families of clouds.
6. Explain the difference between stable and unstable air.
7. List and characterize the various *forms* of precipitation.
8. List and characterize the various types of precipitation in terms of causal mechanisms.
9. Explain the air circulation around high and low pressure centers in both hemispheres.
10. Reproduce an idealized planetary wind and pressure diagram assuming a homogeneous, rotating earth.
11. What is the intertropical convergence zone? What effect does it have on the weather?
12. Discuss the characteristics of cold fronts, warm fronts, and occluded fronts.
13. Why are middle latitudes characterized by frequent weather changes?
14. Explain why winds are important to human beings.
15. What are Rossby waves?
16. Who was Bjerknes?

Selected References

Anthes, R. A.; Panofsky, H. A.; Cahir, J. J.; and A. Rango. 1978. *The atmosphere.* 2d ed. Columbus: Charles E. Merrill Publishing Company.

Barry, R. G., and Chorley, R. J. 1970. *Atmosphere, weather, and climate.* New York: Holt, Rinehart, and Winston.

Battan, L. J. 1983. *Weather in your life.* San Francisco: W. H. Freeman and Company.

Flohn, H. 1969. *Climate and weather.* New York: McGraw-Hill Book Company.

Lutgens, F. K., and Tarbuck, E. J. 1979. *The atmosphere: An introduction to meteorology.* Englewood Cliffs: Prentice-Hall.

Riehl, H. 1965. *Introduction to the atmosphere.* New York: McGraw-Hill Book Company.

Strahler, A. N., and Strahler, A. H. 1987. *Modern physical geography.* New York: John Wiley and Sons.

5

Storms and Weather Forecasting

The tornado is one of the most feared storms in nature.
NOAA.

*T*he ordinary vagaries of weather are punctuated at intervals by storms that add elements of uncertainty and danger to life and property in various environments. A **storm** may be defined as a disturbance of the atmosphere characterized by strong winds, and frequently accompanied by rain, snow, hail, thunder, and lightning. Storms vary markedly in intensity, areal extent, effect on people, and persistence. Some are minor, local, transient, and cause little damage. Indeed, a few are so benign as to be beneficial bringers of rain to parched regions, a breath of cool air in a time of great heat, a purging of air pollution over a big city, or a warming respite from bone-chilling cold. Others are almost unbelievably violent, creating great havoc, but involving only a limited area. Still others may persist literally for weeks and bring widespread devastation to broad areas. Their origins, characteristics, distribution, and spread are essential elements to be considered in an analysis of meteorological and climatological environments. Their contributions to the total physical environment and their effects on each of the constituent environmental elements are significant. They cause suspense and drama in the human experience, and they add an element of catastrophic change to the measured pace of long-term change.

Thunderstorms

Thunderstorms are quite common in warm, humid areas of the tropics and subtropics. Surface heating in unstable or conditionally unstable air causes a strong convectional system to develop. The resulting thunderstorms contain strong updrafts and downdrafts of air, and lightning and resultant thunder occur because of discharges between negatively and positively charged parts of the same cloud, between adjacent clouds, or between clouds and the ground. These charges are induced by frictional contact between water droplets, ice crystals, or between suspended particulate matter. Lightning is a dangerous feature of such storms, and people should get under cover and away from exposed places when a thunderstorm is present. In fact, lightning typically kills more people in the United States annually than are killed in other violent storms. Lightning is an especially dangerous feature in the American Southeast. Of course, lightning is often present in the rainy tropics.

Lightning may be very destructive to property when fires are set by lightning strikes. Occasionally, great financial loss may occur as the result of lightning damage. In March of 1987, for example, an Atlas Centaur rocket and a payload of a navy communications satellite had to be destroyed after proximate lightning strikes caused electrical system malfunctions. The total loss was in excess of $123,000,000. The wisdom of launching a rocket in such marginal conditions is certainly open to question.

In addition to those resulting from convectional circulation because of differential surface heating, thunderstorms are particularly prevalent along and in advance of frontal zones. Lifting, especially along and in advance of cold fronts, may produce thunderstorms of considerable intensity. In fact, thunderstorms may be described as **air mass thunderstorms** if they occur in warm, humid air masses not associated with frontal passage. **Frontal thunderstorms** occur along an active cold or warm frontal passage. Less frequently occluded fronts may produce thunderstorm activity. A line of thunderstorms, known as a **squall line,** may precede an especially vigorous frontal passage, especially a rapidly moving cold front. **Orographic thunderstorms** develop when a warm, humid air mass is forced to rise vigorously over a topographic barrier.

Thunderstorms frequently bring conspicuous quantities of moisture in high intensity showers, sometimes resulting in flash floods. Hail may accompany thunderstorms with violent updrafts. Thunderstorms are also responsible for wind shear—strong updrafts and downdrafts within short distances of each other. Such wind shear areas are hazardous to aviation, creating severe turbulence and, in those cases where an aircraft is just landing or taking off, producing exceptionally dangerous conditions that have led to fatal crashes. Violent downdrafts may literally slam aircraft into the ground in wind shear zones. Although wind shear zones are difficult to predict and not detectable on ordinary radar, it is possible to predict when wind shear may occur, and pilots are usually warned of such conditions.

A great deal of research is being devoted to detection of wind shear, since the effects of such turbulence on civil and military aviation are quite serious. Doppler radar promises greater reliability in predicting wind shear in time to give appropriate warnings to pilots to avoid areas of greatest danger. Because of costs, however, federal and state agencies have been slow to install these systems. The potential to save numerous lives suggests the need for the installation of Doppler radar at the earliest practicable moment.

Typically, the thunderstorm is a localized storm, although it may move over considerable distance on the prevailing winds. Within tropical and humid subtropical environments, however, thunderstorms are omnipresent facts of life, especially during the high-sun season. The **cumulonimbus clouds** may extend upwards of 50,000–60,000 feet (15,000 to more than 18,000 meters). The turbulence in such a storm is monumental, and any aircraft wandering into such a system would be in grave jeopardy.

In a thunderstorm, rapid vertical movement of air causes it to condense and form **cumulus** clouds, which rapidly grow through stages of **cumulus castellatus, cumulus congestus,** and **cumulonimbus** with the towering form and characteristic anvil top. At the freezing level, precipitation in the form of water droplets and ice crystals begins to form within the cloud and fall to the earth; cooler air is drawn down in downdrafts, while the same mechanism that initiated the storm continues to send air upward. In this mature stage of the storm, violent updrafts and downdrafts of air create areas of wind shear that are very turbulent and prospectively dangerous for aircraft that may be flying into them. Fortunately, thunderstorms appear as bright echoes on radar, and pilots can usually take evasive action to avoid areas of greatest turbulence. As a young navigator–radar operator many years ago, the author was among early users of radar in assisting pilots to avoid turbulence. Avoiding the centers of cumulonimbus clouds borne on the winds of the inblowing monsoon in south and southeast Asia was a major task eagerly accepted by the neophyte navigator–radar operator. The turbulent ride in even a modest thunderstorm is an experience to be avoided if at all possible. Eventually, the copious precipitation and associated cooler downdrafts of air may cause the deterioration and ultimate disintegration of the storm.

Lightning

Cumulonimbus clouds are electrically charged, usually with greater positive charges in the upper reaches of the cloud and more negatively charged particles in the basal portion of the cloud. The exact reason for this variation in charge within the cloud is unknown. The prevailing theory is that rising air carries positive charges from near the earth's surface to upper cloud elevations, whereas cooler droplets of moisture move negative charges to lower sections of the cloud, although current research reveals

Figure 5.1
An advancing squall line.

Figure 5.2
Electrical charges in thunderstorms.

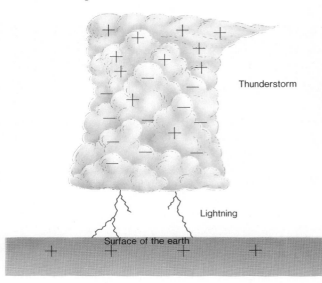

that the pattern may be far more complex than previously believed. When the electrical potential is sufficiently great, there is a discharge of electrical energy, which we call lightning, between cloud and earth, from cloud to cloud, or within the same cloud (figure 5.2). This enormous discharge of energy creates a shock wave, which we hear as thunder. Unless we are very close to a lightning strike, there is considerable delay between the lightning flash, which sends out light at the speed of light, and the thunder clap, which is heard subsequently at the much slower speed of sound. Since light travels at the speed of 186,000 miles per second (about 298,000 kilometers), we see a lightning flash virtually instantaneously, whereas sound travels at somewhat varying rates depending upon the temperature. In most storms, sound would probably travel in the range of 1,100–1,200 feet per second (about 333–363 meters); hence, a lightning flash followed by thunder three seconds later would mean that the observer was 3,300–3,600 feet (almost 1,000–1,100 meters) away from the lightning strike. A lightning flash followed by an almost immediate crack of thunder means that the strike was uncomfortably close to your location. Features that project some distance above the earth's surface, such as trees or man-made structures, are particularly prone to being struck by lightning. In fact, contrary to the phrase "lightning never strikes twice in the same place," lightning frequently strikes certain projecting targets again and again.

Thunderstorms are always present. At any time, there are more than 2,000 storms generating in excess of 6,000 lightning strikes per minute. The obvious problems for aircraft are mirrored by danger at the earth's surface. The number of lightning deaths per year usually exceeds all other weather-related deaths combined in the United States.

Figure 5.3
Cumulomammatus. NOAA

The greatest danger is in the Southeast where the incidence of thunderstorms is highest. People can reduce the danger by some simple, commonsense rules: avoid open spaces, avoid taking shelter under trees (especially isolated trees), and get out of boats. Buildings and automobiles usually provide relatively competent shelter.

Even those people struck by lightning frequently survive if mouth-to-mouth resuscitation and/or cardiopulmonary resuscitation (CPR) are used. As previously indicated, damage from thunderstorms can be quite severe, since wind, lightning strikes, downpours of precipitation, and hail may all contribute to the destruction.

Figure 5.4
Cumulonimbus.
NOAA.

Figure 5.5
Tornadoes are very destructive.
NOAA.

Hail damage is especially significant in the Great Plains and Midwest. Hailstorms are especially damaging to agricultural crops, although large hailstones that are the size of golfballs or larger may damage buildings and vehicles. High wind gusts, often characteristic of thunderstorms, also cause property damage.

As damaging as they are, thunderstorms are exceeded in violence by other types of storms. They are discussed in the next sections of this chapter.

Tornadoes

Tornadoes are the most violent storms in nature. Fortunately, they are usually quite small and localized. Typically, they are extremely low pressure, rapidly rotating funnel clouds that carry winds probably well in excess of 200 miles (300 kilometers) per hour (figure 5.5). The pressures within tornadoes are extremely low—frequently 100 millibars less than pressure in surrounding air. The very steep barometric gradient accounts for the high wind velocities.

The great destruction caused by a tornado is due both to the strong winds and the almost explosive movement of air from buildings into the low pressure column of swirling, rising air. Although tornadoes usually have rather short trajectories over the ground and erratically bounce into and descend from the squall lines that spawn them, they may be totally destructive in the areas where they touch down. Perhaps the only typical thing about a tornado is that each one is atypical. In 1925, for example, a tornado spawned in southeast Missouri, crossed the Mississippi River into southern Illinois, cut a swath across southern Illinois, and crossed the Wabash River before finally dying in southern Indiana. This storm, and many others, effectively debunked the commonly held belief that tornadoes cannot cross rivers. The storm moved along the ground for 125 miles and killed about 625 people in a relatively rural

region at a time when the population was much less dense than it is now. Obviously, this is in sharp contrast to storms that normally are on the ground for only a few miles. Indeed, the "typical" tornado has a destructive width of only a hundred yards or so, and the contact along the ground can usually be measured in a few hundred yards. To repeat what bears repeating, however, tornadoes are noteworthy for their individuality and variability.

Tornadoes usually form when warm, humid air is flowing into a low-pressure cell from the southwest and a cold front is advancing rapidly from the west or northwest. Sharp temperature contrasts are favorable for development of tornadoes along a squall line preceding the cold front or along the cold front itself. Tornadoes occur most frequently during the late afternoon or early evening hours in the Great Plains or Interior Plains of the United States. Although all continents except Antarctica experience tornadoes, about 90 percent of them occur in the United States. The season of greatest frequency for tornadoes is late spring or early summer. At this time, there may be dramatic temperature, pressure, and moisture contrasts between colliding, dissimilar air masses. These contrasts provide the dynamics and energy sources to spawn such storms.

When a tornado approaches, those in its path should take shelter immediately, preferably in a storm cellar or basement. If caught indoors, an interior location is preferable, especially under a heavy piece of furniture, like a table or a desk, because falling or flying debris frequently causes injuries.

If one is caught in the open, it's best to seek shelter in a low area such as a culvert, or a depression, or a ditch. If one is in a car, it's usually possible to get out of the way by moving in a path at right angles to a storm, since the tornado's path along the ground is typically quite narrow. If a person can't outrun it, he or she should take cover in the best available shelter. Unfortunately, tornadoes striking in the middle of the night may announce themselves with a roaring noise only moments before they strike.

Tornadoes cannot be predicted with any degree of certainty, but conditions conducive to tornado formation are well known. From time to time, radio or television may advise listeners and viewers of a **tornado watch.** This means that conditions are right for possible tornado formation—be alert. When a **tornado warning** has been announced, it means that a tornado has been sighted in the vicinity—take cover. It is wise to heed such warnings, for tornadoes are awesomely destructive.

The national center for analysis of severe storms at Norman, Oklahoma is increasing the fund of information about tornadoes with the goal of development of an accurate predictive model and a more precise warning system. The characteristic hook-shape of a tornado on the radarscope allows some time for warning people in areas of imminent danger. Meteorologists continue to seek ways to predict tornadic occurrences with greater accuracy. Many of these intrepid scientists get into vehicles and go hunting tornadoes when atmospheric conditions suggest possible tornado development to learn more about them.

Tornadoes occur over land, although the similar whirling funnel-like shape is characteristic of a **waterspout,** which occurs over water. The pressure gradient within a waterspout is much less than in a tornado, and the winds are much weaker. Waterspouts tend to be very short-lived phenomena. A waterspout is much like the terrestrial **dust devil,** which is a localized whirlwind of minimal destructive capacity occurring in the heat of the day in midsummer primarily in the Plains states.

Although tornadoes are the most intense storms occurring in nature, hurricanes frequently cause greater damage because they are larger, affect a much larger area, and persist for days at a time. A discussion of the salient characteristics of hurricanes follows.

Hurricanes

Tropical cyclones have a variety of names depending on the part of the world where they occur. This tropical and subtropical storm is called **hurricane** in the Western Hemisphere and **typhoon** in most of Asia, although it may be termed **baguio** in the Philippines. They are called simply **cyclones** in the Indian Ocean and Australia, and they are termed **willy-willies** in certain South Pacific islands. All are variants of the same storm type, and develop under the same meteorological conditions, and yield the same legacy of devastation and destruction (figure 5.6).

The typical storm, an intense low pressure, with winds of 74 miles (119 kilometers) per hour or more, typically has 50–100 millibar lower pressure at the center than at the periphery of the storm. Winds circulating into the center of the storm in the normal hemispheric low-pressure circulation (i.e., counterclockwise in the Northern Hemisphere and clockwise in the Southern Hemisphere) are intense and may exceed 150 miles (240 kilometers) per hour in a well-developed storm.

Figure 5.6
Typical tropical cyclone paths.

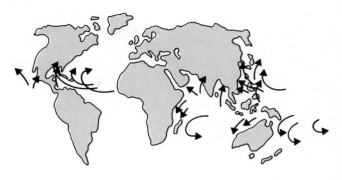

Figure 5.7
Pattern of easterly wave development.
Ralph K. Anderson, et al., Application of Meteorological Satellite Data in Analysis and Forecasting, *ESSA Technical Report NESC 51, NOAA, 1974.*

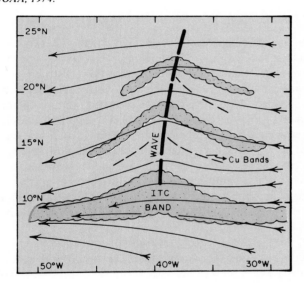

In much of the storm, there are significant updrafts of air that are quite pronounced around the margins of the calm **hurricane eye.** This **eye wall,** around the periphery of the eye, is an area of violent updrafts, whereas the eye is a calm, cloudless area with air slowly descending in the center. The calm eye stands in marked contrast to the seething cauldron of turbulence, cloudiness, and precipitation elsewhere in the storm.

The relatively circular pattern of isobars around a hurricane covers a smaller area, typically about 100 to 600 miles (160 to 965 kilometers) than the ovoid isobaric pattern of a middle-latitude low, which may have a diameter of more than 1,000 miles (1,600 kilometers). Hurricanes normally develop along easterly waves in the subtropics or margins of the tropics. These easterly waves are zones where the isobars bend in an easterly direction and there is some convergence of air (figure 5.7). The exact mechanism for formation is uncertain; a great many easterly waves develop but never become hurricanes. If a circulation develops with wind speeds less than 39 miles (about 62 kilometers) per hour, they are known as tropical depressions. If the circulation continues and wind velocities measure 39 to 74 miles per hour, they are known as tropical storms. Above 74 miles (about 118 kilometers) per hour, they become hurricanes. People in subtropical areas, especially, need to be alert to the possibility of hurricane development, especially during the warm months. In the United States, hurricane development usually occurs from June through October.

The storms develop typically between 5° and 20° latitude over water in the warm months and ride westward on the trades. Such storms often recurve and move northwestward, northward, or northeastward. They last typically for several days and may persist for up to three weeks. Hurricanes move forward at varying speeds, but the average forward movement is 10 to 20 miles (16 to 32 kilometers) per hour.

Mechanisms of Hurricane Destruction

Hurricanes cause destruction in a variety of ways including their high winds that may uproot trees and destroy buildings. Not infrequently, this destruction is augmented by tornadoes, which may develop around the periphery of the storm that has made landfall. These skirting tornadoes add a dimension of uncertainty and destructiveness to a number of hurricanes. In addition, hurricanes are often accompanied by torrential downpours of several inches of rain per hour, which may cause flash flooding. The combination of wind-driven rain may exert considerable hydraulic pressures on exposed surfaces, such as the walls of buildings, adding to a storm's destructive impact.

By far, the greatest destruction along coastal regions occurs because of the **storm surge.** The storm surge is built up because of high winds with a long fetch, or unimpeded trajectory, over open water, and because the surface of the sea actually rises in response to diminishing pressure. Indeed, this rise of water level in response to diminishing pressure amounts to about an inch (2.54 centimeters) for each 2.5 millibar drop in pressure. A 3½-foot rise (about one meter) in sea level in a violent storm, augmented by the high winds, may produce a storm surge of 20 feet (more than 6 meters) or more. The destruction of such surges is almost unbelievable. A wall of water boils across low-lying areas removing everything in its path. Such areas often appear to have been scraped bare by a giant bulldozer blade after the storm surge has receded. Destruction in such a storm surge may be near total.

The movement of people to coastal regions for all the amenities such areas offer has created a potential for great death and destruction during a major storm. A number of relatively small storms and near misses of major

storms in the early 1980s has led many people to a false sense of security. Higher population densities and ill-founded optimism about the destructive power of hurricanes have set the stage for a major disaster during a violent storm. It is almost a safe bet that such a disaster will occur before the twentieth century is completed.

It was such a false sense of security that added to the number of deaths and injuries during Hurricane Camille in 1969. In spite of repeated warnings to evacuate, a number of people decided to ride out the storm along the coastal margin of the Gulf of Mexico. Many of those people perished because of a foolhardy sense of bravado. In fact, a substantial number vanished without a trace as Camille stormed ashore along the Mississippi Gulf Coast.

Fortunately, in spite of their somewhat erratic movements, the tracks of hurricanes can be readily plotted through the use of satellite imagery and weather aircraft penetration. Reasonable predictions as to probable landfall can be made, and those in danger can be asked to evacuate well in advance of the time the storm is likely to strike. Unfortunately, evacuation plans in many areas are

Figure 5.8
Typhoon tip.
NOAA.

inadequate, and evacuation routes may be insufficient to accommodate those requiring evacuation. In addition, an all too complacent citizenry may resist an evacuation order.

When a hurricane approaches, evacuate if you live in a low-lying area and take shelter in authorized Civil Defense shelters, or move far into the interior. Hurricanes quickly lose their source of energy (warm, tropical water) when they move over land, and frictional drag over the surface tends to reduce wind velocities, although they may persist as areas of disturbed weather for some time and frequently bring copious precipitation far into the interior. A few hurricanes carry their destructive powers well into the interior. Camille, for example, reduced some pine forests in its path to confused piles of jackstraws scores of miles inland. Frederic, on the other hand, produced significant timber damage only relatively close to the coast. Major hurricane damage normally tends to be restricted to the coastal margins.

In addition to the obvious danger to life and property, hurricanes can completely alter coastal terrain, making shoreline modifications in a few hours, which haven't been experienced in hundreds—even thousands—of years of less dramatic actions by the usual coastal processes.

Figure 5.9
Radar trace of Frederic's eye.
NOAA.

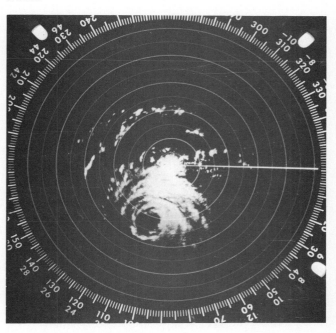

Figure 5.10
Wind trace of Frederic.
NOAA.

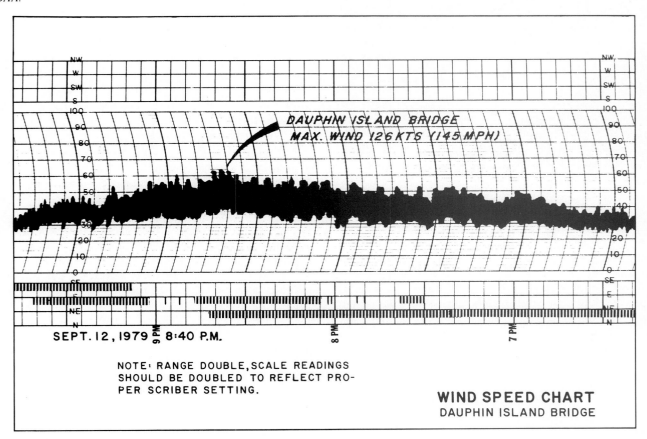

Figure 5.11
Timber damage associated with Hurricane Frederic.

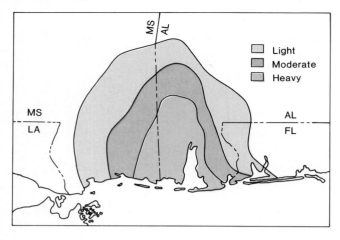

Light
Moderate
Heavy

Figure 5.12
Aerial view of some hurricane damage along a barrier island in northwest Florida.
Jerome Coling.

Extratropical Hurricanes

Although the hurricane that develops in the tropics is the typical storm to achieve hurricane-force wind velocities (greater than 74 miles (119 kilometers) per hour), certain storms developing in Arctic waters may achieve hurricane status. Formation of such storms is somewhat enigmatic, but invasions of warmer waters from lower latitudes may be a significant causal factor.

These storms have received much less attention because they are usually relatively short-lived, and they occur over water or along coastlines that are very sparsely populated. Some comparative data of the extratropical and the tropical hurricanes show their essential similarities and differences.

These Arctic hurricanes typically develop between October and April, whereas the typical Northern Hemisphere hurricane develops between June and October. Air temperatures in Arctic hurricanes are from $-20°$ to $55°$ F ($-29°$ to $13°$ C) compared with the subtropical hurricane air temperature of approximately $80°$ F ($27°$ C).

Arctic hurricanes develop within twelve to twenty-four hours, whereas tropical types require three to seven days. Arctic hurricanes have diameters of 60 to 300 miles (100 to almost 500 kilometers), while tropical types are typically 100 to 600 miles (160 to almost 1,000 kilometers). The eye in each varies from 5 to 100 miles (6.6 to 160 kilometers). Wind velocities in the Arctic storm vary between 50 and 100 miles (80 to 160 kilometers) per hour, and in the tropical storm between 75 and 180 miles (120 to 290 kilometers) per hour. Technically, of course, neither storm is a hurricane unless velocities greater than 74 miles per hour are achieved.

The Arctic storm has a forward speed of 30 miles (almost 50 kilometers) per hour, whereas the tropical variety moves between 10 to 20 miles (16 to 32 kilometers) per hour. Typical lowest pressure in the Arctic storm may be 940 millibars, and in the tropical variety the pressure may drop to 900 millibars. Arctic storms develop to a height of 5 miles (8 kilometers); tropical storms develop to a height of 12 miles (almost 20 kilometers).

Although these Arctic storms are less well known, they are clearly to be reckoned with. They add muscle to the capricious forces of weather and become a potent meteorological, oceanographic, and geomorphological element in certain regions.

As in the case of all storms, a hurricane becomes a geomorphological agent as well as a meteorological event. The effects of storms on landforms stand out in the slower pace of landform change during more normal climatological events. In a sense, these dramatic changes mirror occasional rapid evolutionary changes in the biological world in which long periods of slow evolutionary change are punctuated at intervals by rapid change.

Fortunately, a close network of weather stations, augmented by reports of ships, aircraft, and satellites makes it possible to plot the progress of a hurricane and usually to project probable landfall with some accuracy. Less dramatic weather changes are often more difficult to predict, although a great deal of money, effort, and energy are expended in an attempt to predict weather reliably.

Table 5.1
Deadliest Hurricanes in the United States 1900–1972

Hurricanes	Year	Deaths
Texas (Galveston)	1900	6,000
Florida (Lake Okeechobee)	1928	1,836
Florida (Keys/South Texas)	1919	600–900*
New England	1938	600
Florida (Keys)	1935	408
AUDREY (Louisiana/Texas)	1957	390
Northeast U.S.	1944	390†
Louisiana (Grand Isle)	1909	350
Louisiana (New Orleans)	1915	275
Texas (Galveston)	1915	275
CAMILLE (Mississippi/Louisiana)	1969	256
Florida (Miami)	1926	243
DIANE (Northeast U.S.)	1955	184
Florida (Southeast)	1906	164
Mississippi/Alabama/Pensacola, Florida	1906	134
AGNES (Northeast U.S.)	1972	122
HAZEL (South Carolina/North Carolina)	1954	95
BETSY (Florida/Louisiana)	1965	75
CAROL (Northeast U.S.)	1954	60
Southeast Florida/Louisiana/Mississippi	1947	51

*Over 500 of these lost on ships at sea
† Some 344 of these lost on ships at sea
Source: NOAA Technical Memorandum NWS NHC 7, *National Hurricane Center, Miami, Florida, August 1978.*

Figure 5.13
Aerial view showing sand transport along seaward margins caused by a hurricane.
Jerome Coling.

Weather Forecasting

Weather forecasting is a mixture of art and science and, in some ways, is similar to the diagnosis and prognosis for a disease by a physician. Both fields of science require a number of observations; in both areas, an accurate analysis or diagnosis is necessary prior to weather forecasting or a physician's prognosis. It is, of course, impossible to isolate weather phenomena and conduct controlled experiments on natural processes. Rather, it is necessary to study atmospheric data from thousands of places almost simultaneously. In addition, it is possible to simulate weather conditions in computer programs and to observe certain phenomena in pressure chambers and wind tunnels. Such models, which are becoming increasingly sophisticated, are adding to our knowledge and prediction of weather events, and certain phenomena observable in miniature in experimental circumstances provide new insights into weather characteristics and patterns.

In order to evaluate atmospheric data collected from a network of ground stations, ship and aircraft observations, and satellite telemetry and imagery, weather forecasters need a concise picture of weather patterns,

Figure 5.14
Radar image illustrating the hook form characteristically displayed by tornadoes.
NOAA.

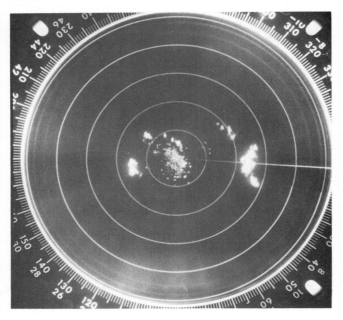

phenomena, and interactions. A synoptic chart provides a map of existing weather conditions, and trend analyses permit the observer to make rational predictions as to future occurrences. At one time weather maps were all laboriously produced by hand plotting, and a very few are still produced in that way, especially in developing countries, but sophisticated computer programs permit the production of maps by automated means, especially in the developed societies. Automated maps are labor-saving interpretive devices that can be produced in a more timely fashion than weather maps drawn by hand.

Such weather maps are very important for geographers as well as meteorologists, since a sequence of three or four maps provides a history of weather change and, simultaneously, provides a vehicle for predicting future weather occurrences. Weather forecasters are able to determine the speed of air mass and frontal movement, to determine whether a given pressure area is building or deteriorating, and whether a particular weather system is gaining or losing intensity (figure 5.15). They can also determine whether an air mass is retaining its physical integrity or whether it is taking on the conditions near the earth's surface and, therefore, being rapidly modified.

Figure 5.15
Weather map showing isobars, pressure cells, and fronts.
NOAA.

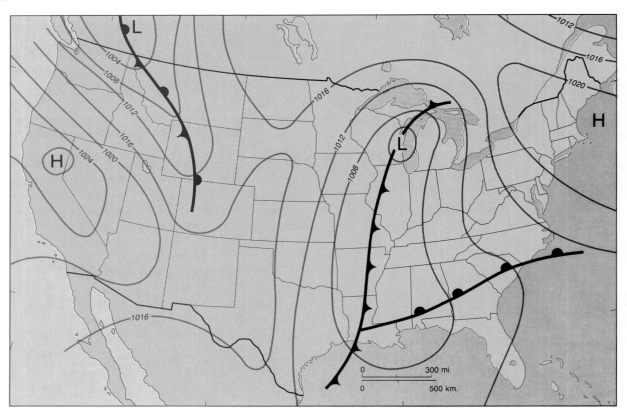

All students of geography should be able to interpret a weather map accurately (figure 5.16). Such maps contain an enormous amount of information about weather conditions existing in a given period of time over a large geographical area. The United States Weather Bureau uses all possible sources of data, not only to provide usual forecasts, but also to issue warnings about approaching storms, cold waves, likelihood of frost, prospects for flooding, heat waves, and the like. Such warnings can save lives and property, and reduce the harmful effects of bad weather. Most people, especially scientists, agree that the more weather data available, the better off we all are.

As a general rule, the accuracy of a forecast diminishes as the time period involved increases. A 24-hour forecast is usually much more accurate than a 3-day forecast, which is typically much more accurate than a 30-day forecast. Forecasters usually hedge their forecasts by reporting the statistical likelihood of the occurrence of a particular event. For example, the forecast may call for a 50% chance of rain. Meteorologists should not be criticized for such efforts, since the dynamic atmosphere is full of surprises where two apparently identical sets of preconditions may yield different weather outcomes.

Considerable effort is being exerted today to achieve more accurate weather prediction. With the use of ground stations more closely spaced than formerly, sophisticated new remote sensing devices, complex computerized simulations, and new scientific theory, more accurate forecasts are a likely outcome. Like the physician's diagnosis and prognosis, however, outcomes are not always as predicted, but both physicians and weather forecasters must use science in the art of diagnosis or forecasting.

Table 5.2
U.S. Natural Hazard Deaths (Yearly Totals)

	1980	1981	1982	1983	1984	1985	1986
Floods	99	90	155	200	125	168	208
Hurricanes	4	0	0	22	4	30	8
Lightning	74	66	77	77	67	73	68
Tornadoes	28	24	64	34	122	94	15

Source: National Weather Service, National Oceanic and Atmospheric Administration.

Figure 5.16
Weather Station Model Symbols.
NOAA.

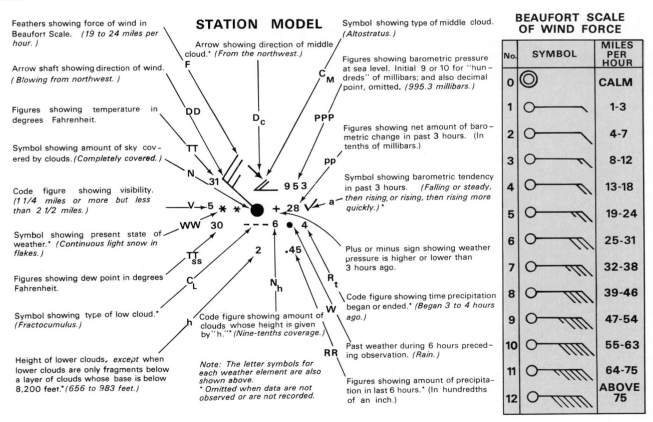

Mathematical models that simulate the atmospheric conditions under numerous scenarios always fall short of perfection because of the almost infinite complexity of actual conditions. Although more precise models are being developed as the mathematics employed is more complex and the computers used are faster, the models will always fall short of perfection in the face of fickle nature. One hundred percent accuracy in weather prediction is probably an unachievable goal.

The long-term aspects of average weather over time produce a climatological pattern that is a fundamental aspect of man's environment. Taken collectively, various elements of the environment, including weather and climate, provide the framework within which people live. These additional elements add to the mosaic that is the earth's environment. Storms add elements of uncertainty and intensity that impact on human adjustments to environment. They are, nevertheless, part of a total meteorological framework, which over time becomes a part of the climatological element of the environment.

Study Questions

1. List the various types of thunderstorms in terms of their origins and development.
2. What is wind shear? What are the principal dangers associated with it?
3. Where are tornadoes most numerous? What is responsible for the great destruction that they bring? What are the meteorological conditions conducive to their formation?
4. What are some appropriate safety tips that apply to tornadoes?
5. What are some regional names applied to the tropical cyclone-type storm?
6. What is a storm surge? Why is it so destructive?
7. What are some appropriate actions to be taken with the approach of a hurricane?
8. Compare and contrast Arctic and subtropical hurricanes.
9. What actions should be taken if you are in the open during thunderstorms?
10. What accounts for increasing accuracy in weather forecasting?
11. Why is weather forecasting not an exact science?
12. Differentiate between weather and climate.

Selected References

Fellows, D. K. 1985. *Our environment: An introduction to physical geography.* 3d ed. New York: John Wiley and Sons.

Flohn, H. 1969. *Climate and weather.* New York: McGraw-Hill Book Company.

Flora, S. 1957. *Hailstorms.* Norman: University of Oklahoma Press.

Flora, S. 1955. *Tornadoes.* Norman: University of Oklahoma Press.

Hare, F. K. 1963. *The restless atmosphere.* New York: Harper and Row, Publishers.

McKnight, T. L. 1984. *Physical geography: A landscape appreciation.* Englewood Cliffs: Prentice-Hall.

Navarra, J. G. 1979. *Atmosphere, weather, and climate: An introduction to meteorology.* Philadelphia: W. B. Saunders Company.

Simpson, R. H., and Riehl, H. 1981. *The hurricane and its impact.* Baton Rouge: Louisiana State University Press.

Weisberg, J. S. 1976. Meteorology: *The earth and its weather.* Boston: Houghton Mifflin Company.

6

Climatic Influences, Controls, and Classifications

Storms are significant climatic controls—this advancing squall line will likely bring rain and perhaps lightning, hail, and high winds.

*C*limate is central to a geographic appreciation of environment, to a humanistic understanding of civilizations, and to the establishment of parameters for future developments. Climate plays a role in determining whether we will eat and, if so, what we will eat; it is a significant factor in influencing the type of shelter required and the clothing that is worn; and it has a bearing on virtually all aspects of civilization. No human being, no landform, and no terrestrial plant or animal can escape the impact of climate. Certain regions are seriously handicapped if there is a deficit of precipitation, if growing seasons are short, or if temperatures are excessively hot or cold. Others may be blessed with a desirable environment of mild temperatures with sufficient and reliable amounts of precipitation from season to season and year to year.

In the past, certain environmental determinists have laid progress (or lack of it) at the doorstep of climatic characteristics. Few would now embrace such deterministic theories, but the influence of climate on man is so pervasive and its role as an aspect of the physical environment so comprehensive that it is essential to understand climatic characteristics and patterns in order to appreciate the physical world.

In addition, climate is a major factor in influencing the biosphere, hydrosphere, and lithosphere. The type of vegetative mantle and dominant animal species, the character of surface and subsurface water supplies and drainage patterns, and the nature and character of agencies of weathering and erosion are principally related to the aspects of a particular climatic regime. Indeed, next to structure and materials, climate is the most important factor in etching the face of the physical landscape.

Several factors, taken together, influence the development of a particular climate in a specific region. These influences act and interact in intricate ways to produce the salient elements of a climatological region.

Climatic Influences

Weather is the state of the atmosphere at any given time. In a sense, it is like a snapshot of temperature, humidity, precipitation, wind, cloudiness, pressure, and other meteorological elements *at a specific time.* Weather maps are graphic depictions of specific meteorological conditions at a particular time. **Climate,** on the other hand, is a composite of weather conditions *over a long period of time.* Climates are produced by the interaction of several major climatic influences and a number of minor ones. The principal influences are (1) sun and latitude; (2) the distribution of land and water bodies; (3) ocean currents; (4) altitude; (5) topographic barriers; (6) winds and air masses; (7) storms; and (8) the attitude, or orientation, of coastlines. These factors, operating in conjunction with minor topographic or atmospheric conditions, combine to produce the array of climatic patterns that characterize the earth. A description of their impact on climates follows.

Sun and Latitude

Our sun is the only important source of heat for the earth, although radioactive decay of certain rocks and volcanic activity add minor amounts of heat to the earth's surface. The amount of radiant energy received at the surface of the earth is called insolation. Since the sun is emitting radiant energy in all directions, the earth is intercepting only a tiny fraction of that radiant energy. In addition, various agents like clouds, dust, and particulate matter reflect substantial quantities of that energy back into space before it can effectively heat the atmosphere or the earth's surface. The amount of insolation received varies in response to changes in the angle at which the sun's rays strike the earth, the length of day and night, and the amount reflected back into space. Sun's rays that strike the earth at an oblique angle are dispersed over a wider surface area and are, consequently, less effective than more nearly direct rays. At high latitudes, the low-angle rays must pass through a greater thickness of atmosphere, losing more heat to the atmosphere in passing through to the earth's surface. There is, in effect, a diminished efficiency in heating as one proceeds from the equator towards the poles.

Since the average length of daylight and darkness for the year is the same everywhere, duration of day and night is not a significant factor in the amount of insolation received during the entire year, but it is important in causing summers to be warmer and winters to be colder in the middle and high latitudes than they are in the low latitudes. The long days of high-latitude summers permit increased warming, whereas the shorter days of a high-latitude winter diminish the effectiveness of the weak, low-angle sun's rays even further. At the equator, the length of daylight and darkness is always nearly equal, and the more nearly direct sun's rays are more effective in warming the earth's surface at low latitudes. The oblique rays received in high latitudes during the winter and the more nearly direct rays received in the summer account for the shift from cold to warm season.

Most meteorologists and geographers assume that the amount of solar energy received at the earth's surface is a constant, although the sun is a variable star. Certain research suggests that climatological events may be influenced by this variability, but no conclusive data have been developed, and no comprehensive theory has yet been advanced to explain such climatological influences. It is clear that sunspots (gigantic magnetic storms) on the surface of the sun influence patterns of precipitation on the earth's surface. Although there appears to be a certain periodicity in sunspot maxima and minima, there is no persuasive theory as to what is responsible for such variations.

Further, the elliptical orbit of the earth about the sun causes its distance to vary from about 91.5 million miles (more than 147,000,000 kilometers) to about 94.5 million miles (more than 152,000,000 kilometers). These distance variations do influence insolation, but apparently the variations are of little significance compared to the angle at which sun's rays strike the earth's surface. It should be recalled that the sun is nearest the earth in January at the height of the cold season in the Northern Hemisphere, and it is furthest away in July in the midst of the Northern Hemisphere summer.

Distribution of Land and Water Bodies

One major characteristic that differentiates the earth from the other planets in our solar system is the great system of interconnected oceans and seas that covers about 71% of the surface. Indeed, the earth as seen from space is very clearly dominated by water contrasted with the relatively small amount of land. Extraterrestrials, observing the earth from space, would undoubtedly give it their name for water rather than earth.

Satellite observations and unmanned probes demonstrate that Mars has no liquid water. Mercury apparently has no atmosphere at all; Venus certainly has no liquid water beneath its poisonous atmosphere; other planets are also almost certainly lacking in liquid water. Certainly, none of them have anything remotely resembling the abundance of water found on the surface of the earth. The oceans and seas of the earth serve as reservoirs of moisture and are major influences in moderating the temperatures at or near the earth's surface. Presumably water was released in a variety of complex geochemical processes attendant to the solidification of the earth from a cloud of cosmic gas and dust. It is assumed that the release of water vapor initiated clouds that precipitated copious quantities of water for centuries until the ocean basins filled and something like the hydrological cycle that exists today was established. The continued cycle of evaporation and precipitation is a prerequisite to the continuation of life on earth.

Water bodies heat and cool more slowly than land because (1) water has a higher specific heat than land (i.e., it takes a greater quantity of heat energy to heat a given volume of water than a like volume of earth material); (2) light and associated heat penetrate transparent or translucent water to some depth, increasing the thickness of material being warmed, whereas there is no comparable penetration of the solid earth; (3) there tends to be a greater reflection of solar energy from water surfaces than from the land; and (4) water moves or circulates, and in the process transfers heat energy to other segments of the water column. Water bodies tend to stabilize temperatures and, as a result, the adjacent atmosphere tends to show a lower temperature range seasonally, or diurnally, than the atmosphere over land areas. Areas at a greater distance from the sea tend to have hotter summers and colder winters than areas near the coast in similar latitudes. Coastal or marine climates tend to experience fewer temperature extremes than comparable continental locations. For example, waters in the Gulf of Mexico along the American Gulf Coast are typically in the 80-degree (F) range in summer and in the 50-degree range in winter.

As a result, temperatures along the coastal margin rarely exceed the 90-degree range in the summer or drop below the 20-degree range in winter, whereas areas a few score miles into the interior might experience highs in the 100-degree range and lows in the teens. The proximity of water and its modifying influence reduce the range of temperatures from season to season. Continentality infers marked temperature contrasts from season to season, whereas maritime conditions are characterized by less significant seasonal swings in temperature.

Ocean Currents

The ocean currents are closely associated with oceans and seas as climatic influences. Currents exist in all substantial bodies of water, but great patterns in oceans and seas transport enormous amounts of cold and warm waters with concomitant effects on climate. The surface currents are related to wind and pressure systems, and the gross patterns may be considered moving in large circulatory patterns, known as **gyres,** around the subtropical highs. In the Northern Hemisphere, this circulation is clockwise due to the Coriolis effect, whereas in the Southern Hemisphere it is counterclockwise. Waters near the equator tend to be driven westward by predominantly easterly winds. Those currents carry a number of local names, but the general flow may be termed the Equatorial Current. These waters are deflected by land masses and the Coriolis effect near their poleward margins, and move into middle latitudes as warm currents. In middle latitudes, the deflective effect of the Coriolis force and the westerly winds drive these currents across the oceans as warm currents or drifts. When they strike continents, much of the water is deflected equatorward. The waters have cooled relative to surrounding water by a trajectory in higher latitudes, and they continue their journey as cold currents until they rejoin the Equatorial Current. Cold currents along the western margins of the continents tend to be reinforced by upwelling cold water brought to the surface by the persistent effect of the Trades blowing surface water towards the west.

Currents are driven not only by wind and pressure belts, but by differences in density caused by surface heating, variations in salinity, and evaporation. Since currents are described as warm or cold when temperatures within the waters are compared to surrounding ocean waters, typically those moving from lower to higher latitudes are warm, whereas those moving from higher to lower latitudes are cold (figure 6.1).

Figure 6.1
Generalized map of surface ocean currents.
From Arthur Getis, Judith Getis, and Jerome Fellmann, Introduction
to Geography, *2d ed. Copyright © 1988 Wm. C. Brown Publishers,*
Dubuque, Iowa. All Rights Reserved. Reprinted by permission.

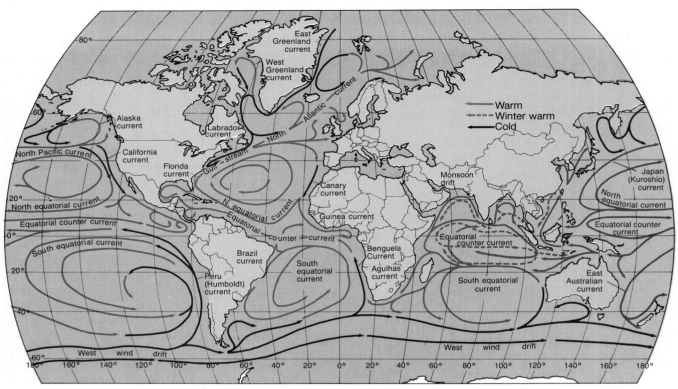

Warm currents tend to warm adjacent land masses, and that effect is most pronounced when augmented by a prevailing onshore wind. For example, climate in northwestern Europe is decidedly mild, especially in winter, when compared to comparable latitudes not directly influenced by the North Atlantic Drift. London, with a temperate climate, is in the same latitude as Labrador and central Canada, where severe continental climates are the rule. On the one hand, London is warmed by air moving onshore across the relatively warm North Atlantic Drift. On the other, Labrador has prevailing offshore winds, and, in any case, the coastal waters are influenced by the cold Labrador Current.

Warm currents also tend to be sources of moisture, since air moving over the currents tends to be warmed, often causing conditions of instability. Cold currents are often associated with semiarid or arid conditions, since air adjacent to and over such currents is chilled, and stability is increased. When cold and warm currents meet, there are often frequent fogs because of the clash of cooler and warmer air. Indeed, some of the world's foggiest regions occur where warm air over the warm current comes in contact with the colder air associated with a cold current.

It is ironic that some of the world's great fishing areas often occur in regions where two contrasting currents meet. The mixing that develops creates a favorable habitat for the planktonic base of the biotic pyramid. These same temperature contrasts, however, often create foggy conditions that make fishing in such areas hazardous.

Basically, the system of ocean currents allows for the surface transport of ocean water from lower to higher latitudes, whereas colder, denser water from high latitudes moves equatorward to replace the waters that have moved to polar positions because of the effects of the heating effect of the sun. This constant movement of waters ensues as nature attempts to bring to equilibrium a shifting spectrum of temperatures, densities, and salinities in the oceans and seas of the world. At the same time, those movements of water affect the temperature and humidity of adjacent air masses, which, in turn, have significant meteorological and climatological ramifications.

A shift of currents may create catastrophic modification of existing climatological norms. The weakening or temporary diversion of the Gulf Stream may contribute to unusually cool summers or cold winters in northwestern Europe. Such shifts may also affect fish, and, in turn, fishermen may experience declines in fish catches in usually

Figure 6.2
Pattern of water temperature during development of El Niño.
Edward Edelson and Mosaic.

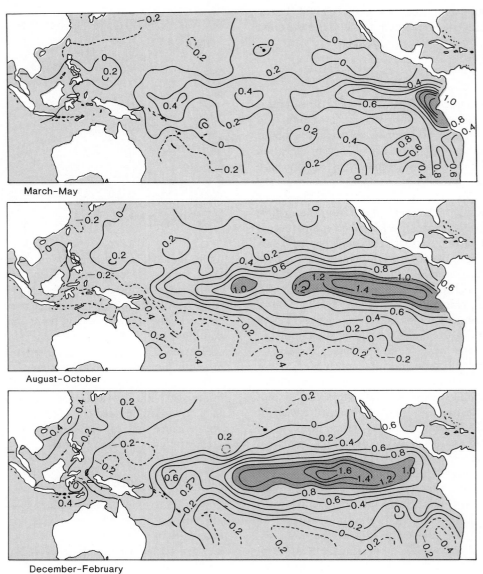

March-May

August-October

December-February

reliable fishing areas. Clearly, such declines can have catastrophic economic effects. Conversely, a strengthening of the current may make for unusually mild winters and warmer than normal summers.

Shifts in the strength and pattern of the Equatorial Current in the Pacific appear to be responsible for the modification of a current called **El Niño.** Significant shifts in the intensity and location of El Niño apparently are responsible for unusual patterns of precipitation and drought, as well as temperature extremes, in low latitudes and middle latitudes. So far, scientific knowledge has not advanced far enough to explain the causes of such strengthening or weakening of currents.

Retrospectively, it is possible to explain *what* happened and the consequences of a shift in position, temperature, or persistence of a current, but it has not been possible to explain *why* a specific event occurred. Strengthening, weakening, or shift of El Niño seem to begin in pressure rises and falls of the southeast Asian archipelagos, which then extend across the Pacific in a series of pulses ultimately affecting El Niño. Meteorologists are learning to recognize the signs for shifts of El Niño, but theories and models have not yet satisfactorily explained *why* the changes occur. Since the climatic changes associated with this shift are often dramatic, widespread, and undesirable, a great deal of additional research will be undertaken to improve forecasting of this event (figure 6.2).

In 1982, for example, the encroachment of warm El Niño waters along the Peruvian coast restricted the upwelling of cold water that resulted in torrential rains along a coast that is normally desert. At the same time, droughts were experienced in many sections of the United States. Both areas were adversely affected, but in quite different ways.

Altitude

Elevation has pronounced effect on temperatures, and it may be influential in affecting precipitation amount and distribution. The average reduction in temperature is about 3½° F/1,000 feet (6° C/kilometer). A rise in elevation causes a significant lowering of temperature from the base to the crest of a mountain. Radiant energy penetrates the thin air of higher elevations without heating it effectively. The principal heating of the atmosphere from the surface of the earth upward is responsible for cooler temperatures at higher elevations and warmer conditions near the surface (figure 6.3). In stable air, the temperature at the base of a 10,000-foot mountain might be 90° F (32° C), whereas at the crest it might be only 55° F (13° C). Elevation provides significant relief from the stifling heat of the lowlands in tropical regions, and sojourners in such lands have been quick to take advantage of such highland areas if they have the means and opportunity to do so.

Tropical highlands, like those of Central Mexico, are often cooler than areas at lower elevations in the middle latitudes. For example, at Mexico City at an elevation of about 7,500 feet, the average annual temperature is about 60° F (15° C). The average annual temperature at El Paso, located 1,000 miles (1,600 kilometers) north of Mexico City and closer to sea level, is 63° F (17° C). The effects of altitude on temperature may be as great in a few thousand feet as that normally seen in hundreds of miles of latitude at low altitudes. Snow clad peaks, like Kilimanjaro in central Africa, which lie almost astride the Equator, testify to the cold conditions at the crest, although the base may be sweltering in tropical heat. Tropical Hawaii may periodically offer snow skiing in the upper reaches of high volcanic peaks, while people are water skiing on the surrounding tropical seas. The vertical zonation of biota and human pursuits on such a tropical mountain is dramatic and significant in the mosaic pattern of life on this planet. It's possible to move from tropical selva at the base through temperate and subarctic forms to permanent snow and ice at the crest of high mountains. For example, the sweltering heat of the Indo-Gangetic Plain is succeeded by permanent snow and ice a few miles away at high elevations in the Himalayas.

As a general rule, increased elevation accentuates precipitation on windward slopes. For example, the amount of precipitation received in the Black Hills is substantially greater than in surrounding sections of the Great Plains. Precipitation in the Ahaggar and Tibesti Domes in the central Sahara is significantly greater (perhaps as much

Figure 6.3
Elevation results in lower temperatures, and mountains may develop a variety of microclimatological conditions.
Daniel Ehrlich.

as 30 inches (76 centimeters) in the mountains and less than 5 inches (12 centimeters) in peripheral areas) than in the surrounding desert. The hilly or mountainous features create enough air turbulence to cause air to be lifted to elevations where the dew point may be reached. The combination of decreased temperature and increasing moisture with increased elevation tends to produce a microcosm of climates and vegetation patterns in a very small area. Broad climatological regions obviously do not adequately explain the variety of microclimates encountered in a mountainous region. Indeed, the heterogeneity of temperature and rainfall regimes in high mountain regions stretches the imagination.

Topographic Barriers

Topographic features may influence climates in very significant ways. Windward slopes may receive copious quantities of moisture because of orographic lift, whereas leeward slopes may be quite dry (figure 6.4). Such barriers are also very important in influencing the tracks of air masses and storms. For example, the locations of the Rockies and the Appalachians tend to channel invasions of cold polar air or warm gulf air into the Great Plains and interior plains of the United States. The Rockies and Sierra Nevada–Cascades tend to shelter the Pacific Coast of the United States from invasions of cold Arctic air.

The effects of the inblowing Indian monsoon are less significant in central Asia, since the Himalayas and associated mountains, are higher than the average thickness of the inblowing air, typically about 18,000 feet (about 5,500 meters). Similarly, India tends to be sheltered from very cold outbreaks of air from the interior of the continent by the same high mountain barriers, although occasional spillover of cold air from the Tibetan Plateau through high mountain passes may bring a chill to the air of northern India.

Figure 6.4
*This precipitation map (in inches for November 1985)
illustrates how moisture-bearing winds from the west are
intercepted and, as a result, central and eastern Washington
experience drier conditions.*
From Arthur Getis, Judith Getis, and Jerome Fellmann, Introduction
to Geography, *2d ed. Copyright © 1988 Wm. C. Brown Publishers,
Dubuque, Iowa. All Rights Reserved. Reprinted by permission.*

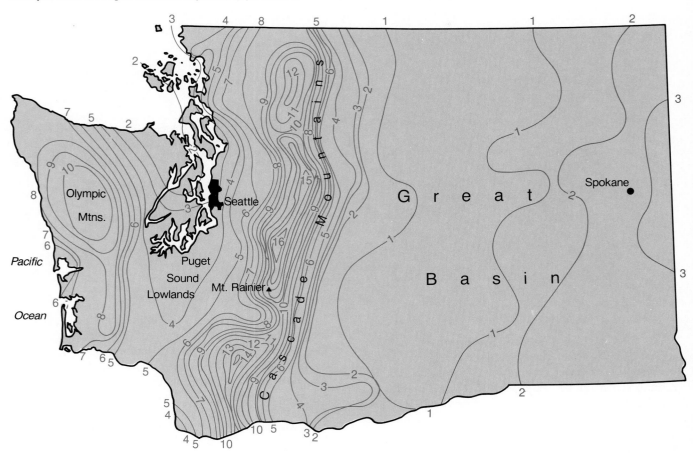

In India, the Western Ghats, which lie at right angles to prevailing winds during the inblowing monsoon may receive well in excess of 80 inches (more than 200 centimeters) of rain a year, whereas the interior of the Deccan Plateau on the lee side of the mountains may receive 30 inches (76 centimeters) or less. Similar contrasts exist between the wet windward slopes of the Olympic Mountains of Washington and the drier city of Seattle. Indeed, these marked contrasts between wet windward and dry leeward sides of pronounced topographic barriers are repeated in many areas of the world.

The pronounced differences between wet windward and dry leeward slopes are explained in adiabatic cooling and heating. Air is cooled at varying rates (i.e., the wet adiabatic rate and the dry adiabatic rate). If condensation occurs on the windward slope, and air is warmed at the dry adiabatic rate on the leeward side, the resultant descending winds are drying. These local descending and drying winds, called **chinooks** in the United States and **foehns** in Europe, have very low relative humidities and may prove to be very trying on the nerves and sensibilities of people if the winds are persistent for several days. Certain areas tend to be extraordinarily dry, because air from almost any direction is descending, warming, and therefore drying. The Dead Sea in Israel and Death Valley in California, both below sea level, are such locales. This descending air contributes to the high temperatures experienced in such areas during the summer season.

Obviously, continuous high barriers have more pronounced influences on climate than discontinuous and lower barriers. It's clear that all topography has some climatological effect because of frictional drag of air moving over the surface. Minor perturbations in atmospheric flow may create interesting localized climatological effects, some of which may be transitory, whereas others are persistent.

Figure 6.5
*Generalized map of the pattern of the inblowing and
outblowing monsoon circulation. (A) represents average lower
level winds for January, the outblowing monsoon, whereas
(B) represents the inblowing monsoon in July.*

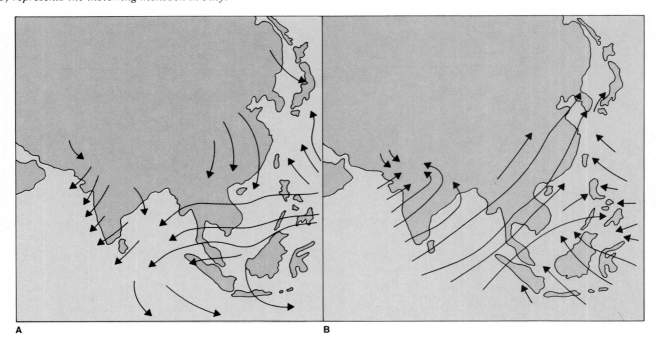

A   B

Winds and Air Masses

Winds blowing from low latitudes towards higher latitudes typically warm the areas over which they blow.
Conversely, air blowing from high to lower latitudes tends
to bring cooler temperatures with it. The air masses are,
in turn, modified by contrasting conditions existing at the
surface of the earth. Warming air tends to pick up moisture from available sources, whereas air that is being
cooled has a lower moisture-holding capacity.

Monsoons. Global circulations, such as those described
in the planetary wind system diagram, are disrupted by
the patterns of land and water distribution and a host of
other factors. **Monsoon** circulations, for example, are seasonal inflows and outflows of air associated with the shift
of pressure cells and the intertropical front. A pronounced
monsoon effect is felt in India when, in the Northern
Hemisphere summer, the intertropical front shifts well
north of the equator moving the southeast trades with it.
The southeasterly winds are deflected to the right by the
Coriolis effect as they cross the geographic equator and
become southwesterly winds. Their flow towards the land
is accentuated by an intense thermally induced low located over northwestern India and Pakistan. Although the
monsoon is a wind circulation that reverses with the season,
it characteristically is moisture bearing as air blows from
sea to land during the summer, and it is drying as air blows
from land to sea during the winter.

A similar flow of air moves over south and east Asia
in response to a large low over Asia. In the winter half of
the year, there is a southward shift of the intertropical front
and the development of a large dome of high pressure over
Asia. Air flows out from this high towards areas of lower
pressure. Depending on slope and exposure, this monsoon
circulation is responsible for a wetter summer half and
drier winter half of the year over much of south and east
Asia (figure 6.5). Pronounced monsoon effects are also felt
in Africa, and monsoon tendencies are encountered in other
continents, including North America.

Diurnal Winds. The land and sea breeze is the diurnal
equivalent of the seasonal monsoon circulation. Especially
in the summer half of the year in middle and low latitudes, the effect of land heating and cooling more rapidly
than water is to create a pressure gradient resulting in air
flow from sea to land at low levels during the day (**sea
breeze**), and from land to sea at low levels at night (**land
breeze**) (figure 6.6). The strength of the breeze depends
principally on temperature contrasts between the land and
adjacent water. Aloft air moves in opposite directions to
the surface flow.

During periods of relative stability in a mountainous
area there is a tendency for warmed air to rise up the slopes
of a mountain during daylight hours (**valley breeze**), while
at night the settling of colder air from higher elevations
to adjacent valleys is known as a **mountain breeze** (figure
6.7). In both of these instances, air moves along a pressure
gradient (i.e., from higher to lower pressure).

Figure 6.6
Land and sea breezes.
From Arthur Getis, Judith Getis, and Jerome Fellmann, Introduction
to Geography, *2d ed. Copyright © 1988 Wm. C. Brown Publishers,
Dubuque, Iowa. All Rights Reserved. Reprinted by permission.*

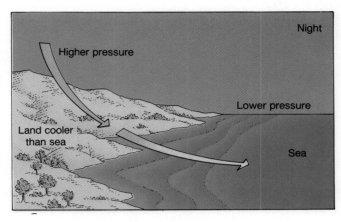

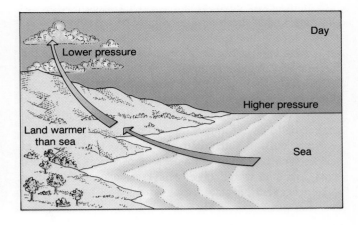

Figure 6.7
Mountain and valley breezes.
From Arthur Getis, Judith Getis, and Jerome Fellmann, Introduction
to Geography, *2d ed. Copyright © 1988 Wm. C. Brown Publishers,
Dubuque, Iowa. All Rights Reserved. Reprinted by permission.*

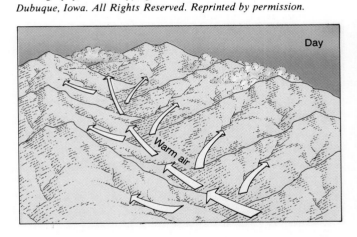

Katabatic Winds. Under calm, clear conditions in mountains and plateaus, cold air may tend to pile up and spill out to adjacent lowlands, especially through natural troughs or valleys. Air from Antarctica and Greenland frequently spills out to the coast through fjorded valleys. These types of winds, sometimes known as **katabatic winds** or **air drainage winds** may acquire local names. In France, such a wind descending from the Massif Central down the Rhone Valley is called the **mistral.** A comparable wind descending out of Yugoslavia into the Adriatic is called the **bora.** A similar wind in southeastern Alaska is termed **taku.** Such winds may acquire significant force if they are channeled down narrow valleys.

The movement of high- and low-pressure cells through mountainous areas may create pressure gradients leading to strong winds flowing through mountain passes.

Since this channelling effect may be pronounced in certain areas, it is not unusual to see highway signs warning of strong and dangerous winds. Such signs are quite common in the high mountain passes of the American West. The presence of snow fences, which are placed at locales where drifting occurs, provides indirect warnings about strong localized winds.

Desiccating winds bearing enormous quantities of dust or sand, and high temperatures blowing out of desert regions into more humid adjacent lands may bear local names like **scirocco** or **harmattan.**

Winds of global or continental scale tend to be steering mechanisms for the movement of weather and storm systems from place to place, and they may bring moisture or increase drying. Generalized circulations, like monsoons, tend to produce alternating periods of wet and dry conditions. Localized winds may produce transient, yet significant, effects on local weather.

In areas where few topographic or vegetative barriers exist to break the wind, it may be an omnipresent fact of the daily environment. In the United States, for example, areas where winds are quite persistent and moderately strong include the Great Plains, the Cape Hatteras area, and the Straits of Juan de Fuca adjacent to the state of Washington. Following the Dust Bowl days of the 1930s, shelter belts of trees were planted at intervals across the Great Plains to modify wind velocities and to mitigate the effects of wind erosion. Although vast segments of the shelter belts have fallen victim to periodic drought, disease, and cutting, some sections still remain and interrupt the vast fetch of the wind as it sweeps across the plains. They do reduce the effects of wind erosion locally, and they have, in general, functioned as intended.

Attitude of Topographic Barriers or Coastlines

The orientation of topographic barriers or coastlines with respect to prevailing winds may have a salient effect on precipitation amounts if the orientation of a barrier or coastline is such that the winds strike the barrier or coast at a high angle (figure 6.8). Apparently the perturbations created by striking a coast at a high angle creates eddies and air currents that may yield additional precipitation. Winds paralleling a coastline rarely yield precipitation on land in any significant amounts. Several areas in the world, especially in the trade wind zones, are dry because the coastline parallels the prevailing wind direction.

Storms

Storms are locally important in producing various weather changes and are, therefore, a climatic influence. Storms are the least predictable of meteorological phenomena and usually produce only transient effects, like flash floods or wind damage, although those effects may be dramatic in a violent storm. Examples include thunderstorms, both those associated with frontal passage and those occurring within warm, humid air masses. Frequently, thunderstorms produce copious precipitation, damaging winds, and dangerous lightning. Thunderstorms may also bear hail, which can cause crop and property damage, and turbulence and wind shear which may create discomfort or danger for airline passengers.

In the humid tropics and subtropics, however, thunderstorms are an ever-present part of the climatological environment, and they are important in determining the amount, intensity, and distribution of precipitation. Cooling showers and attendant winds may break the suffocating heat of the day. Rhythms of life may be influenced by the march of convection and associated cloudiness and precipitation during short periods of time. Frequent,

Figure 6.8
The orientation of a coastline has a significant influence on the pattern and density of proximate cloud cover.
Ralph K. Anderson, et al., Application of Meteorological Satellite Data in Analysis and Forecasting, *ESSA Technical Report NESC 51, NOAA, 1974.*

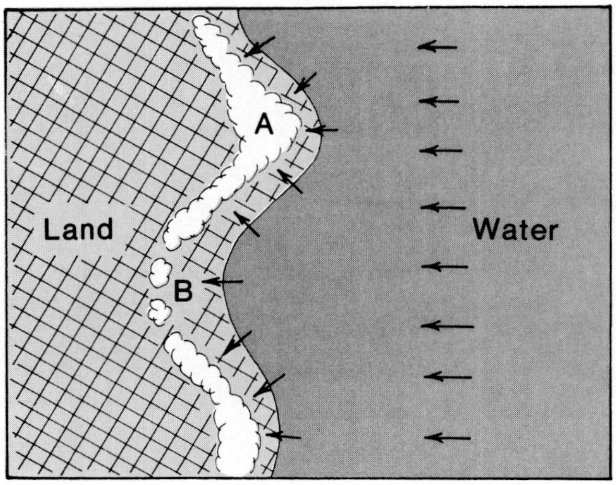

Figure 6.9
The beginning of cumulus buildup in the early morning in the Philippines during the rainy season is a prelude to the development into thunderstorms with afternoon rains.

violent storms increase the rate of erosion and deposition of soil over the rate encountered in areas of lighter rainfall intensity.

Other more violent storms, like tropical cyclones or hurricanes, may be responsible for enormous changes within regions where they occur. The copious precipitation, strong storm surges, and wind and wave action may drastically alter typical climatological patterns in a given season or year. Such storms may also cause dramatic geomorphological changes in a relatively short period of time. Tornadoes may drastically alter the norms of climatological characteristics in regions where they occur.

Figure 6.10
Hail damage can be considerable.
Yearbook of Agriculture, Climate and Man, *U.S. Government*
Printing Office, 1941.

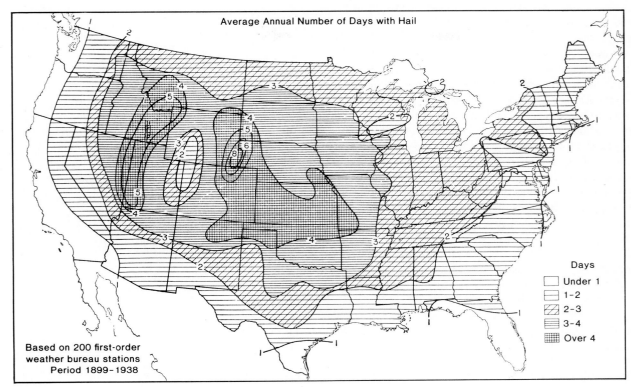

Average Annual Number of Days with Hail

Days
- Under 1
- 1-2
- 2-3
- 3-4
- Over 4

Based on 200 first-order
weather bureau stations
Period 1899-1938

Storms are usually transient events that extend from a few minutes or hours to several days, but tropical cyclones depart radically from climatological norms. They become both a climatic influence and a characteristic of climates in certain regions. The humid tropical and subtropical realms must recognize tropical cyclones as a fact of life; tornadoes must be reckoned with in the interior portions of the United States; and thunderstorms are omnipresent features in humid tropical, subtropical, and many middle latitude realms.

The interaction of the several climatic influences described in preceding sections produces certain areas of temperature, precipitation, wind and pressure, sunshine and cloudiness, and other elements of weather that yield discernible patterns of climate. These patterns form a portion of the fabric of the physical world. Climatological regions depicted depend on the density of the reporting climatological network, the length of the record, and the complexity of the climatic classification scheme used. Macroclimatological and microclimatological regimes are important factors in soil, biological, and geomorphological environments.

The development of classification schemes to characterize climates adds another element of order in the landscape. Examples of such classifications are described in the following section.

Figure 6.11
Tornadoes often punctuate the normal climate with dramatic impact.
NOAA.

Climatic Classifications

The broad outlines of climatic patterns may be deduced from a knowledge of the interaction of climatic influences. Climatic patterns are not accidental, but result from the interplay of climatic influences over time. Climates are classified, for convenience, in a variety of ways, although there is no agreement on the approach to classification or the kind of classification to be used. The core of a climatic region is usually quite well defined and possessed of particular characteristics. The boundaries, or transition zones to other climatic types, tend to shift over time, since, like so many other natural features, there is a gradation into an adjacent type. Typically, climatic classifications have climates grouped with respect to significant aspects of seasonal patterns, departures from certain norms, as well as average annual conditions. All available information about climatic conditions may be used in a classification system, although temperature and precipitation are most frequently used. These two pieces of data are usually readily available, whereas other elements of weather may be missing from the climatological record. Typically variations in precipitation and temperature are the significant weather elements for human activities.

More complex classification systems may be developed when more climatic elements are included (e.g., evapotranspiration, relative humidity, sunshine and cloudiness, wind direction and velocity, and so on). Mean conditions, frequency data, and departures from long-term norms are all elements that could add precision to a climate classification system.

Climatic classifications are especially significant to geographers, since they enable scientists to plot distribution patterns and reveal spatial relationships. These patterns, when combined with other forms or elements of the physical environment, provide powerful tools of inference about environmental opportunities or restraints, and human response to environmental conditions. Certainly climate, even with the mitigating opportunities of modern technology, continues to be a major factor influencing human occupance of a region. For example, although several "permanent" scientific stations exist in Antarctica, it seems certain that that inhospitable region will resist large scale colonization. Only the presence of a particularly vital or scarce mineral, and international agreements permitting the exploitation of such a resource would alter the existing sparse and transient settlement pattern.

The problem of finding a satisfactory classification of climates is a difficult one. Many schemes for classifying climates have been proposed, but none is satisfactory for all regions and all purposes. The shifting nature of boundaries between climatic regions produces difficulties that should be addressed by the classification scheme. A fixed boundary between climatic regions, as established by any classification system, is an arbitrary one. To be generally functional, the classification system must focus on mean conditions over a period of time rather than extremes that may occur from time to time. At certain times, it may be useful to use frequency data rather than mean data, especially in a climate characterized by large fluctuations from year to year.

The climatic classification used in this book is a system established by the late Wladimir Köppen. It has the advantage of being developed from rational premises relating principally to vegetation patterns (although some of those premises have been challenged subsequently); it proceeds in a logical step-by-step manner in developing the classification; it is easily understood; and it is widely used. The Köppen system has been widely imitated and modified by other climatologists for a variety of reasons. An American climatologist, C. Warren Thornthwaite, developed a popular climatic classification system that is widely used, but the Köppen system suits the purposes of this book better. Students should become adept in their understanding and use of the Köppen system. It not only provides a framework of understanding, but also a platform for further study and analysis by the serious student of climatology.

In the Köppen system, six major climatic groups are identified. Four of these are determined largely by temperature, a fifth by rainfall (or rather, the lack of it), and the sixth is determined by elevation. The six groups are listed:

1. Tropical rainy (A): all mean monthly temperatures exceeding 64.4° F
2. Dry (B): deficient rainfall and evaporation exceeding precipitation received
3. Humid mesothermal (C): at least one month less than 64.4° F, coldest month greater than 32° F, and the warmest month greater than 50° F
4. Humid microthermal (D): coldest month less than 32° F, but warmest month greater than 50° F
5. Polar (E): with warmest month less than 50° F and
6. Undifferentiated highland (H): with great climatological variations in mountainous regions.

These primary groups are further subdivided based on precipitation distribution and temperature refinement. In using the classification or key in the last pages of this chapter, the process is much like using a "go/no-go" gauge. For example, if one has determined that a particular climate is humid, then the procedure is to test the temperature regime. If all months have a mean temperature greater than 64.4° F, the first letter of the classification is A. When that has been determined, one checks in the precipitation column to the right of A. If, in our example, the rainfall of every month is greater than 2.4 inches, the second letter is f. One then checks back under A for the tertiary type of temperature refinement. If the variation between the cold month and hot month is less than 9° F, the letter i applies; hence, we would have a classification of Afi.

Figure 6.12
The sea–land interface tends to moderate temperature extremes on land areas.
Daniel Ehrlich.

If, in the primary type of a humid station one determines that A does not apply, a determination is made to ascertain whether C applies. If it does, the secondary type is determined by testing secondary characteristics opposite C in the chart. And, finally, the tertiary characteristic is determined under C in the left column.

Practicing by using a variety of climatological data will quickly enable the student to classify stations correctly. Obviously, when a large number of stations have been catalogued, it is possible to plot them on a map and insert lines separating different climatic regions. In many instances, using frequency rather than average data will produce modified patterns, especially in regions with wide variations in temperature and precipitation data.

Also, from time to time, it is useful to produce climatic maps on an annual basis. When this is done for a large number of years, areas of particular risk may be identified in comparing overlying plots. It is certainly true that departures from mean conditions are especially prevalent in many climatological environments. This is particularly characteristic of climates situated near the humid-dry boundaries.

Further, it is always important to recognize that certain important elements of the climatological environment may be missing from a particular classification scheme. It is important to know, for example, that the Köppen symbol of BSk for the Oklahoma Panhandle does not really address the persistence and velocity of wind and the resulting high evapotranspiration rates. In addition, the mean conditions mask extremes of temperature, which frequently exceed 100° F in the summer and drop to below zero in winter. The blazing heat of the sun in a desert region is somewhat masked by mean shade temperatures, but shade is almost absent. The biting cold of a cold air mass invasion of the American Southeast is obscured by a mean winter temperature for a cold month, which might be in the 40°s or 50°s F. Numerous other examples of mean conditions hiding vast departures of temperature or precipitation from the norm could be cited.

It should be recognized in the following chapters that the average conditions that mark climatological regions hide an amazing variety of conditions when detailed analysis is employed. Sometimes these variations from mean conditions are extraordinarily important in human adjustments to regions, in the development of natural vegetation regimes, in edaphic soil expression, and in geomorphological processes and results. Students should be alert to the fact that generalizations, including climatological generalizations, often obscure great variety when a situation is examined in detail.

Primary types *Temperature*	Secondary types *Precipitation*	Tertiary types
A: All months 64.4° F (18° C) or more, and not "B".	Af: All months 2.4 inches (6 centimeters) or more Am and Aw: At least one month < 2.4 inches Boundary between Am and Aw is determined by following equation. I = computed Index r = rainfall of driest month $$I = \frac{98.4 \text{ inches} - \text{annual precipitation**}}{25}$$ If r > I, then it is an Am climate If r < I, then it is an Aw climate	i: Annual range < 9° F (5° C) g: Hottest summer month comes before the solstice and the summer rainy season
C: Coldest month < 64.4° F (18° C), but > 32° F (0° C); and not "B"	Cw: Winter dry; precipitation in the driest month of winter < one-tenth of the amount of wettest month of summer Cs: Summer dry; precipitation in the driest month of summer < one-third wettest month of winter and less than 1.2 inches (3 centimeters) for driest month. Cf: No dry season.	a: Warmest month > 71.6° F (22° C) b: Warmest month < 71.6° F, but at least 4 months > 50° F (10° C) c: Warmest month < 71.6° F, and only 1–3 months >50° F
D: Coldest month < 32° F (0° C); Warmest month > 50° F (10° C); and not "B" E: Warmest month < 50° F (10° C)	Dw: Same as for "C" climates above Ds: Same as for "C" climates above Df: Same as for "C" climates above (Precipitation *not* used for E) EM: Coldest month > 20° F (−7° C) ET: Warmest month > 32° F (0° C) EF: Warmest month < 32° F (0° C)	a, b, and c: Same as for "C" d: Coldest month < −36.4° F (−38° C)

Note: This key is organized so that the user classifies climatic data by a process of elimination. If the data from a weather station fit the prescribed limits of a type and its subtypes, then there is no need to go to the next category. However, if the given data do not fit the particular limits, move on through the key until a type is found that meets all requirements.
**Köppen used the Celsius scale and metric system in devising the system, but those data have been translated to the Fahrenheit scale and English system of measure to accommodate the array of climatological data available in the United States.*
***Köppen used a table, but this formula yields the same result.*

Study Questions

1. Explain the pervasive influence of climate on human activities.
2. Distinguish between weather and climate.
3. List and describe how each of the several climatic influences affects climatic patterns.
4. Explain why land heats and cools more rapidly than water.
5. What is El Niño? What kind of climatic influences does it exert?
6. What factors influence the development of monsoon circulations?
7. List some regional examples of katabatic winds.

8. What were the principal factors used in developing various iterations of the Köppen classification of climates?
9. Who was C. Warren Thornthwaite?
10. Why does one use different formulae in determining humid-dry boundaries based on seasonality of precipitation?
11. Explain why mean conditions may mask great climatological variations within a particular climatic region.
12. If data were available and you were developing or modifying a climatic classification scheme for your area of the country, what additional element(s) of climate would you use in your system?

Primary types *Temperature*	Secondary types *Precipitation*	Tertiary types
B: The boundary between Dry (B) and Humid (A,C,D) climates is determined by the following equations. I = computed Index t = mean annual temperature R = mean annual precipitation	The boundary between BS (Steppe) and BW (Desert) is determined by the following equations. BS: if $R > \frac{1}{2}I$ For each of the three seasonal distributions BW: if $R < \frac{1}{2}I$ of precipitation.	h: Coldest month $> 32°$ F k: Coldest month $< 32°$ F For more detailed classification, the B climates are sometimes further classified and labeled with the following symbols: s: Summer drought; at least three times as much rain in the wettest winter month as in the driest summer month. w: Winter drought; at least ten times as much rain in the wettest summer month as in the driest winter month. n: Frequent fog.

If R < I, then it is Dry, B.
If R > I, then it is Humid;
   A, C, or D, *NOT* B.

1. $\dfrac{Precipitation\ mainly\ in\ summer}{(70\ percent\ in\ summer\ 6\ months)}$

   I − 0.44 (t − 7)

2. $\dfrac{Precipitation,\ mainly\ in\ winter}{(70\ percent\ in\ winter\ 6\ months)}$

   I = 0.44 (t − 32)

3. $\dfrac{Precipitation\ evenly\ distributed}{(less\ than\ 70\ percent\ in\ either}$
   summer or winter 6-month period)

   I = 0.44 (t − 19.5)

Selected References

Critchfield, H. J. 1974. *General climatology.* 3d ed. Englewood Cliffs: Prentice-Hall.

Doerr, A. H., and Guernsey, J. L. 1976. *Principles of physical geography.* 2d ed. Woodbury: Barron's Educational Series.

Kendrew, W. C. 1961. *Climates of the continents.* 5th ed. London: Oxford University Press.

McKnight, T. L. 1984. *Physical geography: A landscape appreciation.* Englewood Cliffs: Prentice-Hall.

Mather, J. R. 1974. *Climatology: Fundamentals and applications.* New York: McGraw-Hill Book Company.

Oberlander, T. M., and Muller, R. A. 1989. *Essentials of physical geography today.* 3d ed. New York: Random House.

Scott, R. C. 1989. *Physical geography.* St. Paul: West Publishing Company.

7

Climatic Regions—The Humid and Polar Climates

Many marine west coast (Cfb) environments support luxurious forest vegetation.
Daniel Ehrlich.

> *There is a sound of abundance of rain.*
> *I Kings 18:41, The Bible*

The geographical distribution of distinct climatic regions is produced by the interplay of the various climatic influences. The connections and interactions of climatic influences combine to yield climatic characteristics including the march of temperature, the amount and distribution of precipitation, as well as other elements of climate. As we have seen, climatic classifications typically use some combination of temperature and rainfall to develop the scheme. In humid climates, precipitation exceeds evaporation. The humid climatic types, their characteristics, regional distribution, and factors that have caused them to develop are discussed in the pages that follow. The characteristics and climate patterns are essential pieces of information for an understanding of and an appreciation for the vegetative, geomorphological, and edaphic associations encountered in different regions. Climates are, in fact, a part of the environmental mosaic that creates the physical face of the earth. That physical environment is our only home, and climate is an omnipresent facet of daily living.

Essential characteristics, distribution patterns, and major controls of the principal climatic types are elucidated in the sections that follow.

Tropical Rainy Climates (A)

Tropical rainy climates are frost-free throughout the year. Indeed, most tropical climates have temperatures far above freezing at all times. Seasons are based principally upon distribution and amounts of rainfall rather than temperature, since there are no large swings of temperature from season to season. Some tropical areas have large amounts of fairly evenly distributed rainfall, whereas others exhibit distinct wet and dry seasons. When Köppen established the A regime in his classification, he believed that the 64.4° F (18° C) isotherm for the cold month established the limit for certain tropical species of plants. Of course, Köppen used the Celsius scale when he developed his system of climatic classification because of his European origins. In light of present knowledge, it seems fairly clear that absolute minimum temperatures establish the climatological limits for certain tropical species of plants, rather than mean conditions, but Köppen didn't know that, and he can't be faulted for acting on a presumed scientific "truth" of the time. Indeed, science constantly uncovers new truths with new insights, new observations, and new technologies. One of the fallacies in establishing climatic classifications on the basis of vegetation associations is that intricate interplay between climatic characteristics, including extremes, and soil, drainage, and slope conditions. Together, these conditions seem to be of greater significance in plant associations than are mean climatic conditions. In sum, Köppen's boundary was established on the basis of an imperfect understanding of temperature and biological relationships. Nevertheless, the 64.4° F cold month isotherm embraces tropical conditions as most people perceive them and is a useful parameter to use for the boundary of A climates.

Similarly, the 2.4 inches (6 centimeters) isohyet for the dry month was used by Köppen to separate tropical wet and dry environments. Again, Köppen believed 2.4 inches of precipitation was the monthly mean necessary to sustain tropical evergreen species. This is not specifically correct because of the variations in evaporation rates, slope, exposure, and soil conditions, but it is a reasonable boundary to be used in the climatic classification scheme (figure 7.2). The tropical wet (Af), or tropical rainforest climate, is somewhat more restricted in area than the tropical wet and dry (Aw), or tropical savanna, and the tropical monsoon (Am) climates together. The efforts to tie climates to vegetation are evident in the names attached to various climatic types, but it must be stressed that many factors other than mean and monthly conditions of rainfall and temperature are important in the ecological matrix that creates vegetation regions. Soil textures and structures, slopes and exposure, and evapotranspiration rates are all meaningful, but this is a **climatic** classification and many data other than temperature and precipitation are not readily available, especially in many of the less developed areas of the world.

Figure 7.1
The tropical rainy climates support dense forest vegetation.

Tropical Rainforest (Af)

The tropical rainforest climate by definition receives 2.4 inches or more of rainfall every month, and it has uniformly high temperatures throughout the year. This climatic type is located astride the equator, and usually extends from 5° to 10° north and south latitude, except on the eastern side of continents where onshore trade winds striking the coast at high angle, buttressed by orographic lift, may extend a tongue of Af climate as far as 20° to 25° from the equator. This is especially notable along the Atlantic coast of Brazil and in Madagascar where there are extensions of Af climates into nominally subtropical realms.

Monthly and annual temperatures average about 80° F (about 26° C) for the typical tropical rainforest environment. Frequently, there is very little range in temperature from month to month and season to season. These universally high temperatures are maintained by high-angle sun throughout the year along with days that are always about twelve hours long. Diurnal ranges in temperatures are often significantly greater than annual ranges. Daytime highs are typically in the 90° F range, while night temperatures frequently drop into the low 70° range and occasionally into the high 60° range. It has been said that "night is the winter of the tropics," since absolute lower temperatures are experienced at night than mean temperatures between seasons. High humidities contribute to the enervating heat of the region. They are said to have high **sensible temperatures** (i.e., the combination of heat and humidity make the environment uncomfortable to most people). High humidities reduce the efficiency of the perspiration–evaporation mechanism employed by the human body.

Figure 7.2
General locations of the tropical rainy and tropical wet and dry climates.
From Arthur Getis, Judith Getis, and Jerome Fellmann, Introduction to Geography, *2d ed. Copyright © 1988 Wm. C. Brown Publishers, Dubuque, Iowa. All Rights Reserved. Reprinted by permission.*

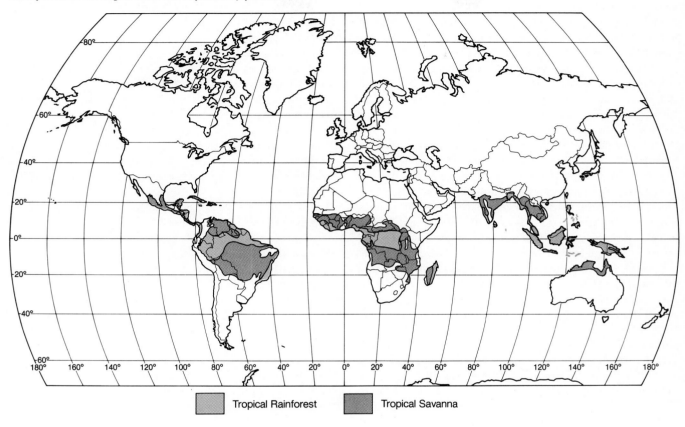

Tropical Rainforest   Tropical Savanna

Table 7.1										
Apparent Temperature										
Relative Humidity (%)	**Air Temperature (Fahrenheit)**									
	70	**75**	**80**	**85**	**90**	**95**	**100**	**105**	**110**	**115**	**120**
0	64	69	73	78	83	87	91	95	99	103	107
10	65	70	75	80	85	90	95	100	105	111	116
20	66	72	77	82	87	93	99	105	112	120	130
30	67	73	78	84	90	96	104	113	123	135	148
40	68	74	79	86	93	101	110	123	137	151	
50	69	75	81	88	96	107	120	135	150		
60	70	76	82	90	100	114	132	149			
70	70	77	85	93	106	124	144				
80	71	78	86	97	113	136					
90	71	79	88	102	122						
100	72	80	91	108							

Source: NOAA.

Clearly, areas of hilly terrain might have significantly cooler temperatures induced by elevation and still have every month above 64.4° F. In the grand scheme of things, such cooler regions might occupy a small area, and yet they might be favored areas for human settlement. Indeed, the hill station was a constant place of respite for European colonials in tropical regions. The British made regular pilgrimages to stations like Simla or Srinagar in northern India to escape the summer heat of lowland regions during the heyday of the British raj.

Areas of tropical rainforest are either under the constant influence of the equatorial area of low pressure and the intertropical convergence zone, or they are influenced by orographic lift along trade wind coasts. The absence of a dry season is accounted for by convectional uplift along the intertropical convergence zone, the development of air mass thunderstorms, or by orographic lift. Those areas located at any distance from the equator generally exhibit high-sun maxima and low-sun minima in precipitation, but there is no distinctly dry season, and no month receives less than 2.4 inches of precipitation. Convectional lift may drive saturated air to 50,000 feet (more than 15,000 meters) or more in enormous thunderstorm cells. The attendant downpours of rain and the pyrotechnics of lightning and thunder are regular features of the rainforest environment. Rainfall is usually well in excess of 60 inches (more than 150 centimeters) per year, frequently exceeds 100 inches (250 centimeters), and may be more than 200 inches (500 centimeters) per year, especially on exposed coasts with some orographic lift. Mountain barriers lying perpendicular to onshore winds in such regions are among the wettest places on earth.

The daily march of temperature, cloudiness, and precipitation are closely related. The day typically will dawn clear with temperatures in the 70° range. The air quickly warms, and by midmorning, air temperatures are into the 80° range and cumulus clouds have sprouted everywhere. The heating of the day slows as the clouds continue to grow, since the growing clouds block some of the insolation. By midafternoon, as the temperatures rise into the 90° range and the cumulus have grown through a series of towering forms into cumulonimbus, a downpour of rain comes amidst the crackling of lightning and the roar of thunder. Copious quantities of precipitation may fall in a short period of time temporarily stilling the cacophony emanating from insects and native animals, and substituting the drumbeat of volumes of water and the rumble of thunder, but typically the clouds have cleared by sunset. The thunderstorm breezes slacken, and a calm night sees the temperatures descend into the 70° range before dawn. Nights are clear to partly cloudy. The sun

rises, and the whole process begins again in a kind of regularized natural dance. This daily monotony of heat, humidity, and rainfall tends to be enervating for the sojourner from the middle latitudes. In past eras, visitors sought hill stations in the hottest season. Modern refrigerated air conditioning is a great boon during the heat of the day. Native peoples, except for the very affluent, continue to cope with high heat and humidity in time-honored ways. Not surprisingly, the midday siesta continues to be a fact of life in such regions. Closed businesses reopen later in the day, and the evening meal may be served quite late. Humid subtropical regions, like the southeastern United States, experience many of the same conditions in the summer season.

Some have suggested that slavery in the Americas was a Caucasian response to the dreadful heat and humidity of the summers. Would such an institution have developed in a society with today's technological advantages?

Heavy rainfall provides adequate and, at times, excessive moisture throughout the year. Water tables are near the surface; swamps and marshes are found in low-lying areas; and some of the great rivers of the world (e.g., the Amazon and Congo) drain the runoff from the convectional storms, which make their daily contributions to the water budget. High temperatures, moisture, and high relative humidities foster the growth of molds and fungi of all kinds. Mildew is a constant problem and must be fought vigorously.

The major land areas of the world having a tropical rainforest climate are (1) the Amazon Basin of South America, (2) the Congo Basin of Middle Africa, (3) the east coast of Madagascar and portions of Brazil, and (4) the mainland and islands of Southeast Asia. The climatological data for Singapore are characteristic of this type of climate.

Singapore—Lat. 1°14'N.; Long. 103°55'E.

	Jan.	Feb.	Mar.	Apr.	May	June
T.	78.3	79.0	80.2	81.5	81.5	81.1
P.	8.5	6.1	6.5	6.9	7.2	6.7

July	Aug.	Sept.	Oct.	Nov.	Dec.	Year
81.6	80.6	80.4	80.1	79.3	78.6	80.1
6.8	8.5	7.1	8.2	10.0	10.4	92.9

Classifying Singapore according to the Köppen system would yield the symbols Afi. High temperatures, copious precipitation, and small annual temperature range are all exhibited by the climatic data. It would be apt to describe the climate as monotonous, at least by the standards of residents of the middle latitudes, and most people from such latitudes would find it difficult to function there without artificial cooling devices.

Figure 7.3
Colombo, Sri Lanka.

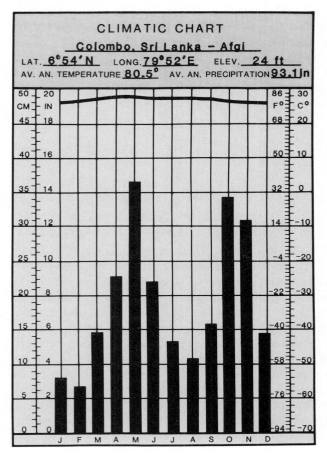

CLIMATIC CHART
Colombo, Sri Lanka – Afgi
LAT. __6°54'N__ LONG. __79°52'E__ ELEV. __24 ft__
AV. AN. TEMPERATURE __80.5°__ AV. AN. PRECIPITATION __93.1__ in

Figure 7.4
Papeete, Tahiti.

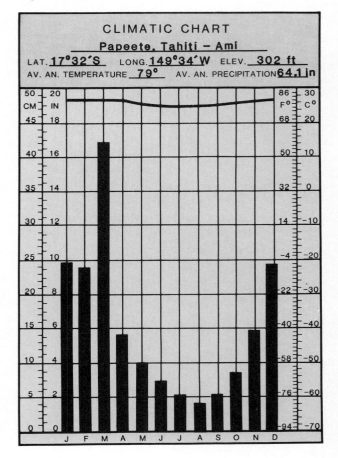

CLIMATIC CHART
Papeete, Tahiti – Ami
LAT. __17°32'S__ LONG. __149°34'W__ ELEV. __302 ft__
AV. AN. TEMPERATURE __79°__ AV. AN. PRECIPITATION __64.1__ in

Tropical Monsoon (Am)

Tropical monsoon climates have at least one month (frequently several) with precipitation less than 2.4 inches. The high-sun periods are wet, and the low-sun periods are dry. Low latitudes insure reasonably high temperatures year around, but there is typically a significantly greater range of temperature than in the Af environment. High temperatures tend to occur before the summer solstice, because cloudiness subsequent to the onset of the rainy season mitigates the temperature somewhat and modifies the "typical" pattern of postsolstice temperature maxima. When this situation exists, the symbol g becomes a part of the classification. This pattern of the warm month preceding the summer solstice is especially noticeable in certain sections of India and adjacent areas of Bangladesh where the effects of the inblowing monsoon and attendant cloudiness are especially pronounced.

When the monsoon indraft of air during the summer half of the year brings air heavy with moisture, orographic or convectional uplift produces precipitation. There is a significant peak of rainfall during the summer, while the winter season, or time of outblowing air, is almost rainless. The time of the outblowing monsoon is notably less humid, and the combination of cooler temperatures and lowered humidities provides blessed relief from the high temperatures of the wet season. Indeed, the winter season can be quite pleasant in such environments.

The monsoon is said to "break" (i.e., the rainy season begins after long periods of increased cloudiness) at later and later dates as distance from the equator increases. The monsoon may break in southern India, for example, well before precipitation begins at the latitude of Calcutta.

Figure 7.5
The idyllic South Sea Island of Moorēa has a tropical monsoon climate.
Daniel Ehrlich.

Major areas of the world with a tropical monsoon climate include portions of southern and southeastern Asia, the Guinea coast area of Africa, and certain islands of the West Indies. Monsoon-like indrafts of air augment the precipitation of the southeastern United States during the summer season, especially in late summer.

Akyab, Burma, exhibits noteworthy characteristics of the tropical monsoon climate. A cooler dry season is followed by an enervating period of copious precipitation, high humidity, and high temperatures.

Akyab, Burma—Lat. 20°08′N.; Long. 92°55′E.

	Jan.	Feb.	Mar.	Apr.	May	June
T.	70.0	72.5	78.0	82.5	84.0	81.5
P.	0.1	0.2	0.4	2.0	15.4	45.3

July	Aug.	Sept.	Oct.	Nov.	Dec.	Year
80.5	80.5	81.5	81.5	78.0	72.0	79.0
55.1	44.6	22.7	11.3	5.1	0.7	202.9

The Köppen classification for Akyab is Amg. There is a distinct wet and dry season and the warm month (May) comes before the summer solstice. Sun's rays are nearly vertical during that month, and maximum cloudiness at the height of the monsoon rains in subsequent months mitigates the temperatures somewhat. Copious precipitation during the high-sun period is augmented by a coastal location with a hilly hinterland. The whole area is literally saturated with moisture and humidity during the wet season. It is very difficult to dry anything, and molds, fungi, and hydrophytic vegetation thrive at that season. It is a kind of natural hothouse in which plants thrive and people wilt.

Tropical Savanna (Aw)

The wet and dry tropical savannas are located poleward from tropical rainforests from 5° to 15° from the equator. Typically, tropical savannas have less precipitation than tropical rainforest and tropical monsoon environments, and precipitation is asymmetrically distributed in a wet high-sun and a dry low-sun pattern. The equatorial low pressure and the intertropical convergence zone move over savanna regions in the summer, bringing periods of instability and convectional rain. During the low-sun period, trade winds and higher pressure dominance produce drier conditions.

Northwest India receives the precipitation even later, and the dry season onset there precedes its beginning further south and east. As a general rule, too, precipitation reliability diminishes as one proceeds north and northwest in India and on into Pakistan.

The daily march of temperature, cloudiness, and precipitation during the summer is very similar to that experienced in tropical rainy environments, except that monsoon regions may have more protracted periods of rain. Some regions may record enormous totals of precipitation during the wet season. During the dry season in tropical monsoon environments, there may be protracted periods without clouds when relative humidities drop and temperatures are somewhat more comfortable.

Average temperatures in tropical savanna regions tend to be similar to those in the tropical monsoon environment. The hot month may precede the summer solstice, because of increasing cloudiness in the summer wet period.

Annual precipitation typically ranges between 30 and 60 inches (76 and 152 centimeters) per year. Not only is the amount less than for the tropical rainforest, but the reliability is less. Also, the beginning and end of the rainy periods are quite unpredictable. The length of the rainy season varies from five to eight months depending primarily upon proximity to or distance from the equator. Those areas closest to the equator tend to have the longest wet season and the larger amount of precipitation. Obviously, those at somewhat higher latitudes tend to have a shorter rainy period and a lower total amount of precipitation. Also, increasing distance from the equator tends to reduce precipitation reliability. During the wet high-sun period, precipitation is usually at least ten times that of the low-sun period. At Darwin, Australia, for example, precipitation ranges from 15.3 inches in January (high-sun period for a Southern Hemisphere station) to 0.1 in the three low-sun months. A sere, brown landscape in the winter season is succeeded by a brilliant green in the wet period.

Again, the high-sun period is characterized by a sequence of temperature, humidity, cloudiness, and precipitation very similar to the tropical rainy environments. The low-sun period is marked by cloudless skies and low relative humidities. These times of low sun are much more pleasant for human occupants of the area.

Darwin, Australia—Lat. 12°28'S.; Long. 130°51'E.

	Jan.	Feb.	Mar.	Apr.	May	June
T.	84	83	84	84	82	79
P.	15.3	13.5	9.6	4.1	0.6	0.1

	July	Aug.	Sept.	Oct.	Nov.	Dec.	Year
	77	79	83	85	86	85	83
	0.1	0.1	0.5	2.0	4.7	9.8	60.4

Darwin is classified as Awgi according to the Köppen classification. The warm month, November, comes before the summer solstice, and the period of maximum cloudiness occurs during the peak of the wet season. The annual range of mean monthly temperatures is less than nine degrees. Darwin's coastal location helps to mitigate the range of temperature from season to season.

Increasing latitude diminishes temperatures sufficiently to provide for a transition from the tropical to subtropical or mesothermal environments. Those climates will be discussed in the sections that follow.

Figure 7.6
Ho Chi Minh City, Vietnam.

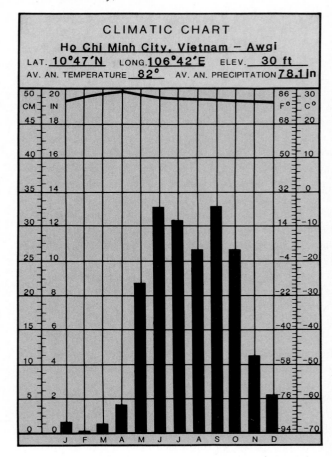

Humid Mesothermal Climates (C)

Humid mesothermal climates are divided into three well-defined climatic regions: (1) humid subtropical, (2) Mediterranean (dry subtropical), and (3) marine west coast. Of the three types, the marine west coast environment extends to higher latitudes. The latitudinal extent of these climates is quite broad, extending from the edge of the tropical latitudes almost to the Arctic Circle in the Northern Hemisphere and to about 50°S. in the Southern Hemisphere. Essentially, however, these are middle latitude climates with discrete seasons marked by heat and cold as major distinguishing characteristics from tropical climates, rather than precipitation amounts and distribution, although asymmetrical precipitation patterns are characteristic of several of the climatic types.

Humid Subtropical (Cfa or Cwa)

Humid subtropical climates typically extend from 25° to 40° latitude along the eastern margins of continents, although most such climates are limited to the area between 28° and 38° latitude. In such areas, except in portions of east Asia and east Africa, the spring and summer months are wetter, although no season is dry. In certain portions of east Africa and east Asia, there is a distinct dry season in winter because of the monsoon effect. Rainfall typically ranges between 30 and 65 inches (76 and 165 centimeters) annually. Mobile, Alabama, a humid subtropical station with 65 inches (165 centimeters) of precipitation annually, is the wettest large city in America. Pensacola, the author's home, sixty miles east of Mobile, essentially mirrors Mobile's climatic characteristics.

Summer temperatures are warm to hot with monthly means in the 70° or 80° F range, although daily maximum temperatures often exceed those encountered in the humid tropics. Monthly means in the winter are typically in the 40° or 50° F range, although somewhat lower temperatures may be encountered as distance from the sea increases. Average annual temperatures are typically in the 50° or 60° F range. Residents find both the heat and cold to be penetrating because of relatively high humidities at almost all seasons. Diurnal ranges in temperature are not usually great because of high humidities and significant amounts of cloudiness.

Humid subtropical climates are dominated by maritime tropical air during the summer. In winter, there is a frequent succession of middle latitude cyclones in North America; whereas in Asia, there tends to be minimum rainfall because of the effects of the outblowing monsoon. Indeed, the Cwa climate varies from the Aw climate only in having a short cool season. These somewhat cooler conditions occur either because of slightly higher latitudes or increasing elevation. Nashville, Tennessee illustrates the characteristics of a typical humid subtropical station where there is no distinct dry season. Nashville is Cfa in the Köppen scheme.

Nashville, Tenn.—Lat. 36°10'N.; Long. 86°46'W.

	Jan.	Feb.	Mar.	Apr.	May	June
T.	39	41	50	59	68	76
P.	4.8	4.2	4.1	4.4	3.8	4.2

	July	Aug.	Sept.	Oct.	Nov.	Dec.	Year
	79	78	72	69	49	41	59
	4.1	3.5	3.5	2.4	3.5	3.9	47.4

The dry season variant of this climate, showing the influence of the monsoon circulation, is illustrated by Hanoi. The summer maximum of precipitation during the inblowing monsoon and the dry winter season are clearly

Figure 7.7
Wilmington, North Carolina.

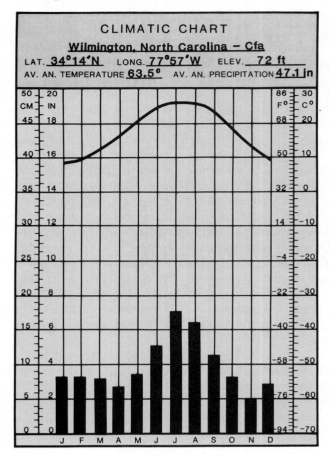

Figure 7.7
Wilmington, North Carolina.

discernible. The relatively low latitude of this station almost qualifies it as a tropical savanna environment. Because Hanoi has two months with a mean temperature below 64.4°, it is a Cwa environment rather than an Aw.

Hanoi, Vietnam—Lat. 21°02'N.; Long. 105°52'E.

	Jan.	Feb.	Mar.	Apr.	May	June
T.	62.0	63.5	68.5	75.5	82.0	85.0
P.	0.7	1.1	1.5	3.2	7.7	9.4

July	Aug.	Sept.	Oct.	Nov.	Dec.	Year
84.5	84.0	82.0	77.5	71.0	65.5	75.5
12.7	13.5	10.0	3.9	1.7	0.8	66.2

The total amount of precipitation at Hanoi is almost the same as for certain Gulf Coast cities (e.g., Mobile and Pensacola), but the Gulf Coast experiences no distinct dry season under mean conditions. Even with areas that are classified as Cfa, however, periodic dry spells may punctuate the pattern of fairly evenly distributed precipitation.

Figure 7.8
Forest-clad slopes of the humid subtropical climatic region of southeastern Oklahoma.

Figure 7.9
Lashio, Burma.

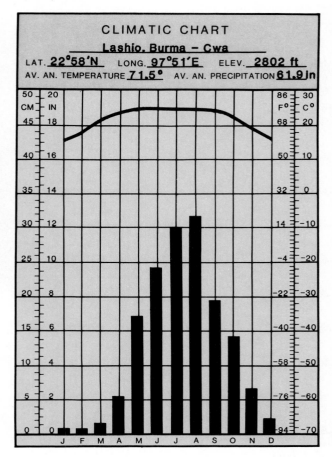

Mediterranean (Csa, Csb)

As the name implies, the type sample for this climatic regime is to be found around the periphery of the Mediterranean Sea. This climatic regime differs in three significant ways from the humid subtropical climate: (1) it receives less total precipitation; (2) drier summers are characteristic; and (3) the regions are typically located on western sides of continents except where large bodies of water permit penetration further eastward, as in the case of the area around the Mediterranean Sea. Mediterranean climates are almost subhumid with total amounts of precipitation typically ranging between about 15 and 25 inches (38 and 65 centimeters) per year, and the climatic regime is sometimes termed dry subtropical. The winters are relatively wet when a parade of cyclonic cells carried by the westerlies dominate the region. During the almost rainless summer, subtropical high-pressure cells dominate the Mediterranean regions, and cyclonic storms usually move well poleward of the areas.

Hot temperatures are characteristic of Mediterranean climates in the summer, and mild temperatures are the rule in the winter. A latitudinal position on the west side of continents, comparable to that of the humid subtropics on the east coasts of continents, makes for modest ranges between summer and winter temperature conditions. Prevailing westerly winds bear marine influences that ameliorate temperature extremes during the winter season. The major areas with a Mediterranean climate include the margins of the Mediterranean Sea, southern California, central Chile, southwestern and southern Australia, and southern Africa. Occasional freezing temperatures are encountered in winter, but frosts are rare.

Largely frost-free conditions, high percentage of sunshine, and availability of irrigation water in the mountains that bound several of these regions have combined to create major horticultural areas, especially in the United States and middle Chile. The ability to bring fruits and vegetables to market in the "off" season gives these regions major competitive advantages. The summer is sunny almost all of the time, and even winters experience a high percentage of sunshine. Thunderstorms are uncommon. Winter rains tend to come in gentle showers, although occasional downpours may create severe problems. Relative humidities are quite low in the summer season and even in the winter wet period, they tend to average less than humidity levels in humid subtropical areas. Most people find Mediterranean climatic regions to have especially pleasant temperatures and precipitation conditions for daily living.

Figure 7.10
Adelaide, Australia.

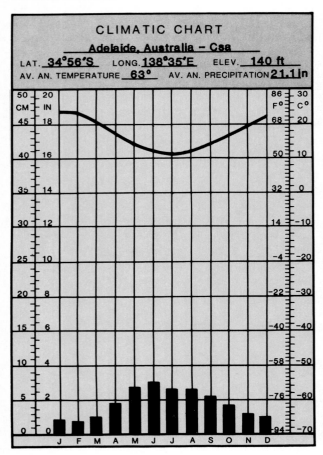

CLIMATIC CHART

Adelaide, Australia – Csa

LAT. **34°56'S**   LONG. **138°35'E**   ELEV. **140 ft**

AV. AN. TEMPERATURE **63°**   AV. AN. PRECIPITATION **21.1 in**

Los Angeles, California is a typical Mediterranean station, although it is on the margin of semiaridity. Availability of water, especially in the summer season, is a problem in such an environment. Los Angeles is Csb.

Los Angeles, Calif.—Lat. 34°00'N.; Long. 118°15'W.

	Jan.	Feb.	Mar.	Apr.	May	June
T.	54.5	55.5	57.3	59.7	62.1	65.2
P.	3.3	3.2	2.9	0.9	0.4	0.1

July	Aug.	Sept.	Oct.	Nov.	Dec.	Year
70.2	71.1	69.4	65.1	60.9	55.3	62.2
0.0	0.0	0.2	0.7	1.2	2.7	15.6

Marine West Coast (Cfb, Cfc)

Marine west coast climates typically occur poleward of 40° on the west sides of continents. In areas where there are few topographic barriers, marine influences may penetrate some distance into land areas. Where barriers are close to the coast, marine influences may be restricted to a very narrow coastal fringe. In Europe, the absence of significant coastal barriers allows penetration of the climate some distance away from the coast. In southern Chile, in contrast, the climate is restricted to a narrow coastal strip, because the Andes restrict penetration of maritime influences into the interior.

Figure 7.12
Auckland, New Zealand.

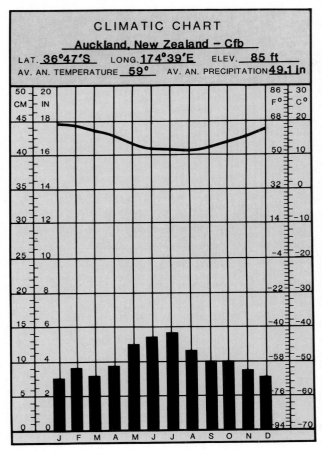

CLIMATIC CHART

Auckland, New Zealand – Cfb

LAT. **36°47'S**   LONG. **174°39'E**   ELEV. **85 ft**

AV. AN. TEMPERATURE **59°**   AV. AN. PRECIPITATION **49.1 in**

Marine west coast climates are characterized by relatively cool summers, and, for the latitude, quite mild winters. Monthly averages during the winters are frequently in the 40° and 50° F range, whereas summer monthly averages are typically in the 50° and 60° F range. Diurnal and annual temperature ranges are moderate, and extremes of heat or cold are rare. Annual averages range between 45° and 55° F (about 7° and 13° C).

Moderate precipitation occurs throughout the year, but winter precipitation is heavier than summer. Frequent cyclonic storms march across the region borne by prevailing westerly winds. The higher frequency of these storms in the winter season accounts for the peaking of precipitation at that time. In some areas, cyclonic storms are augmented by orographic lift, and when that occurs, copious quantities of moisture may fall. During winter,

light showers, gentle drizzles, and cloudy and foggy weather are common. In summer, sunny days are more frequent, relative humidities are lower, and rainfall is lighter, although there is a high percentage of cloudiness throughout the year. Fogs are not uncommon even in summer, especially in northwestern Europe.

Although winter temperatures are rarely severe, the high humidities and cloudiness cause the cold to be penetrating. The high percentage of cloudiness and the shorter winter days add a measure of gloom, and psychological depression of a mild kind is quite commonplace in winter. Severe depression in a few has been linked to shortened periods of daylight. In summers, the temperatures are cool to mild.

Certain areas illustrate the contrasts that can occur within short distances depending upon slope and exposure. On Mt. Olympus in the state of Washington, precipitation exceeds 200 inches per year (more than 500 centimeters). In Seattle, less than a hundred miles away, the rainfall is only about 33 inches (more than 80 centimeters). Both stations are in the marine west coast environment. Sequim, located in a peculiar rainshadow pocket on the Olympic Peninsula, receives only 17 inches (about 43 centimeters) per year. The great disparity in rainfall amounts illustrates the fact that considerable differences may exist from place to place within the broad climatic regions delineated, generated especially by variation in slope and exposure.

The climatic data for Bordeaux provides a useful example of a marine west coast environment. There is somewhat more precipitation in the winter half of the year, but the summer is by no means rainless. Bordeaux is Cfb.

Bordeaux, France—Lat. 44°50'N.; Long. 0°43'W.

	Jan.	Feb.	Mar.	Apr.	May	June
T.	41.5	44.0	49.0	53.5	59.0	64.5
P.	2.7	2.8	2.9	2.6	2.5	2.3

July	Aug.	Sept.	Oct.	Nov.	Dec.	Year
69.0	68.5	64.5	56.5	48.0	43.0	55.0
2.0	1.9	2.2	3.0	3.9	3.9	32.7

Clearly, the C climates show greater ranges in temperature than the A climates. The microthermal (D) climates, which will be discussed subsequently, exhibit even greater temperature ranges primarily because of greater continental influence. The latitudes where D climates are located are dominated by colder air masses, and rainfall amounts tend to be less because of distance from major sources of moisture.

Figure 7.13
Distribution of the C climates.
From Arthur Getis, Judith Getis, and Jerome Fellmann, Introduction
to Geography, *2d ed. Copyright © 1988 Wm. C. Brown Publishers,
Dubuque, Iowa. All Rights Reserved. Reprinted by permission.*

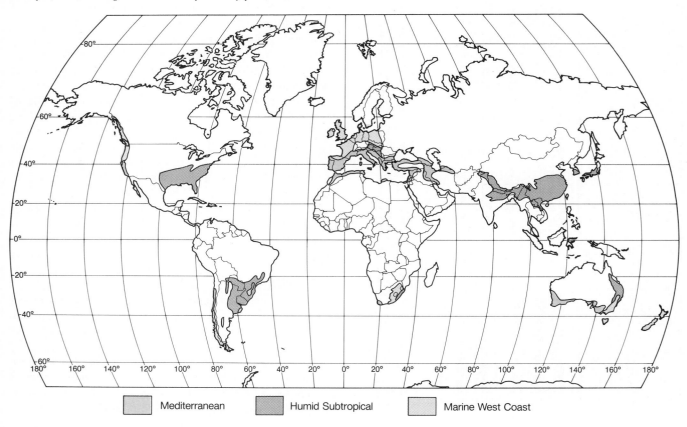

Mediterranean	Humid Subtropical	Marine West Coast

Humid Microthermal Climates (D)

Humid microthermal climates are typically continental in character and are best represented in North America and Eurasia. They have lower winter temperatures, shorter growing seasons, and greater temperature ranges on the average and, in the extremes, than is characteristic of the mesothermal environments. Humid microthermal climates are located predominantly in the interior and eastern coast of the higher middle latitudes in the Northern Hemisphere. They are most significantly influenced by large land areas, relatively high latitudes, and the interaction of contrasting air masses. The absence of large land masses at appropriate latitudes in the Southern Hemisphere limits the occurrence of the climatic types south of the equator. Three types of microthermal climates are recognized (i.e., humid continental warm or long summer, humid continental cool or short summer, and subarctic or subpolar).

The humid microthermal climates are marked by sharp temperature contrasts from season to season, and extremes of heat and cold for absolute maxima and minima are also characteristic. Typically, too, rainfall diminishes with increasing latitude and distance from the sea.

Humid Continental Long Summer or Warm Summer (Dfa, Dwa)

Humid continental long summer climates extend from about 38° N. latitude to about 48° N. along the eastern coasts and well into the interior of North America and Eurasia. These latitudes are under the influence of the westerlies almost all of the time in North America, and there is a steady progression of cyclonic storms across the region augmented by convectional storms, especially in spring and summer. In Asia, there is a pronounced influence from the monsoon circulation accounting for the fact that there is a pronounced winter dry season.

Temperatures are highly variable both seasonally and diurnally. Insolation exceeds that of the tropics in the summer because of the length of the daylight period. Winter temperatures are low because of low-sun angle, short period of daylight, the large land masses involved, and the frequent encroachment of cold, polar air. Latitudinal position restricts invasions of warming tropical air. There may be sharp temperature contrasts at all seasons because of the passage of fronts. Winter temperature means are in the 20° and 30° F range, with at least one month averaging below freezing. Cold waves may cause temperatures to plunge well below zero. In fact, these cold snaps occur with considerable frequency. Cold, often accompanied by snow and long days under dull, lead-gray, stratus skies, adds a somber note to all the microthermal environments in the winter season. The short daylight hours and the often persistent periods of cloudiness have been linked to a high incidence of chronic depression in winter among residents of these regions. It's becoming increasingly clear that human beings are much more photosensitive than was previously believed. Some treatments for depression apparently induced by a shorter photoperiod have included exposing depressed patients to quite bright lights for substantial periods. Such treatment appears to be effective.

The summers are warm to hot with mean temperatures into the 70° F range. Occasional heat waves may cause summer maxima to soar above 100° F (37° C).

Precipitation is also variable, but, in general, it decreases both in amount and reliability from the humid margins near the coast toward continental interiors and toward higher latitudes. Maximum rainfall occurs during summer and, in Asia, there is a distinct summer maximum and winter dry period. The impact of the inblowing and outblowing monsoon is pronounced in Asia, resulting in the same seasonal pattern of precipitation as that which occurs in tropical monsoon regimes. The slight peaking of precipitation amounts in the Western Hemisphere during summer can be attributed to convectional augmentation of cyclonic precipitation in the summer half of the year. In North America, there is a more equal rainfall regime than that which exists in Asia, since the monsoon is not as significant a phenomenon in North America.

Chicago illustrates quite effectively the climatic characteristics of a humid continental long summer or warm summer climate. Chicago is classified as Dfa.

Chicago, Illinois—Lat. 41°53′N.; Long. 87°38′W.

	Jan.	Feb.	Mar.	Apr.	May	June
T.	26	27	37	47	58	68
P.	2.1	2.1	2.6	2.9	3.6	3.3

	July	Aug.	Sept.	Oct.	Nov.	Dec.	Year
	74	73	66	55	42	30	50
	3.4	3.0	3.1	2.6	2.4	2.1	33.2

Mukden, China illustrates the pronounced wet and dry periods encountered in Asia because of the monsoon effect. It also illustrates the wider temperature swings associated with the large Asian land mass. Mukden is Dwa.

Mukden, China—Lat. 41°48′N.; Long. 123°23′E.

	Jan.	Feb.	Mar.	Apr.	May	June
T.	11.5	17.0	32.0	49.0	62.0	72.5
P.	0.3	0.3	0.7	1.1	2.7	3.3

	July	Aug.	Sept.	Oct.	Nov.	Dec.	Year
	78.0	76.0	68.5	49.5	31.5	15.0	46.5
	7.2	6.7	2.5	1.4	1.1	0.6	27.9

Figure 7.14
Lincoln, Nebraska.

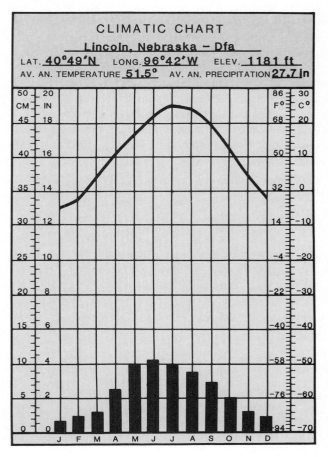

The transition to the short summer phase of the humid continental environment occurs near the poleward margins of the long summer phase. Lines between the two are arbitrary and tend to blur from year to year.

Humid Continental Cool Summer or Short Summer (Dfb, Dwb)

The humid continental cool or short summer is at a slightly higher latitude and tends to be further to the interior than the humid continental long summer type. Precipitation tends to be somewhat less, summers are shorter and not quite so hot, and winters are longer and more severe. Extremes of temperature and rapid change are also hallmarks of this climate. Bone chilling outbreaks of cold Arctic air occur frequently during the winter half of the year. Occasional surges of hot air encroach upon the region in the summer season, but such invasions are limited in scope and duration. The same climatic influences operate as in the humid continental warm summer, and variations from it may be attributed largely to higher latitude and greater continentality.

Toronto and Vladivostok illustrate some of the contrasts between the North American regime with no distinct dry season and the Asian station where winters are dry. Obviously, the Asian monsoon accounts for the contrast in precipitation regimes, and the continentality of the Eurasian landmass explains the different temperature characteristics. Although the two cities are located at approximately the same latitude, it can be observed that Vladivostok has a substantially colder winter. It also should be noted that the outblowing monsoon and the offshore winds ensure that continental influences predominate, although Vladivostok is a port city. Toronto is Dfb and Vladivostok is Dwb.

Toronto, Canada—Lat. 43°40′N.; Long. 79°24′W.

	Jan.	Feb.	Mar.	Apr.	May	June
T.	23.0	22.5	25.0	42.0	53.5	63.5
P.	2.7	2.4	2.6	2.5	2.9	2.7

July	Aug.	Sept.	Oct.	Nov.	Dec.	Year
69.0	57.5	60.0	48.0	37.0	27.0	45.0
2.9	2.7	2.9	2.4	2.8	2.6	32.1

Vladivostok, U.S.S.R.—Lat. 43°7′N.; Long. 131°55′E.

	Jan.	Feb.	Mar.	Apr.	May	June
T.	6.5	14.0	26.0	40.0	49.0	57.5
P.	0.3	0.4	0.7	1.2	2.1	2.9

July	Aug.	Sept.	Oct.	Nov.	Dec.	Year
65.5	69.5	66.5	48.0	30.0	14.0	40.0
3.3	4.7	4.3	1.9	1.2	0.6	23.6

Progression to higher and higher latitudes produces more severe cold temperatures during the winter season and reduced amounts of moisture. This is a harsh environment, especially in winter, but it is not as severe as the subarctic.

Subarctic or Subpolar (Dfc, Dwc/d)

The subarctic or subpolar climate is the climate most continental in character. Typically, it extends from about 50° to 65° latitude. This climate exists only in the Northern Hemisphere, because large landmasses are absent in appropriate latitudes in the Southern Hemisphere. Temperatures are so severe that at the northern limits only one month may have a mean temperature above 50° F, while six or more months have mean temperatures well below freezing. Southern reaches border on humid continental short summer or semiarid climatic regimes. Long, bitterly cold winters with very short summers are the rule. The winters are gloomy times with the sun making very short appearances only slightly above the horizon. In summer, there are long periods of daylight, but the sun is never very high in the heavens. Maximum temperatures in the summer frequently go above 80° F (26° C). As a rule, subarctic climates have the greatest temperature ranges

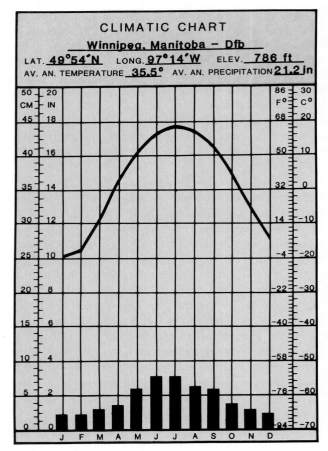

Figure 7.15
Winnipeg, Canada.

of all climates. Ranges of 120° to 150° F between coldest and warmest days are not uncommon. At Verkhoyansk in the Soviet Union, extremes between the all-time high and all-time low have exceeded 180° F (82° C). Oimekon, near Verkhoyansk, has recorded the lowest official temperature in the Northern Hemisphere of −90° F (−67° C), although that same station recorded an unofficial −108° F (−77° C). The lowest temperature in the Western Hemisphere, −87° F (−65° C), was recorded in interior Greenland, and the coldest temperature in Canada was recorded in the Yukon Territory, −81° F (−62° C), until the winter of 1989 when temperatures in interior Alaska plunged to −83° F (−63° C). This bone-chilling cold can have significant effects on people and shelter. Brief exposures by people not properly clad in cold outbreaks may produce severe cases of frostbite. Prolonged exposures to such cold, even by those who are properly attired, may produce cases of hypothermia, which can lead to death. Working outdoors in such regions during the winter season is particularly difficult. Not surprisingly, a number of Soviet Gulags (concentration camps) are located in such regions to supply forced labor, and because remote locations make escape difficult.

Figure 7.16
Churchill, Canada.

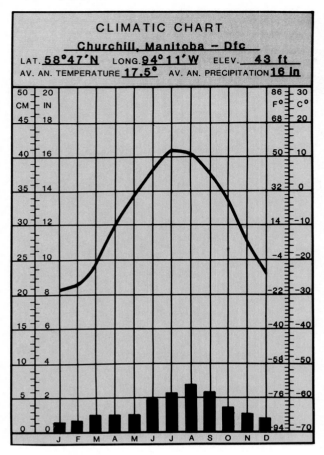

CLIMATIC CHART
Churchill, Manitoba – Dfc
LAT. 58°47′N   LONG. 94°11′W   ELEV. 43 ft
AV. AN. TEMPERATURE 17.5°   AV. AN. PRECIPITATION 16 In

Figure 7.17
Verkhoyansk, U.S.S.R.

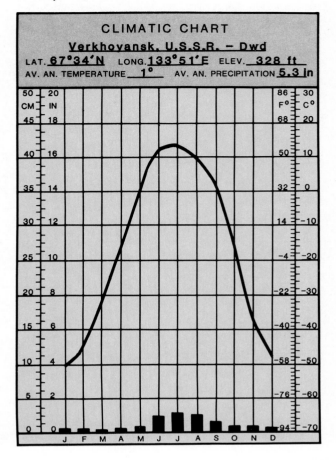

CLIMATIC CHART
Verkhoyansk, U.S.S.R. – Dwd
LAT. 67°34′N   LONG. 133°51′E   ELEV. 328 ft
AV. AN. TEMPERATURE 1°   AV. AN. PRECIPITATION 5.3 In

Subarctic regions generally receive small amounts of precipitation, typically between 10 and 20 inches (25 and 51 centimeters) per year. The cold air contains small amounts of moisture to begin with, and movement towards lower latitude tends to increase moisture-holding capacities. A summer maximum of precipitation occurs, which is quite pronounced in Asian stations. Winter snows are usually light, although there may be significant accumulations, since there is negligible melting or sublimation between successive storms. Frequently, too, high winds may create blizzard conditions either during snowfalls or from the powdery snow picked up from the ground.

Dawson illustrates the tendency towards a summer maximum of precipitation, whereas Okhotsk dramatizes a winter-dry situation where the monsoon effect is quite pronounced. Both illustrate the severity of temperatures during the winter half of the year as well as the short, cool nature of the summer and the low total amount of precipitation. Okhotsk has a somewhat lower range of temperature reflecting its coastal location compared to

Dawson's location well in the interior. Dawson is Dfc and Okhotsk is Dwc.

Dawson, Alaska.—Lat. 64°03′N.; Long. 139°25′W.

	Jan.	Feb.	Mar.	Apr.	May	June
T.	−23	−11	4	29	46	57
P.	0.8	0.8	0.5	0.7	0.9	1.3

July	Aug.	Sept.	Oct.	Nov.	Dec.	Year
59	54	42	25	1	−13	23
1.6	1.6	1.7	1.3	1.3	1.1	13.6

Okhotsk, U.S.S.R.—Lat. 59°21′N.; Long. 143°17′E.

	Jan.	Feb.	Mar.	Apr.	May	June
T.	−11.5	−8.0	10.0	19.5	32.5	42.0
P.	0.1	0.1	0.2	0.4	0.9	1.6

July	Aug.	Sept.	Oct.	Nov.	Dec.	Year
52.5	54.0	46.0	27.0	5.0	−6.0	21.9
2.2	2.6	2.4	1.0	0.2	0.1	11.8

Figure 7.18
Arctic and subarctic climates.
From Arthur Getis, Judith Getis, and Jerome Fellmann, Introduction
to Geography, *2d ed. Copyright © 1988 Wm. C. Brown Publishers,*
Dubuque, Iowa. All Rights Reserved. Reprinted by permission.

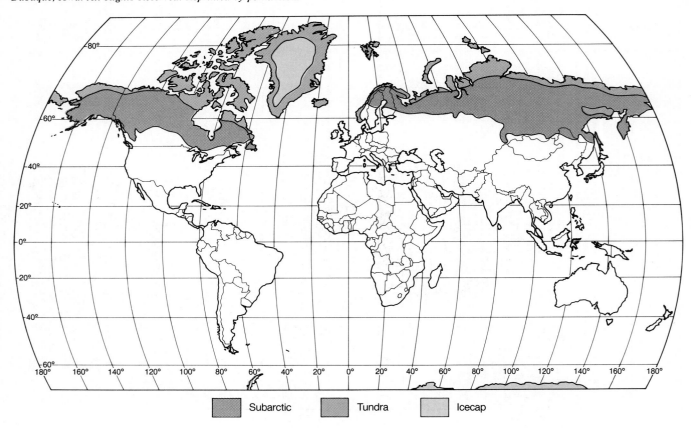

| | Subarctic | | Tundra | | Icecap |

Table 7.2 **Windchill Dangers**												
Wind Speed (MPH)	**Air Temperature (Fahrenheit)**											
	50	*40*	*30*	*20*	*10*	*0*	*−10*	*−20*	*−30*	*−40*	*−50*	*−60*
Calm	50	40	30	20	10	0	−10	−20	−30	−40	−50	−60
5	48	37	27	16	6	−5	−15	−26	−36	−47	−57	−68
10	40	28	16	4	−9	−21	−33	−46	−58	−70	−83	−95
15	36	22	9	−5	−18	−36	−45	−58	−72	−85	−99	−112
20	32	18	4	−10	−25	−39	−53	−67	−82	−96	−110	−121
30	28	13	−2	−18	−33	−48	−63	−79	−94	−109	−125	−140
40	26	10	−6	−21	−37	−53	−69	−85	−100	−116	−132	−148

Source: National Safety Council.

Note: Windchill factor shows the temperature that would have the same cooling effect on exposed human skin as a given combination of temperature and wind speed.

Obviously, at lower latitudes with higher temperatures and higher evaporation rates such stations would be semiarid. Under very high temperatures, such precipitation amounts might be characteristic of deserts.

The polar climates are even less comfortable than the subarctic, although both can be described as hostile to human occupance. Their distribution, controls, and characteristics are discussed in the following section.

Polar Climates (E)

Polar climates are generally quite inhospitable to human occupance, and settlements occur typically in response to military necessity, scientific inquiry, or because of the availability of mineral or animal resources. There are, of course, small native settlements of people who have adjusted to harsh nature and who derive their living from the local environment. These climates, located on the fringes of the Arctic Ocean and including the Antarctic continent, are characterized by penetrating cold, no warm season, and limited precipitation. The two principal types of polar climates recognized in the Köppen system include the tundra (ET) and ice cap (EF). EM climates exist in certain high latitudes where oceanic influences keep the ranges in cold- and warm-month temperatures relatively low. The EM variant exists largely in certain high-latitude islands such as the Falklands (Malvinas), South Georgia, and the South Shetlands, especially, in the vast oceanic expanses of the Southern Hemisphere, although portions of coastal Greenland also experience this climate.

Tundra (ET)

Tundra climates are at such high latitudes that they are dominated by cold air masses. They are found principally along the margins of the Arctic Ocean poleward of the subarctic climates. Average temperatures of the warm month are, by definition, greater than 32° F (0° C), but less than 50° F (10° C). Köppen chose the 50° F warm-month isotherm because he believed that was the effective limit of tree growth. Obviously, other factors are involved including soil, slope, and exposure conditions, but the choice hinged on a presumed vegetation limit. As in other Köppen climatic zones, vegetation extends across lines he established in response to microclimatological differences or because of special soil or slope conditions.

Mean annual temperatures are below freezing, and summers are short and quite cool. Cool to cold temperatures and attendant chilled air masses contribute to the fact that annual precipitation is quite low. Normally, the precipitation total is less than 10 inches (25 centimeters) per year. The permanently frozen ground beneath the surface (**permafrost**) may prevent or slow water percolation, and low evaporation rates may produce swampy conditions in certain locales during the summer months. This

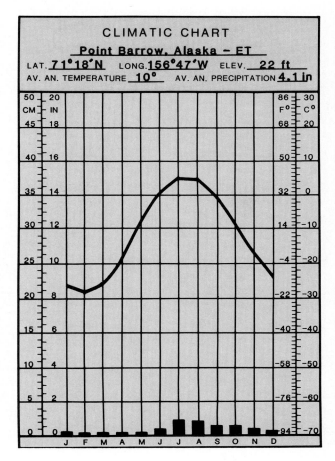

Figure 7.19
Point Barrow, Alaska.

has helped to contribute to the misconception that tundra areas receive more precipitation than is the case. The summer half of the year receives most of the moisture. The amount of moisture that falls as snow is quite low, but snow accumulates over the winter with almost no melting and limited sublimation, giving the illusion of precipitation abundance.

Barrow, Alaska illustrates the characteristics of a tundra environment. A cold winter is succeeded by a short, very cool summer. The amount of precipitation received is quite modest. Barrow is ET.

Barrow, Alaska—Lat. 71°23'N.; Long. 156°30'W.

	Jan.	Feb.	Mar.	Apr.	May	June
T.	−19	−13	−14	−2	21	35
P.	0.3	0.2	0.2	0.3	0.3	0.3

July	Aug.	Sept.	Oct.	Nov.	Dec.	Year
40	39	31	16	0	−15	9.9
1.1	0.8	0.5	0.8	0.4	0.4	5.6

Figure 7.20
Ivigtut, Greenland.

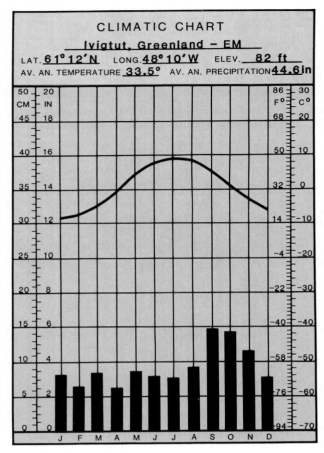

```
CLIMATIC CHART
Ivigtut, Greenland – EM
LAT. 61°12'N   LONG. 48°10'W   ELEV.   82 ft
AV. AN. TEMPERATURE 33.5°   AV. AN. PRECIPITATION 44.6 in
```

Figure 7.21
Eismitte, Greenland.

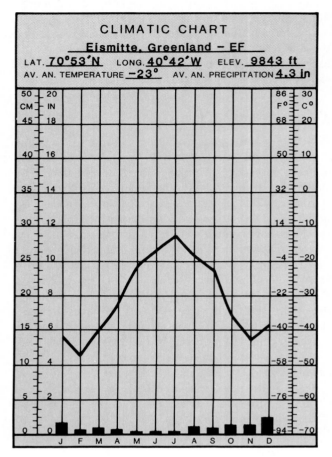

```
CLIMATIC CHART
Eismitte, Greenland – EF
LAT. 70°53'N   LONG. 40°42'W   ELEV. 9843 ft
AV. AN. TEMPERATURE –23°   AV. AN. PRECIPITATION 4.3 in
```

At lower latitudes, such precipitation amounts would produce semiarid conditions at best, and if temperatures were moderately high, such precipitation amounts would probably yield desert conditions.

Polar Ice Cap (EF)

The only land areas with a polar ice cap environment are interior Greenland and Antarctica. No month has an average temperature above 32° F (0° C); hence, a permanent cover of ice and snow accumulates. Although the total amount of precipitation is quite small, there is a steady accumulation of snow and ice because there is essentially no melting. At the Soviet Vostok scientific station in Antarctica, the lowest temperature ever recorded on the earth, −127° F (−87° C), was measured.

Eismitte, Greenland, illustrates the severity of these climates. Below freezing temperatures through the year and frequent high wind velocities make such a climate almost totally inhospitable for man. Clearly, such environments will continue to resist large scale human colonization. Eismitte is an EF station.

Eismitte, Greenland—Lat. 70°53'N.; Long. 40°42'W.

	Jan.	Feb.	Mar.	Apr.	May	June
T.	−43	−53	−40	−25.5	−6	2
P.	0.6	0.2	0.3	0.2	0.1	0.1

	July	Aug.	Sept.	Oct.	Nov.	Dec.	Year
	10	−1	−8	−32.5	−45	−37	23
	0.1	0.4	0.3	0.5	0.5	1.0	4.3

The severity of the E climates compares with the heterogeneity characteristic of the Highland (H) climates.

Figure 7.22
High mountains show distinct climatic zonations. The Canadian Rockies extend well above the tree line, although they are not extraordinarily high.
Department of Regional Industrial Expansion photo, Government of Canada.

Figure 7.23
In Utah, mountains exhibit a heterogeneity of microclimates. Utah Travel Council.

Highland (H)

The H symbol in the Köppen climate is designed to cover the great diversity of climatic conditions encountered in a mountainous region. There are, depending on elevation, slope, and exposure, almost limitless varieties of climatological regimes in mountainous areas. Areas on windward slopes might receive an abundance of moisture, whereas comparable elevations on leeward slopes might be almost rainless. Exposure or protection relative to prevailing winds might produce mild or severe perceived temperatures. Variations in elevation might create temperatures varying from tropical to polar. The H symbol is used to mark this great variety on small-scale maps. On large-scale maps of an area with a close network of stations, it would be possible to provide a detailed map of other Köppen types that might well extend from an A climate near the base of a mountain in a tropical area to an E at the crest of a high summit. Great heterogeneity of climate characterizes each level of elevation, each exposure of slope, and each intervening valley and isolated peak.

In some highland regions, especially in tropical latitudes, it is commonplace to attach names to vertical zones characterized by temperature differences. In tropical Latin America, for example, **tierra caliente** (hot land) extends from sea level to approximately 3,500 feet (more than 1,000 meters); **tierra templada** (temperate land) extends from about 3,500 feet to about 8,000 feet (almost 2,500 meters); **tierra fria** (cold land) from 8,000 feet to about 12,000 feet (more than 3,600 meters); and **tierra helada**

(ice land) above 12,000 feet. The elevations are arbitrary, but they are indicative of the significant diminution of temperature with increased elevation. The elevations vary with latitude, so the zone of permanent cold temperatures would be at a significantly lower elevation in middle and higher latitudes.

Basically, and above all, the H symbol masks a bewildering array of microclimates that could be depicted on only the most detailed large-scale maps. Microclimates are hidden in all climatic regions, however, because the simplified maps are based on generalizations that are not borne out when elements of climate are examined in great detail. Indeed, most generalizations, including this one, tend to break down or be modified when phenomena are examined in great detail.

Microclimatological Regions

Hidden within the broad climatological regimes described by Köppen or any other climatologist are microenvironments that might be at variance to the broad climatic regime. Protected valleys in humid environments might experience dry conditions although, in other circumstances, a valley in a dry area might have higher relative humidity as the result of protection from the wind.

Other elements of climate not used in Köppen, such as wind directions and velocities, amount of available sunshine received, and relative humidity may be significant features to human occupance of a region and to the ecological succession of vegetation patterns in an area. The incidence of sunshine and cloudiness are certainly worth noting as are the frequency of violent storms. Clearly, if additional climatic elements are considered, it's possible to develop more sophisticated classification schemes.

Figure 7.24
Köppen's world climates.

Köppen's World Climates

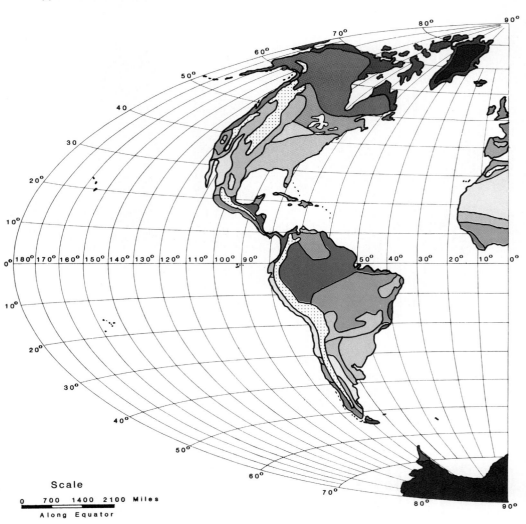

Scale

0   700   1400   2100   Miles

Along Equator

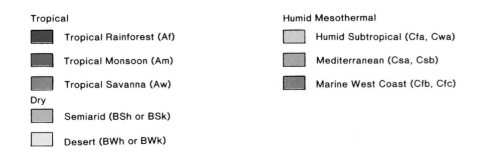

Tropical

- Tropical Rainforest (Af)
- Tropical Monsoon (Am)
- Tropical Savanna (Aw)

Dry

- Semiarid (BSh or BSk)
- Desert (BWh or BWk)

Humid Mesothermal

- Humid Subtropical (Cfa, Cwa)
- Mediterranean (Csa, Csb)
- Marine West Coast (Cfb, Cfc)

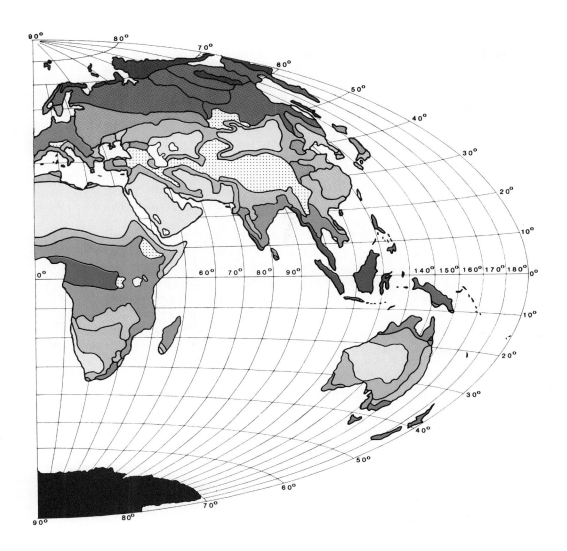

Aitoff's Equal Area Projection

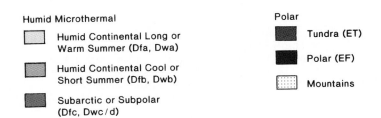

Humid Microthermal

| | Humid Continental Long or Warm Summer (Dfa, Dwa) |

| | Humid Continental Cool or Short Summer (Dfb, Dwb) |

| | Subarctic or Subpolar (Dfc, Dwc/d) |

Polar

| | Tundra (ET) |

| | Polar (EF) |

| | Mountains |

Similarly, the variations from mean conditions may be the most salient features of a particular climatological regime. It should be remembered that there is an almost infinite variation in the microclimatological patterns on the earth.

Microclimatological elements might establish parameters that would permit or preclude certain agricultural pursuits. For example, in an area where peach trees might experience frosts while they are in blossom or after fruit is set, a location on the intermediate north slopes (in the Northern Hemisphere) might be preferred, since such a position would inhibit blossoming until after the danger of frost is past, and air drainage would tend to protect such orchards after blossoming and fruiting.

In contrast, orange growers in central Florida might situate their groves on south-facing slopes adjacent to a lake. Such a location would protect trees bearing fruit from errant invasions from the north of cold, polar air.

The ragged edge of the tree line in the Colorado Front Range is apparently related to microclimatological variations. The relict stands of trees in isolated canyons of western Oklahoma seem to exist because of a more favorable microclimate. A dense and long-term net of climatological stations would doubtless add variety to the patterns of earth climates. The generalized patterns are, however, a useful addendum to an understanding of world environments.

As indicated in an earlier section, there may be vast differences in climatic classification from year to year. Variations in Oklahoma's climate, as depicted in several illustrations, reflects its position at the arid–humid margin on the one hand and its situation relative to the sharply contrasting continental polar and maritime tropical air masses on the other.

An oscillating jet stream can create dramatic contrasts from the long-term means, which tend to modulate the extremes. Obviously, such dramatic swings from "normal" conditions greatly increase the climatological risks in regions that experience them.

The general patterns depicted on small-scale maps, however, provide us with another element in the basic order in the landscape. The essential climatological environment establishes the parameters for plant colonization and growth while creating the conditions for weathering and erosion as part of the cycle of erosion.

Humid climates vary sharply from those that are moisture deficient. In the following chapter, the austere environments created when evaporation exceeds precipitation will be considered. The dry climates, although generally sparsely populated, occupy a large portion of the earth's surface.

Study Questions

1. What is the general location of the Af climates?
2. What accounts for the extension of Af climates into higher latitudes, notably Brazil and Madagascar?
3. What accounts for the wet and dry season in the Am environments? In the Aw environments?
4. What accounts for the g symbol being employed with some frequency in Am or Aw environments?
5. Why do most American humid subtropical environments have a Cfa symbol while a number of those in southeast Asia have a symbol of Cwa?
6. Account for the wet and dry periods in the Mediterranean climate.
7. What is a typical temperature regime for a Marine West Coast station in northwestern Europe?
8. Account for the relatively mild conditions in England in winter compared to the harsh environment in southern Labrador when both areas are located at comparable latitudes.
9. Explain why most North American humid continental stations have a symbol of Df while many of those in Asia have a symbol of Dw.
10. Where have the coldest surface temperatures on earth been encountered?
11. What is permafrost?
12. Explain why boundaries between climatic regions tend to be zones of transition.
13. Explain climatological variations in a given year from mean conditions.
14. What are microclimates?
15. Under what conditions might frequency data be more useful than mean conditions in producing a map of climates?

Selected References

Bryson, R. A., and Hare, F. K. 1974. *Climates of North America.* Amsterdam: Elsevier Scientific Publishing.

Fellows, D. K. 1985. *Our environment: An introduction to physical geography.* 3d ed. New York: John Wiley and Sons.

Flohn, H. 1969. *Climate and weather.* New York: McGraw-Hill Book Company.

Kendrew, W. C. 1961. *Climates of the continents.* 5th ed. London: Oxford University Press.

Mather, J. R. 1974. *Climatology: Fundamentals and applications.* New York: McGraw-Hill Book Company.

Miller, A., and Thompson, J. C. 1970. *Elements of meteorology.* Columbus: Charles E. Merrill Publishing Company.

Riehl, H. 1965. *Introduction to the atmosphere.* New York: McGraw-Hill Book Company.

Strahler, A. N., and Stahler, A. H. 1987. *Modern physical geography.* 3d ed. New York: John Wiley and Sons.

8

Climatic Regions—The Dry Climates and Climate Anomalies

The dry conditions in the Sinai Desert are revealed in these barren hills.

My soul thirsteth for thee, my flesh also longeth after thee; in a barren land where no water is.

Prayer Book, 1662

*T*he semiarid and arid climates of the world occupy an enormous portion of the earth's surface, perhaps as much as one-third of the land area, and there is a slow, but inexorable expansion of such dry regions, apparently at a more rapid rate because of man's destructive modification of the physical environment. There is every reason to fear that the process of desertification will continue and, as population continues to grow, accelerate. The dry climates continue to challenge the ingenuity of men and, except in unusual circumstances and restricted locales, resist efforts at permanent, close settlement. In the dry climates, evaporation exceeds precipitation, and the entire ecosystem is in a precarious balance because of the vagaries of precipitation, in terms of amount, distribution, reliability, and intensity. Water supplies are precious and exotic rivers (i.e., those that rise in humid areas and flow across dry regions), and underground sources of water are widely used to enable settlements to exist in areas where they would not otherwise be possible.

Dry climates occupy large areas in low latitudes from the interior to the western coasts of continents. In middle latitudes, they tend to occupy areas to the east of high mountain barriers and in the interior of continents.

121

Figure 8.1
*Precipitation variability. As a general rule, areas with low
mean precipitation have a lower probability of receiving
precipitation close to the mean in a given year.*
From Arthur Getis, Judith Getis, and Jerome Fellmann, Introduction
to Geography, *2d ed. Copyright © 1988 Wm. C. Brown Publishers,
Dubuque, Iowa. All Rights Reserved. Reprinted by permission.*

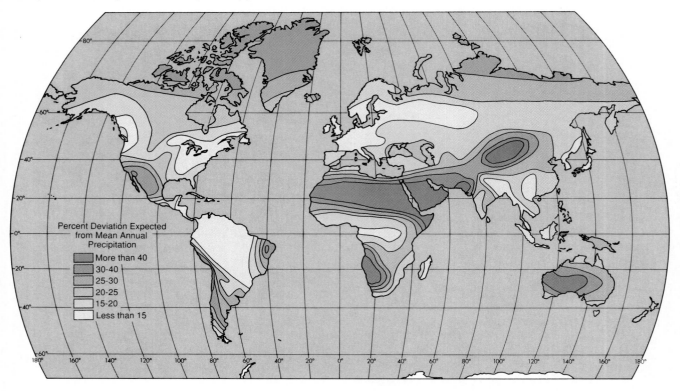

The aridity of a region is based not only on the
amount of rainfall, but also on the distribution and intensity of precipitation, as well as the type of soil and rate of
evapotranspiration. For example, two areas receiving the
same amount of precipitation might be vastly different in
terms of the rate of drying if, in one area, the rain comes
in a few intense showers where moisture runs off quickly,
compared to another region where slow soaking rains are
experienced more frequently. Obviously, too, a hard impervious soil will shed moisture quickly, whereas a permeable soil will make it available for replenishment of groundwater and to plants. High winds and temperatures might
accentuate aridity in one area while cooler temperatures
and moderate winds might lessen dry conditions in another. We have seen, for example, that tundra and ice caps
receive very small amounts of precipitation, but they aren't
deserts in the usual sense because evaporation rates are
low, and permafrost or ice prevents percolation of water
to any great depth. Frequently, dry areas experience high
and quite persistent winds. In part, this situation exists
because vegetation is sparse and provides less frictional
drag.

As a result, it is essentially impossible to establish
rainfall limits to demarcate the boundaries of the dry regions. Generally, however, dry climates are characterized
by (1) low precipitation amounts; (2) erratic distribution
of precipitation from month to month and year to year;
(3) high evaporation rates that normally exceed precipitation; (4) precipitation that does come may fall in hard,
intense, and short-lived showers; and (5) the reliability of
precipitation amounts tends to decrease as aridity increases. Because of infrequent hard showers, runoff may
be accelerated and groundwater recharge may be slowed.
All this is a kind of vicious circle that restricts vegetative
growth and leaves the rock and soil bare to the direct rays
of a merciless sun, which increases evaporation rates. As
devegetation occurs, usually as the result of overgrazing
or the gathering of desert plants for fuel, there is a tendency for relative humidities to be lowered and for drying
to accelerate.

Figure 8.2
The world's dry climates.
From Arthur Getis, Judith Getis, and Jerome Fellmann, Introduction
to Geography, *2d ed. Copyright © 1988 Wm. C. Brown Publishers,
Dubuque, Iowa. All Rights Reserved. Reprinted by permission.*

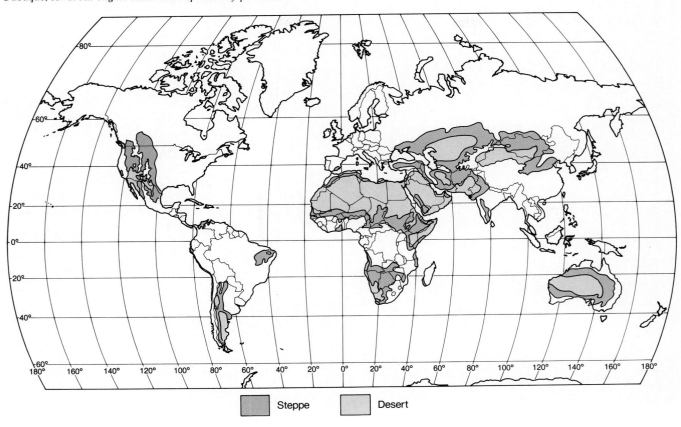

Steppe　　Desert

Deserts (BWh or BWk)

The biggest expanse of desert in the world is found between 15° and 30° north and south latitude. These areas are characterized by wide temperature variations, especially between daily maxima and minima, low relative humidity, and limited precipitation. Violent convectional showers, which are very limited in distribution and frequency, sometimes occur, but skies are generally almost cloudless, and sunshine is abundant at all seasons. Indeed, as a general rule, such areas are among the sunniest on earth. The unrelenting sun, which now adds to high temperatures and accelerated desiccation, may ultimately prove to be an important energy resource in such areas.

Regions between 15° and 30° tend to be dominated by subsiding air associated with the subtropical highs or, at other times, by trade winds that are blowing towards warmer latitudes picking up moisture rather than delivering it, except on certain high, exposed mountain barriers. The descending air of the subtropical high is being warmed by compression, and, as a result, tends to be drying. Indeed, these areas of descending air are some of the driest places on earth. Desertification tends to intensify in areas where evaporation exceeds precipitation. In desert areas, clouds sometimes release precipitation that is evaporated by extremely low relative humidities before it reaches the ground.

Low-Latitude Deserts

The low-latitude deserts are distinguished by the tertiary letter "h" in the Köppen classification, which testifies to the high mean temperatures encountered there during the high-sun season and the generally high annual temperature. The clear skies and minimum vegetation also account for the exceptionally high maximum temperatures encountered in desert regions. The generally clear air permits the receipt of maximum insolation. Rapid radiation through this clear air at night accounts for the sharp contrasts between daytime highs and nighttime lows. The

Figure 8.3
The dry conditions at Massada in Israel are obvious.

highest shade temperature, in excess of 136° F (more than 57° C), encountered on earth was recorded in the Sahara Desert in Libya. Death Valley in California has recorded a temperature of 134° F (56° C). In summer, temperatures frequently exceed 100° F (37°C), in the shade, and temperatures in the sun may be very high indeed. Such highs in the sun may be warmer than the typical thermostat setting about 140° F (60° C) on home hot water heaters. Rocky surfaces often become hot enough to burn bare skin with prolonged exposure. It is literally possible to fry an egg on these surfaces.

Prolonged exposure to summer desert heat carries the very real danger of heatstroke for the unwary. High dosages of ultraviolet light, which comes from prolonged exposure to the sun, increase the likelihood of skin cancer developing in middle or later life for desert inhabitants. Those living an outdoor life in desert areas show a much higher incidence of such cancers on exposed skin surfaces than do residents of humid regions. Aficionados of the sun are warned that the lovely bronze tan of youth may be a precursor to numerous wrinkles and cancerous skin lesions later on. Reduced exposure to sun is the prudent course when practicable, and the use of a good sunscreen is important when prolonged exposure to the sun is unavoidable. Sunlight is known to aggravate the conditions of those with systemic lupus erthymatosis (a chronic arthritic disease affecting skin and mucosa), and ultraviolet light can cause retinal damage, especially in those people with light-colored irises. Sun worshippers should exercise moderation in their exposure to the sun.

Because of the rapid radiation of heat from the earth after the sun has gone down, there may be considerable variations between day and night temperatures in desert areas. The searing heat of the day may well be followed by the chill of a desert night. Those unacquainted with this rapid cooling may experience uncomfortably chilly nights if they are not properly clothed or sheltered. Clearly the cooling from daytime highs well over 100° F to lows at night in the 70° range, represents a significant diurnal range. It is the relentless aridity, however, that is the significant aspect of desert environments.

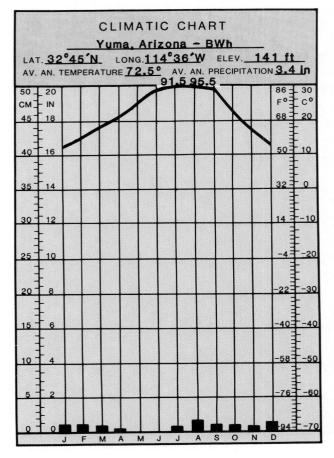

Figure 8.4
Yuma, Arizona.

CLIMATIC CHART
Yuma, Arizona – BWh
LAT. 32°45′N LONG. 114°36′W ELEV. 141 ft
AV. AN. TEMPERATURE 72.5° AV. AN. PRECIPITATION 3.4 in
91.5 95.5

Some of the low-latitude deserts are almost rainless. Some stations, especially in the Atacama Desert of Chile, have been rainless for several years at a stretch. The driest known place on earth is at Arica, in northern Chile, which has an annual precipitation of 0.03 inch. Wadi Halfa, Sudan receives only an average of 0.1 inch of precipitation annually. Stations such as these may go years without receiving any measurable amounts of rainfall. In the case of Chile, the arid conditions have permitted the commercial extraction of natural nitrate. The dry conditions have precluded the dissolving of the natural deposits. In humid regions, such mineral salts would long since have been leached out or washed away.

Extremely low relative humidities cause rapid drying of exposed skin or any wet surface. Such dry air may be quite conducive to the development of static electricity in any application of friction. The very dry air has also been responsible for the preservation of many organic and inorganic materials. What we have learned about many ancient civilizations, such as the Egyptian or Persian, has been enhanced by the preservative characteristics of a dry environment. Some mummies from ancient Egypt have been remarkably preserved by the very low relative humidities. Many of the accouterments of daily living or dress have been similarly preserved. Recent rediscovery (1988) of foods buried with the Pharaoh Tutankhamen gives interesting insights into the domestication of certain plants as well as the nature of agriculture several millennia ago.

Because of the high percentage of sunshine and the low incidence of cloudiness, deserts provide almost optimum flying conditions. It's no accident that Edwards Air Force Base in the desert of California, for example, has become a test facility for exotic aircraft and the site for important training in air-to-air combat. The general reliability of its clear weather also makes it the primary site for the space shuttle to land. The Soviet air base near Baikonur in Soviet Central Asia also makes use of the good flying weather characteristic of a dry environment.

In certain deserts, such as the Atacama in South America and the Kalahari in Africa, where desert conditions extend to the coast in regions that have cold ocean currents, there may be frequent fog at the coastal margin. The lower portions of the atmosphere may be chilled to the dew point by contact with the cold current. Indeed, the Köppen classification includes a symbol (n) to denote frequent fog wherever it may occur. These fogs may create curious microbiotic zones with interesting plants and animals subsisting in an environment where the amount of precipitation would not support them, but the moisture available in the fog supports essential metabolic processes. Such an intricate web of life in dunes, which at first appear devoid of life, is encountered at the coastal fringe of the Kalahari Desert of Africa known as the Namib.

Deserts, like other climatological regions, exhibit variations, but the central imperative of aridity is omnipresent. Cairo illustrates the essential characteristics of the low-latitude desert. The Köppen symbol for Cairo is BWh. When one considers these conditions, it should be recognized that temperature means may be fairly reliable indicators of what will be encountered in a given year, but departures from norms in precipitation may be radical in amplitude and frequent in occurrence.

Cairo, Egypt—Lat. 30°31′N.; Long 31°15′E.

	Jan.	Feb.	Mar.	Apr.	May	June
T.	55	57	63	70	76	80
P.	0.4	0.2	0.2	0.2	0	0

July	Aug.	Sept.	Oct.	Nov.	Dec.	Year
82	82	78	74	65	58	70
0	0	0	0	0.1	0.2	1.3

Middle-Latitude Deserts

Middle-latitude deserts are frequently located on the lee side of topographic barriers or are situated deep within the interior of large landmasses far from possible sources of moisture. The rainfall is meager and unreliable, there is greater range in temperature from hot to cold season, and the mean annual temperature is less than for that of low-latitude deserts.

Sufu, China illustrates the effect of higher latitude and continental location. The march of temperature shows a greater amplitude than that for Cairo, and the amount of precipitation is somewhat greater reflecting the influence of the occasional cyclonic storm that affects Sufu. Certainly the cold winters of Sufu contrast sharply with the relatively mild winters of Cairo. The Köppen symbol for Sufu is BWk.

Sufu, China—Lat. 39°24'N.; Long. 76°07'E.

	Jan.	*Feb.*	*Mar.*	*Apr.*	*May*	*June*
T.	22.5	31.0	45.5	59.5	69.5	76.5
P.	0.6	0.1	0.5	0.2	0.3	0.2

July	*Aug.*	*Sept.*	*Oct.*	*Nov.*	*Dec.*	*Year*
80.0	78.0	70.0	57.0	41.5	27.5	55.0
0.4	0.3	0.1	0.1	0.2	0.3	3.2

Because there are varying degrees of aridity in deserts, efforts have been made to differentiate further the inherent characteristics of arid lands. One such effort has been made by the American scholar, Peverell Meigs. He has used a fairly elaborate scheme to classify different desert regions. Essentially, the scheme provides a kind of index of the hospitability or inhospitability of desert areas. The varying levels and intensities of arid conditions show the Peruvian-Atacama, portions of the Sahara, and the Namib-Kalahari as the driest and most hostile of desert environments. Those deserts where conditions are not quite so austere include the Arabian, Turkestan, Patagonian, Gobi, Thar, Australian, and North American Deserts. In spite of the fact that certain deserts are less hostile than others, few reflect close settlement except in response to economic or strategic advantage.

Those not accustomed to desert environments may make extraordinary efforts to mitigate the heat of the summer. The author of this work spent a protracted period of time living and working in Iran. Cooling in the summer was very important to maintain general health and well-being. Since he was employed by a government agency, application was made for an appropriate cooling device. Either an evaporative or refrigerated air conditioner was offered. He chose, wisely as it turned out, the evaporative cooler. It worked very effectively using a principal described earlier in this book (i.e., heat is taken up when evaporation occurs). As air is drawn across a water-saturated pad, the latent heat of vaporization cools the air drawn through the pad. Since relative humidities are quite

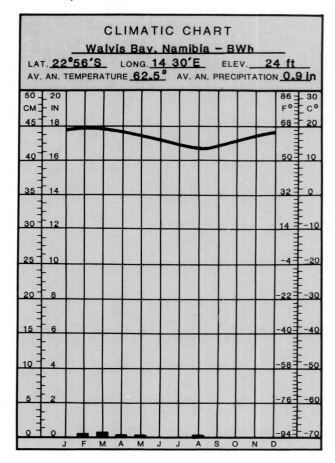

Figure 8.5
Walvis Bay, Namibia.

CLIMATIC CHART
Walvis Bay, Namibia — BWh
LAT. **22°56'S** LONG. **14 30'E** ELEV. **24 ft**
AV. AN. TEMPERATURE **62.5°** AV. AN. PRECIPITATION **0.9 in**

low, the cooling is considerable. The air drawn into the house was also humidified, mitigating the very low relative humidities somewhat. Other employees who chose the more familiar refrigerated air conditioner were less comfortable.

In contrast, in a humid environment, like the humid southeast where the author now lives, an evaporative cooler would be very inefficient. Indeed, the refrigerated air conditioner is a must to reduce humidities for personal comfort and to reduce the ravages of fungi such as mildew. We idly speculate from time to time how people lived and functioned effectively in such a climatic environment before the days of air conditioning. In our musings, the answer is probably—not very well.

It seems clear that deserts are expanding in size, especially along the Sahel fringe in Africa. Increased use of subsurface water, drastic overgrazing, the cutting of existing vegetation for fuel, and close human settlement all seem to be contributing to an encroachment of the desert on less arid regions. Further, there is the growing suspicion that this is an irreversible process, at least during

Figure 8.6
The desert of Iran is a hostile and forbidding place.

Figure 8.7
Certain midlatitude deserts can support fairly close growing vegetation, especially after errant rainstorms.

It is difficult to teach an ethic of conservation where survival is a constant battle. A few sprigs of grass or the branches from a bush may feed a goat, and twigs and dung provide fuel for cooking. In the long term, such activities hasten desertification. In the short run, hungry stomachs win out over the need for conservation.

Semiarid Climates (BSh or BSk)

Tropical semiarid or steppe (BSh) climates surround deserts except on the western side of continents where deserts extend to the sea. The climatic regions differ from deserts in having a somewhat greater amount of precipitation, which is usually a bit more reliable. In low-latitude steppes, the wet season occurs at the high-sun season when certain unstable conditions associated with the Intertropical Convergence Zone spill over infrequently from adjacent savannas. The dominance of the subtropical highs during the winter season accounts for the dry conditions that exist then. These enormous domes of high pressure tend to suppress convectional activity and deflect the occasional low-pressure storm to higher latitudes.

The wet season typically lasts from three to five months, but the average time may vary considerably from year to year. Amounts of precipitation are also unreliable, and the unreliability increases towards the dry margins of the steppe. Many such semiarid regions are in danger of slipping over the line to desert conditions primarily because of the depredations of people. Devegetation of a region may speed desiccation of the topsoil, and it tends to reduce relative humidities and chances for dew or rain. Drawdown of water supplies may have similar effects, and cultivation or overgrazing will hasten the deterioration of

any reasonable length of geologic time. What is more troublesome is the fact that this expansion may disturb generalized atmospheric circulations with negative climatological consequences. The exact nature of such dislocations cannot be predicted, but it is that very uncertainty that potentially makes them so ominous. This process of desertification seems to be continuing at a rapid pace in Africa, where food production capacities are inadequate and declining, and fuel sources are few and far between. Outlook for the future in an area where there is rapid destruction of land are less than happy. Indeed, Africa is the only continent where food production per capita is less than it was two decades ago.

Nevertheless, some agencies are attempting to revegetate areas denuded by overgrazing, cutting of wood for firewood, or poor agricultural practices. This battle between the subsistence farmer and the conservationist may determine whether the deserts expand or whether the extension of bare earth and sand will be halted by a stand of green vegetation.

Figure 8.8
Tehran, Iran.

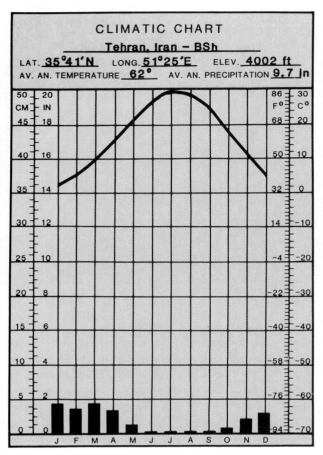

CLIMATIC CHART

Tehran, Iran — BSh

LAT. **35°41'N** LONG. **51°25'E** ELEV. **4002 ft**

AV. AN. TEMPERATURE **62°** AV. AN. PRECIPITATION **9.7** in

Figure 8.9
Semiarid landscapes, such as this one in western Oklahoma, suffer from frequent periodic droughts.

the environment. It seems likely that steppes will yield to desertification on the dry margins, and humid areas may degenerate to semiaridity. This is an immediate problem for people who live in such regions; it is a persistent problem for the affluent who supply aid to such people; and prospectively, it is an ominous problem for the world at large because of the destruction of at least marginally productive land and the resulting burden on productive regions elsewhere.

Typically, precipitation amounts range from 10 to 20 inches (25 to 51 centimeters) per year, although higher amounts may be received when the evaporation rates are very high. High evapotranspiration rates reduce the effectiveness of the precipitation that does fall. The data for Monterrey are typical of the low-latitude semiarid environment. Monterrey is classified as BSh.

Monterrey, Mexico—Lat. 25°40'N.; Long. 100°18'W.

	Jan.	Feb.	Mar.	Apr.	May	June
T.	58	62	68	73	79	82
P.	0.4	0.5	0.7	1.1	1.2	2.3

	July	Aug.	Sept.	Oct.	Nov.	Dec.	Year
	82	83	78	71	64	57	71.4
	2.1	2.0	4.4	2.4	1.3	1.0	19.4

Middle-Latitude Steppe

Middle-latitude semiarid (BSk) environments are also transitional between arid and humid climates. Like low-latitude steppes, these regions are characterized by low rainfall amounts and unreliability of the precipitation that falls. These climates fringe middle-latitude deserts, which tend to be situated further into the interior of continents. Precipitation tends to be somewhat more evenly distributed, except in areas experiencing monsoon regimes, such as those in Asia, where there is a decided summer maximum of precipitation. This monsoon tendency is especially pronounced in Asia and in certain portions of Africa. It is less noteworthy in North America, although monsoon-like indrafts do affect certain portions of the American Southwest.

Denver exhibits the characteristics of the middle-latitude steppe. Temperature ranges between seasons are greater, and, in the case of Denver, the summer temperatures are somewhat cooler than might be expected because of the elevation of the mile-high city. Denver is BSk.

Denver, Colorado—Lat. 39°41'N.; Long. 104°57'W.

	Jan.	Feb.	Mar.	Apr.	May	June
T.	30	32	39	47	57	67
P.	0.4	0.5	1.0	2.1	2.4	1.4

	July	Aug.	Sept.	Oct.	Nov.	Dec.	Year
	72	71	62	51	39	32	50
	1.8	1.4	1.0	1.0	0.6	0.7	14.3

Figure 8.10
Bourke, Australia.

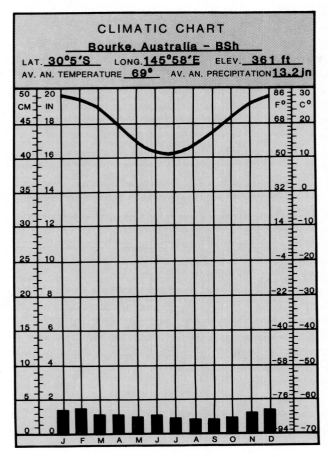

CLIMATIC CHART
Bourke, Australia – BSh
LAT. **30°5′S**   LONG. **145°58′E**   ELEV. **361 ft**
AV. AN. TEMPERATURE **69°**   AV. AN. PRECIPITATION **13.2 in**

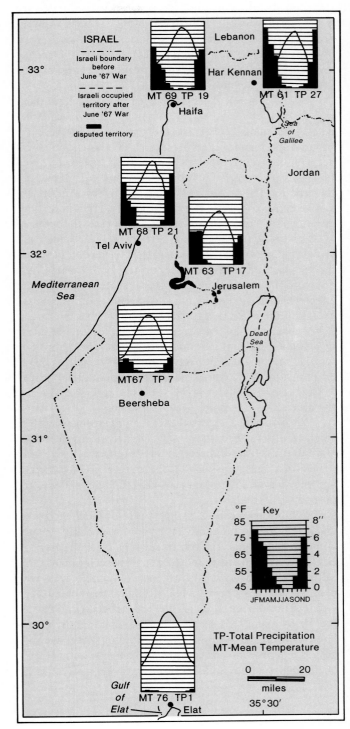

Although Denver is situated about a mile above sea level, it is located in a kind of topographic trench between the margins of the high plains and the edge of the Rockies. Such a topographic basin and a burgeoning population (with all of its industrial and transportation growth) have contributed to a steady reduction in air quality. Air pollution has, in fact, become a major problem for the city.

These changes caused by people remind us that climate has shown significant variations in the past, and logic suggests that the future will be marked by similar changes. It behooves us to have some appreciation for and understanding of the nature of such changes.

Climate Variations and Climatic Cycles

There are more and more data to suggest that we are living in a relatively benign climatological era. Average temperatures during the past 1,000 years have been significantly higher than during the previous 100,000 years, and there have apparently been a series of peaks and troughs during the past 1,000,000 years. During a number of those troughs, there have been buildups of continental ice sheets that have produced numerous glaciated landforms in middle- and high-latitude landscapes.

Climatologists are watching the increasing level of CO_2 in the atmosphere as the result of the rising consumption of hydrocarbon fuels. An increase in the CO_2 will enhance the "greenhouse effect" of the earth, resulting in warming. Many postulate that such warming will cause a more pronounced rising of air in equatorial locations resulting in that air moving further poleward before settling in the subtropical highs. If such an event occurs, there will be a tendency for drier conditions to creep towards higher latitudes, and the equatorward margins of low-latitude steppes and deserts would perhaps experience higher precipitation.

It's clear that there have been massive climatological dislocations in the past—many of them occurring before the genus *Homo* became *sapiens,* or in the early stages of prehistory. It seems likely that those swings will occur in the future. Various explanations have been given for past variations from "normal." It has been suggested that the burden of particulate matter in the atmosphere has been much greater during times of contemporaneous eruptions of numerous volcanoes. Historical accounts of massive eruptions, including several in the recent past, show that there may be a considerable burden of atmospheric dust for some time after an eruption. Clearly, dust affecting the atmosphere on a global scale would have an effect on insolation and radiation of heat from the earth. The eruption of Krakatau (Krakatoa) in what is now Indonesia during the last century had a measurable effect in reducing ambient temperatures around the globe. The total climatological impact is uncertain. Recent large eruptions in North and Central America have appeared to have minimal influence on global climatic patterns, but the precise impact is not yet understood.

Others have suggested that the precessional cycle of earth rotation may affect insolation at various latitudes. Data derived from indirect sources are inconclusive, and other climatologists assert that the precessional arc is small and the change so gradual that climatological effects are probably minimal.

Still other scientists suggest that cyclical variations in sunspot activity may affect insolation, and, as a result, climatic conditions. In certain sections of the American Great Plains there appears to be some association of rainfall peaks and troughs with the approximate eleven-year cycle of sunspot activity. The droughts in the southern Great Plains during the Dust Bowl of the 1930s and during the 1950s and 1970s seem to have been generally correlated with sunspot minima. Some climatologists have suggested that peaking and troughing of sunspot activity may be related to tidal influences on the sun generated by changes in planetary configurations. Most, however, are uncertain as to the causes of sunspot variations, and specific climatic relationships to such variations remain to be worked out.

Some paleogeologists postulate a 26,000,000-year periodicity of animal extinctions, perhaps climatologically induced. There are suggestions that an undiscovered planet or an invisible solar twin may be responsible. Others scoff at such extinctions and unseen solar system celestial bodies. The broad implications of such relationships are difficult to trace, and even indirect evidence allows us to assess only a very small period of earth history.

Others consider the variations of sea temperature, sometimes almost convulsive in nature with the upwelling of cold water in a new area, or the slowing of upwelling in an area where upwelling has been prevalent, as being of major climatological significance. This does create short-term dislocations in such features as El Niño. The changes are uncertain and so far unpredictable. In any case, they tend to be short-lived in effect. It is certain, however, that climates do undergo short- and long-range changes. Uniformitarianism strongly infers that past changes suggest future changes. Since the direction and timing of such changes are uncertain, they are probably not worth losing sleep over. Further, even if the changes could be predicted with some certainty it might be impossible to affect the outcome or satisfactorily adjust to the consequences.

Other causal factors for climatic change will doubtless be advanced, and a more complete understanding of climatic variations will almost surely be forthcoming borne on the wings of better data, more sophisticated measuring devices, a closer net of observations, and more sophisticated theoretical considerations. Because climate has such dramatic influences on human occupance of the earth, it is imperative that our understandings and predictions improve.

The earth's atmosphere is a giant heat engine. The more we can know about the atmosphere and its response to natural and human change, the better the chances are that man will be able to cope with departures from perceived normal conditions. In chapter thirteen of this book, some more detailed suggestions will be made attendant to possible climatological changes involved in the spread of glaciers at various times in earth's history. Cyclical events, like glaciation, hint at continuing climatic changes. Just as the dimensions and directions of future climatic oscillations cannot now be predicted with any degree of accuracy, neither can we fully explain some apparent extant climatic anomalies.

Figure 8.12
The greenhouse effect. Various gases are trapping heat in the lower atmosphere and increasing the average temperature.
Electric Power Research Institute Journal.

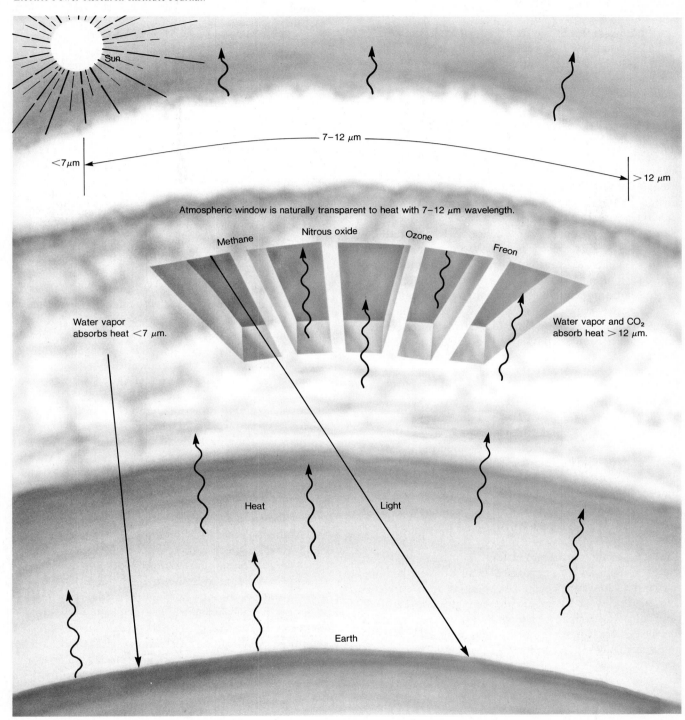

Climatic Anomalies

In certain areas, there seem to be regions where climatic patterns do not fit the apparent circumstances, or where variations are quite large within a restricted location. These areas do, of course, obey natural laws, but a brief description of a few examples of these apparent anomalies will suffice to show that generalized patterns as depicted on small-scale climate maps are often misleading, and the mechanisms creating anomalies are neither completely understood nor appreciated.

Guadeloupe. To illustrate, the island of Guadeloupe in the French West Indies is only 688 square miles (1,107 square kilometers) in area, and yet the contrasts in climate from one place to another are significant. The island is divided roughly into two equal halves: a low-lying limestone platform in the east, known as Grand Terre; and a volcanic mountain range in the west, known as Basse Terre. Although the eastern half is struck directly by the trades, it is so low-lying that there is negligible orographic lift, and rainfall is generally less than 40 inches (102 centimeters) per year (figure 8.13). The porous limestone material and the constant trades add an element of physiological aridity to the relative scarcity of rainfall, and that portion of the compound island exhibits almost semiarid characteristics. Vegetation is semixerophytic compared to the lush rainforest in the mountains of Basse Terre. The mountainous western half lies athwart the trades and orographic lift may yield more than 180 inches (more than 450 centimeters) of rain per year. An almost semiarid environment *on the windward side,* is contrasted with an Af environment less than 20 miles (32 kilometers) away. Interior valleys on the lee slope of the western mountains may be characterized by peculiar microclimatological aberrations, some of which exhibit subhumid characteristics. Indeed, on this leeward side of Basse Terre, the rainfall is again less than 40 inches. Similar apparent anomalies exist elsewhere.

Yucatan. In the northern Yucatan Peninsula of Mexico, there is a semiarid zone primarily because of wind tending to parallel a low-lying coast. Elsewhere, the peninsula has a tropical rainy environment. Interestingly, the interior of the peninsula seems to benefit from sea breezes from different segments of the peninsula converging to force uplift and accompanying precipitation. Ironically, the coastal fringe in the north does not have its precipitation augmented in this way. Under normal circumstances, it would be considered anomalous for a low-lying interior to receive greater precipitation than sections along the coast in similar latitudes. Other regions also exhibit certain climatological peculiarities that are not fully understood.

Northeast Brazil. An area in Northeast Brazil, known as the Dry Zone, has semiarid conditions, occasionally deteriorating to arid, although it is surrounded by tropical savanna. Two major factors apparently contribute to uncertain climatological conditions in this unhappy area. First, it is somewhat lower than surrounding regions, and air moving in tends to be warming by compression. Further, the Intertropical Convergence Zone bends away from the region in the normal wet season, but the cause of the bend is not fully understood. In truth, all the reasons for the dry conditions are not understood, but the anomaly remains to create a specter of recurrent drought and poverty for people of the region, because occasional good years lead to closer settlement than bad years will support.

Ahaggar and Tibesti. Just as there are areas of drier climate than would be expected with a cursory examination of a climatological map, so there are areas with a wetter environment than otherwise would be expected. For example, the Ahaggar and Tibesti Domes deep within the heart of the Sahara, are significantly wetter than surrounding areas because their elevation intercepts air masses, and orographic lift yields significantly larger amounts of rainfall than is characteristic for the adjacent desert. Exposed sediments along those slopes become aquifers for fringing oases located some distance away. Indeed, a number of the oases along established caravan routes owe their existence to the fact that mountains have pierced moisture-bearing winds that nurture aquifers, which supply oases with life-giving water. A similar pattern of greater precipitation in the highlands of Yemen on the Arabian Peninsula supports oases in the deserts of Saudi Arabia and the adjacent United Arab Emirates.

There are enormous variations of climate within the regions depicted on a small-scale map, and there may be great variations of climatic conditions from year to year. We are apparently living in a period of benign climate that was preceded by and may be succeeded by more austere conditions. Cyclical variations do apparently occur; but the reasons for the changes are imperfectly understood in some instances and are almost totally unknown in others.

Figure 8.13
Average annual precipitation of Guadeloupe (in inches).
Data from: Climatological Data: West Indies and Carribean Section,
and weather records at Ralzet, Guadeloupe.
Modified from a map by Don R. Hoy.

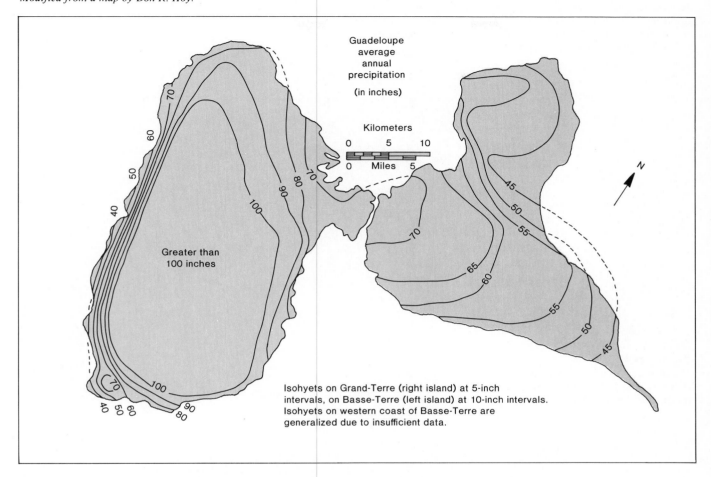

Guadeloupe
average
annual
precipitation

(in inches)

Kilometers

0   5   10

0   Miles   5

Greater than
100 inches

Isohyets on Grand-Terre (right island) at 5-inch
intervals, on Basse-Terre (left island) at 10-inch intervals.
Isohyets on western coast of Basse-Terre are
generalized due to insufficient data.

Climate will continue to remain a dominant element in shaping the physical landscape of the earth and in human affairs. All of the other spheres (i.e., lithosphere, biosphere, and hydrosphere) are affected greatly by the pulsing changes in the atmosphere. Even with all the knowledge and technology at our command, we remain at the mercy of the elements of climate, which affect many aspects of our lives and activities.

Elements of climate are constantly at work in sculpting the face of the land. The cataclysmic forces that uplift the surface of the earth are contesting with the climatologically driven elements of weathering and erosion, which are constantly attempting to reduce the earth to base level.

The earth's lithosphere will be examined in subsequent chapters of the book. Modifications of the face of the land occur because of original uplift and subsequent sculpting largely driven by climatic forces.

Figure 8.14

De Martonne's Index of Aridity is an early scheme of climatic classification. Iar = P/(T+10) where P is precipitation in millimeters and T is temperature in degrees Celsius or Iar = 45.72P/(T − 14) where P is precipitation in inches and T is temperature in degrees Fahrenheit. For mean conditions, the index of 22 is approximately the humid-semiarid boundary. Note that only the Oklahoma panhandle is semiarid under mean conditions.

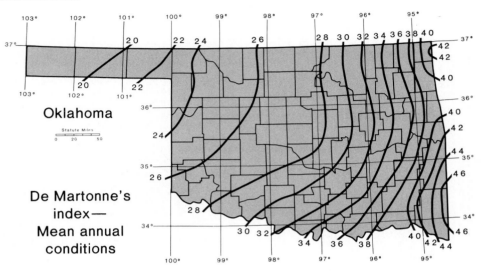

Oklahoma

De Martonne's index— Mean annual conditions

Figure 8.15

De Martonne's Index of Aridity for a dry year. Note that more than half the state is semiarid or arid using the 22 index as a line of demarcation.

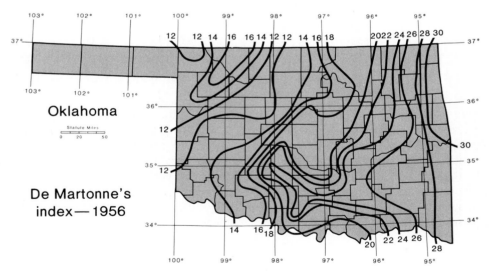

Oklahoma

De Martonne's index— 1956

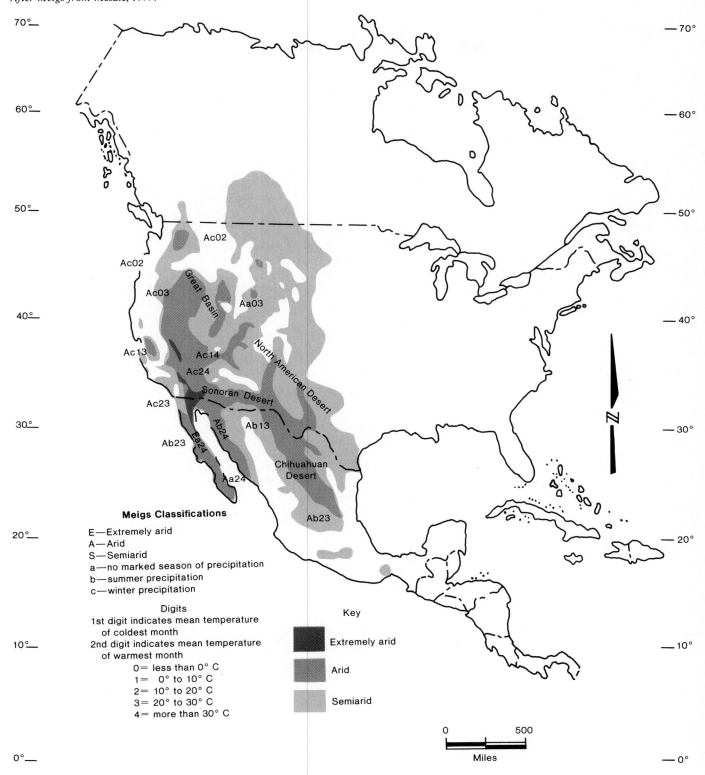

Figure 8.16
North America exhibits varying dry conditions.
After Meigs from Mosaic, *1977.*

Ac02

Ac02

Ac03

Great Basin

Aa03

Ac13

Ac14

North American Desert

Ac24

Ac23

Sonoran Desert

Ab24

Ab13

Ab23

Ea24

Aa24

Chihuahuan
Desert

Ab23

Meigs Classifications

E—Extremely arid
A—Arid
S—Semiarid
a—no marked season of precipitation
b—summer precipitation
c—winter precipitation

Digits
1st digit indicates mean temperature
of coldest month
2nd digit indicates mean temperature
of warmest month
    0= less than 0° C
    1=　0° to 10° C
    2= 10° to 20° C
    3= 20° to 30° C
    4= more than 30° C

Key

Extremely arid

Arid

Semiarid

0　　　　500

Miles

Study Questions

1. What is meant by physiological aridity?
2. What are the principal climatic controls producing aridity in low-latitude deserts? In middle-latitude deserts?
3. Where are the driest places on earth?
4. Why are many tundra stations with an annual precipitation between 5 and 10 inches not classified as semiarid or arid climates?
5. Explain why the symbol n is frequently applicable to low-latitude coastal deserts.
6. What accounts for the fact that deserts are apparently expanding at the present time?
7. What is the wet season for most low-latitude steppes? What accounts for this pattern?
8. What are the likely consequences of an increasing burden of CO_2 in the air?
9. List and comment upon some of the theories that have been proposed to explain long-term climatic changes.
10. Describe the climatic conditions that exist in various parts of the Island of Guadeloupe.
11. Explain why the Brazilian Dry Zone is an area of incipient poverty.
12. What suggestions have been advanced as to possible causes of recurrent droughts in the Great Plains?
13. If the "greenhouse effect" results in more vigorous equational uplift of air and settling of air in high pressure ridges at higher latitudes than is now the case, speculate about the effects upon climate in Florida and California.
14. What accounts for the extreme aridity encountered in certain low-latitude deserts along the west coasts of continents?

Selected References

Bryson, R. A., and Hare, F. K. 1974. *Climates of North America.* Amsterdam: Elsevier Scientific Publishing.

Flohn, H. 1969. *Climate and weather.* New York: McGraw-Hill Book Company.

Kendrew, W. C. 1961. *Climates of the continents.* 5th ed. London: Oxford University Press.

Scott, R. C. 1989. *Physical geography.* St. Paul: West Publishing Company.

Trewartha, G. T. 1981. *The earth's problem climates.* 2d ed. Madison: University of Wisconsin Press.

———. 1941. *Climate and man: Yearbook of agriculture.* Washington: United States Department of Agriculture.

9

The Earth's Crust

Some segments of the earth's crust have remained above sea level for protracted periods of earth history like this portion of the Canadian shield. Department of Regional Industrial Expansion photo, Government of Canada.

*"This earth is not the steadfast place
We landsmen build upon;
From deep to deep she varies pace,
And while she comes is gone."*
William Vaughn Moody, *Gloucester Moors*

When the earth was formed from cosmic debris of gas, dust, and other materials more than 4,000,000,000 years ago, it began a never-ending process of changing its surface form, which has not yet concluded and which will doubtless continue until the earth is charred by a bulging sun billions of years from now. Even then, as a piece of cosmic debris, it may be incorporated into some new coalescence of matter, which will preface a new beginning of a cycle of universe formation.

Indeed, change is a constant and continuing cosmic fact of life. Presumably such change will continue until an expanding sun boils the water away billions of years from now. Most current thinking holds that the final paroxysms of the death of the sun will result in the vaporization of the inner planets, including the earth, about 5,000,000,000 years into the future. Before this event, the world will have revealed a constantly changing face for about 10,000,000,000 years.

At its birth and during its early development, the heat of compression augmented by the radioactive decay of many of its constituent elements caused earth materials to sort themselves out in a rough approximation of their density as the cosmic dust and gas coalesced into planetary form. Although the outer surface of the earth is perceived as solid, it should be recognized that many materials beneath the surface are in a plastic (semifluid) state, or, at least, they act like very viscous fluids in the way they move and in their deformation characteristics. The movement of this plastic material beneath the crust and the solid materials above it is responsible for much of the pattern of landforms on the earth's surface. These structural forms, as modified by elements of weathering and erosion, produced the landforms that are our geomorphological legacy. The throbbing cadence of convection-like movement in both shallow and deep zones of the earth means that the crust is constantly being contorted, broken, reduced, and subjected to the injection and intrusion of the products of volcanic fire. These inexorable forces are in a constant state of warfare with the agents of degradation, which work to reduce uplifts to grade.

To understand the nature of aggradational forces, it is necessary to know something of the interior structure of the earth beneath our feet. A brief description of what is known and inferred about the earth's interior follows.

The Earth's Interior

Since the deepest wells drilled in search of minerals or for general exploratory purposes go down only about seven miles (more than 11 kilometers), much of our knowledge of the interior of the earth must be gained from indirect methods. Projects to bore through the earth's crust to the mantle have not yet been completely funded, but geophysicists and other earth scientists are anxiously awaiting such a project. Drilling technology for such a program exists, and it is hoped that funding, perhaps through an international consortium, will be forthcoming to permit such drilling before the end of the century. Since the earth's crust is not as thick in the ocean basins, such a drilling project will probably occur at sea.

The action of and reaction to seismic waves created by earthquakes or explosive percussion provide significant pieces of evidence as to the character of the earth's interior. Further, gravitational relationships and magnetic fields provide additional pieces of data that allow us to make quite reasonable inferences about the nature of the earth's interior. As more and more sophisticated technology is developed, scientists are gaining better insights into the interior of the earth, which has not yet been probed directly.

Deep mines and wells show gradual increases in temperature with greater depth. Rates of increase vary from about 1° F per 60 feet to about 1° F per 100 feet. This temperature gradient is doubtless greater near the surface than at extreme depths, for if continued at the same rate to the center of the earth, a temperature of several hundred thousand degrees Fahrenheit would result. Available data suggest that although the interior of the earth is quite hot, it is not nearly as hot as suggested by extrapolation of near-the-surface increases of temperature to the center of the earth. It is generally believed that considerable heat is generated by radioactive decay of rocks at a depth of 25 to 30 miles (40 to 48 kilometers). Presumably, also, the compression of material and residual heat from the time of earth formation contribute to the extreme heat encountered at great depths. In any case, there is considerable heat flux in the interior of the earth, and the products of that heating are often brought to the earth's surface in quiet or explosive fashion.

The earth will doubtless continue to cool over time, and eventually the restless crust may cease to be as active as it now is. It seems clear, however, that millions of years will elapse before such cooling occurs. In the meantime, we can anticipate that landforms will continue to develop as they have for hundreds of millions of years, and we can also anticipate that the agents of weathering and erosion will attack exposed surfaces attempting to reduce them to grade. The deposition of these products of erosion will tend to fill declivities and smooth profiles while providing new products for lithification as part of an ongoing rock cycle.

Rocks at a depth of about 30 miles have a temperature of about 2,000° F (1,100° C), which is above the melting point of ordinary rocks at the earth's surface; however, the melting point of rocks at great depth is elevated because of pressure of overlying materials. By the time the outer core of the earth is reached, temperatures of about 4,000° F (2,200° C) would be encountered. It is assumed that the inner core may be even hotter. It may be as hot as 10,000° F (5,500° C). As a matter of fact, experimental modelling in 1987 suggested that the core may be as hot as 12,000° F (6,600° C). Absolute precision in determining temperatures in the deep interior of the earth may be impossible to achieve. Enough about the interior of the earth can be inferred, however, to permit us to describe its important characteristics.

The earth may be divided into a series of zones in profile (figure 9.1). The outer skin of the earth, composed of relatively low-density materials and known as the crust, is from about 5 to 40 miles (8 to 65 kilometers) in thickness. There is considerable variation in thickness from place to place. The crust is typically thicker under the continents and thinner beneath the seas. The specific gravity of the upper reaches of the earth's mantle is about 3.0 to 3.5, whereas it may increase to about 4.5 in its lower portions. The earth's crust averages about 2.65, although it may approach 3.0 near the mantle margins. In contrast, the earth's core may have a specific gravity as high as 13.0. Increasing density towards the core should be anticipated, since it reflects the arrangement of materials in approximate relationship to varying specific gravities.

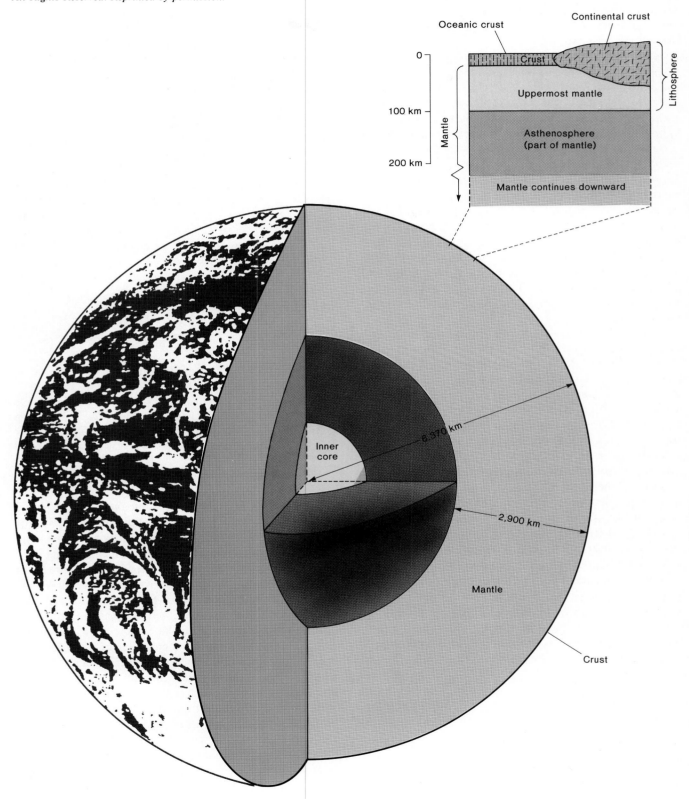

Figure 9.1
Cross section of presumed physical structure of the earth.
From Charles C. Plummer and David McGeary, Physical Geology,
4th ed. Copyright © 1988 Wm. C. Brown Publishers, Dubuque, Iowa.
All Rights Reserved. Reprinted by permission.

At the base of the crust is a zone known as the **Mohorovicic discontinuity,** or simply **Moho,** which is the contact line between the lower reaches of the earth's crust and the mantle, which lies immediately below the crust. The mantle is about 1,800 miles (2,900 kilometers) thick and appears to be more rigid near its upper reaches to less rigid at a depth where it comes in contact with the core. The core of the earth, with a radius of about 2,100 miles (about 3,400 kilometers) behaves like a liquid in its outer margins, yet there appears to be a solid inner core of about 600 miles (965 kilometers) in thickness. Recent information suggests that irregular projections of the core may extend some distance into the mantle. Further speculation suggests that this irregular boundary between mantle and core may be responsible, at least in part, for the small wobble that characterizes earth rotation.

Our description of the core of the earth must remain tentative until such time as reliable data make a more precise description possible. The high-density core is apparently composed largely of nickel and iron. The mantle, on the other hand, is probably composed in significant part of **olivine,** an iron magnesium silicate. The spinning of an earth with an iron–nickel core is probably responsible for the magnetic lines of force. This magnetic field may be observed in a variety of ways including magnetic compass observations.

The mantle is further divided into an upper portion that is solid and seems to merge with the crust as part of the **lithosphere.** Below the lithosphere is the **asthenosphere,** which has plastic-like characteristics. Indeed, convectional movement within the asthenosphere and associated shifting of the surface rocks is principally responsible for the development of most first-order landforms. The **plate tectonics** theory, which will be discussed in a subsequent section of this chapter, hinges on the movement of solid rocky plates generated by circulations within the asthenosphere. These convectional patterns in the asthenosphere are apparently slow motion analogues to those in the atmosphere.

Direct observation of the outer skin of the earth permits us to know only about one percent of the total radius of the earth. This outer crust is the home of man, however, and it is the portion of the earth that sustains all life on the planet. As previously indicated, the earth's crust is quite thin. Although we perceive the crust to be solid, it is subjected to repeated movements, which have their origins at greater depths. Those deep-seated movements contort, bend, and break the crust to produce an array of landform features.

Figure 9.2
Basalt does make its way to the surface and is a significant component of the earth's crust, as in this exposure near Sweet Home, Oregon.
Daniel Ehrlich.

The crust is roughly divided into two layers. A lower, essentially continuous layer, is composed principally of basaltic rock. This zone of **simatic rocks** takes its name from the principal rock elements, silicon and magnesium.

Above the simatic rocks is a zone of less dense rocks called **sialic rocks.** This term derives from the silicon (Si) and aluminum (Al) compounds found in these rocks. The sialic layer is discontinuous, but generally overlies the simatic rocks, especially on the continents. Rocks in the earth's crust have sorted themselves out roughly according to their specific gravity, with the lighter sialic rocks usually found at the surface, generally resting on the simatic rocks. These granitic rocks and their derivatives are quite common at and near the earth's surface. As they are broken down, their weathered products yield the raw materials, which are subsequently deposited to form some of the great series of sedimentary rocks.

Materials of the Earth's Crust

The earth's rocky crust is composed of rocks that are combinations of minerals. A **mineral** is a natural inorganic substance possessing definite physical and chemical characteristics. Most minerals have a crystalline structure, although certain crystals may be detectable only by microscopic examination. The lattice-work connections in minerals may be simple or quite complex. The chemical and physical bonds may be weak or strong, and this bonding determines the rock hardness and resistance to weathering and erosion. In addition, minerals are recognizable by a series of other characteristics including color, hardness, luster, fracture, and streak (i.e., the color left by rubbing the mineral against a white ceramic plate). These characteristics are used by geologists to make first approximations of identification, although more detailed chemical and physical tests, including x-ray diffraction, are necessary for precise classifications of certain rocks. Experienced geologists can recognize most rocks encountered in the field readily with nothing more than the naked eye or a small hand lens. Esoteric specimens must be subjected to microscopic, crystallographic, and chemical analyses in laboratories to make precise identification possible.

The elements, which are the building blocks of minerals, are present in greatly varying amounts. Oxygen, the most abundant, makes up about 47 percent of exposed rocks. Silicon accounts for about 28 percent, whereas aluminum, iron, potassium, calcium, sodium, and magnesium collectively account for about 24 percent. Together, all other elements make up only a little more than 1 percent of the earth's crust. Clearly, these elements are combined in an intricate array of chemical bonding and crystal structures, which yield many different minerals. The circumstances of formation and the presence or absence of impurities add to the complexity of mineral types and forms, and of the rocks that they combine to produce. These rocks exhibit quite different responses to weathering in different climatological realms. Some may be quite resistant in almost all realms. Others may be weak under almost all conditions. Some may be resistant in certain environments, but quite weak in others.

Common minerals include the silicates, oxides, and carbonates, obviously deriving from combinations of silicon, oxygen, and carbon with other elements. Common rock-forming minerals include the aluminosilicates [i.e., feldspar (plagioclase and orthoclase), biotite mica, hornblende, olivine, and quartz]. Oxides are less common rock forms, but hematite is a notable exception. Common carbonate minerals are calcite and dolomite. There are, of course, other mineral families, but the ones mentioned are the common rock formers.

Figure 9.3
The rock cycle.
From Charles C. Plummer and David McGeary, Physical Geology, *4th ed. Copyright © 1988 Wm. C. Brown Publishers, Dubuque, Iowa. All Rights Reserved. Reprinted by permission.*

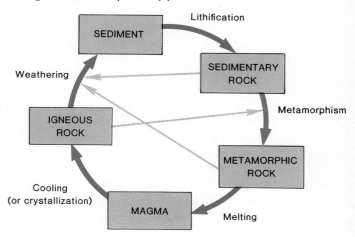

These minerals in various combinations and circumstances are the building blocks of many rocks. Just as the chemical bonding in minerals is a significant determinant in the rate of chemical weathering, so is the combination of minerals in rocks significant in their resistance or susceptibility to physical weathering and erosion. Similarly, the interplay of varying climatological conditions affects the rate and character of weathering.

Igneous Rocks

Rocks that have solidified directly from molten materials are called **igneous rocks** (figure 9.3). To a certain extent, all other rocks derive from igneous rocks; therefore, igneous rocks are commonly referred to as **primary rocks.** Igneous rocks make up the largest percentage of the earth's crust, but they are generally covered at the earth's surface by sedimentary or metamorphic rocks. As soon as igneous and other rocks are exposed they are subjected to agents of weathering and erosion. As primary rock materials break down, they provide the raw materials for secondary rocks. In the subduction zones at the margins of certain tectonic plates, rocks of the crust are carried downward to be remelted and perhaps extruded subsequently as additional igneous rock, and the rock cycle begins anew. So long as internal heat causes the convectional forces that drive plate tectonics, new igneous rock will be intruded into and extruded upon the surface of the earth's crust.

Igneous rocks are distinguished by the following characteristics: (1) they contain no fossils; (2) they usually manifest a quite uniform appearance; (3) except in very unusual circumstances, they are not layered; and (4) they are crystalline in structure, although those crystals may be microscopic in size in certain igneous rocks.

Igneous rocks may be **intrusive** (solidified beneath the surface of the earth) or **extrusive** (solidified at the earth's surface) in origin. Because they cool more slowly, intrusive rocks typically show a better developed and larger crystalline structure. Some common intrusive rocks include granite, diorite, gabbro, peridotite, and diabase. Some common extrusive rocks are obsidian (volcanic glass), rhyolite (the extrusive equivalent of granite), andesite (which corresponds to diorite), and basalt (the extrusive analogue of diabase). There are literally hundreds of variations and combinations of igneous rocks. Except in peculiar circumstances, the more esoteric specimens have less to do with landform or soil development. The esoteric forms may, however, have important economic ramifications as sources of ores or precious stones. Sophisticated searches for mineral resources using the latest technology are augmented by individual prospectors seeking the mother lode in time-honored ways.

The chemical composition of igneous rocks ranges from **acidic** (those rocks that are rich in light minerals, notably silica), to **basic** (rocks high in the so-called ferromagnesium compounds). The acidic rocks tend to disintegrate or break up around the mineral crystals as they weather, and many basic rocks tend to decompose (i.e., the crystalline structure tends to be broken up into the principal mineral constituents). Such acidic rocks often weather into coarse-textured, infertile soils that are subject to physiological aridity. Frequently, too, basic rocks are composed of a higher percentage of the mineral elements essential to the production of a friable and fertile agricultural soil. As a general rule, acidic rocks weather into less productive agricultural soil than do the basic rocks.

Under certain circumstances, gas is trapped in igneous rock as it erupts. If these gases escape to the outside atmosphere as the rocks are cooling, the cavities are left to produce a porous rock known as **scoria**. Bits of volcanic ash may be solidified into light, porous, and fine-grained rock called **pumice**. Ash falls may develop as layers, which form deposits of **tuff** when hardened.

The variety of igneous rocks, far exceeding the few examples described in this text, has been and are the raw materials that lead to the subsequent production of sedimentary and metamorphic rocks (figure 9.5). Existing igneous activity continues to bring new rock materials to the surface as old material is subjected to weathering and erosion. As igneous materials are weathered and eroded away,

Figure 9.4
Magma (A) *has solidified into igneous rock. The land* (B) *is uplifted and subjected to erosion. Material is transported to the sea where it is deposited and becomes sedimentary rock.* From Charles C. Plummer and David McGeary, Physical Geology, *4th ed. Copyright © 1988 Wm. C. Brown Publishers, Dubuque, Iowa. All Rights Reserved. Reprinted by permission.*

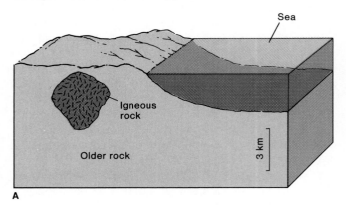

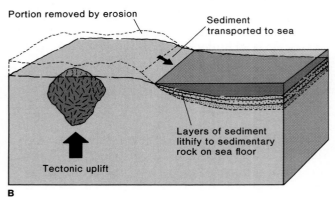

they become the raw material for reconsolidation into sedimentary rocks as they are deposited by running water or wind. Subsequently, both sedimentary and igneous rocks are subject to changes in character producing metamorphic rocks.

Sedimentary Rocks

Sedimentary rocks are composed of cemented fragments of existing rocks, or they are deposits of organic materials or chemical precipitates. **Clastic** sedimentary rocks are those that are formed when weathered products of existing rocks are cemented together in a process of **lithification** to form new rocks (figure 9.6). These clastic sediments are composed of layers of such sediments as clay, sand, gravel, or cobbles where they are deposited primarily by water, but occasionally by the wind. The size of the particles depends upon the source materials, the power

Figure 9.5
Light-colored rhyolite has been exposed along the Yellowstone River canyon.

Figure 9.6
Larger particles are deposited near shore in clastic sediments.
From Carla W. Montgomery, Physical Geology. Copyright © 1987 Wm. C. Brown Publishers, Dubuque, Iowa. All Rights Reserved. Reprinted by permission.

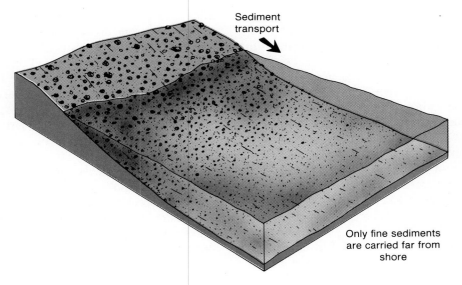

Sediment transport

Only fine sediments are carried far from shore

of the eroding agent, and distance of the deposits from the source material. Typically, coarser fragments are deposited at the base of a rock section, whereas the finer particles are deposited last, and at greatest distance from the eroded source, leaving fine-grained rocks. Sedimentary rocks are deposited in layers, or **strata,** which reflect seasonality of flow into the sea or a lake, or which reflect the variability in eroding power of the streams that contribute to the deposited load. Wind-blown deposits may also be lithified and exhibit stratification (figure 9.7). Succeeding lava flows may also be stratified, but, of course, lava is igneous rock. Indeed, the superposition of layers of rock is useful in establishing geologic chronology and in making geomorphological connections of similar rocks, often over great distances. This concept of **superposition** (i.e., older rocks being overlain by younger materials) is a basic underpinning of stratigraphy and geochronology (figure 9.8). Of course, subsequent tectonic forces may overturn or disturb an orderly process of deposition. Geologists and geomorphologists must be alert to disturbances of original deposits, or stratigraphic and landform analyses may be flawed. Since original sediments were derived from igneous rocks and continue to be derived from other rocks, they are termed **secondary rocks.**

When clastic sediments (i.e., sediments containing individual particles or pieces of rock from which they were derived) harden into sedimentary rocks, the name applied to them changes to reflect the change in physical state. Cemented gravels, cobbles, or boulders, which have been rounded by the action of running water, bound together in a fine matrix of materials are called **conglomerates.** When angular pieces of broken rock are cemented together by a finer matrix the resulting rock is termed **breccia.** Sand that is cemented together becomes **sandstone;** and compacted and hardened clay becomes **shale.** Other intermediates may include **siltstones,** or **mudstones,** reflecting the difference in particle size of the materials that have been cemented together (figure 9.9). Because of heterogeneity of source materials and circumstances of deposition, clastic sedimentaries have great variations in particle size of constituent parts.

In comparison, certain other sedimentary rocks may be chemical precipitates resulting from the deposition of materials formerly held in solution. **Limestone,** which is composed of calcium carbonate, is such a rock, as is its cousin, **dolomite,** which is a calcium magnesium carbonate. Certain kinds of limestone may be considered clastic in character. When shells of marine organisms (composed of limestone extracted from sea water) are cemented together by other calcium carbonate from solution, the resulting **coquina** is part chemical precipitate and part clastic sediment.

Quartz deposited from solution in an amorphous form frequently occurring as nodules in limestone deposits is called **chert.** Particular varieties that fractured in

conchoidal form were the flints used to produce arrow points or cutting tools worked by our ancestors in the Stone Age. Edges produced in such flint and in volcanic glass, or obsidian, were very sharp. As a matter of fact, glass (quartz) edges are used in precise microtomes to produce thin sections to be investigated by sophisticated instruments like the electron microscope.

Coal is a significant sedimentary rock composed of the compacted semidecayed organic remains of plants growing in primordial swamps. Hot, humid conditions prevailing in the area enhanced vegetative growth and organic accumulation. Acidic waters inhibited complete decay of vegetable materials falling into such waters. As such organic remains accumulated, a soft fibrous material called **peat** was first produced. Over time, the accumulation of materials above caused this material to become **lignite,** and, finally, with lithification, **bituminous coal** was produced. If subsequent folding occurred, the bituminous coal was hardened further to produce **anthracite.** Pressures caused by an overlying burden of sediment or subsequent earth movements increased the lithification process.

Certain chemical sediments, called **evaporites,** have evolved in mineral-rich enclosed water bodies where evaporation is especially rapid. Rocks of this type include **halite** (rock salt), **anhydrite,** and **gypsum.** The presence of evaporites indicates an arid or semiarid environment at the time of deposition. Such rocks fall prey to rapid destruction in humid environments, although they retain their integrity for long periods in dry regions.

Figure 9.8
Under certain circumstances, cross-bedding of rocks occurs.

Figure 9.9
*Mud cracks may become lithified and a part of the
sedimentary rock like these in eastern Oklahoma.*

Figure 9.10
Navajo sandstone in Zion National Park, Utah.

Figure 9.11
The concept of superposition is effectively illustrated in this cross section in Utah. The light-colored limestone is younger than the darker shale at the bottom of this exposure.

Metamorphic Rocks

Both igneous and sedimentary rocks may be changed by heat, pressure, solution, or cementing action to produce a new rock type distinct from the types altered. This change, or metamorphosis, produces the third principal type of rock, **metamorphic**. Metamorphics, themselves, may be further modified to produce other metamorphic classes, and their weathered particles become the stuff of new sedimentaries.

Granite, an igneous rock, may be changed to **gneiss**, a metamorphic rock. The heterogeneous arrangement of crystals in granite may be changed to a banded arrangement as the granitic materials are heated and subjected to pressure by subsequent igneous activity or crustal movements. The resulting banding gives an almost layered appearance, and rocks, such as gneiss, are said to be **banded**. Further compaction of the banded rock into a plate-like arrangement of minerals produces **schist**. Coarse-grained rocks typically produce gneiss, whereas fine-grained rocks often become schists.

Shale that has been metamorphosed by considerable pressure yields **slate**. Slate will cleave along planes, and its fine-grained structure often gives it significant possibilities for human use. A slate-like rock that has developed a silky sheen in the metamorphic process is called **phyllite**.

Rocks that are composed dominantly of a single mineral are not banded by metamorphosis. Limestone is crystallized into **marble** when it is subjected to substantial heat and pressure attendant to compressional forces. When the interstices in sandstone are replaced by quartz in solution, the resultant metamorphic rock is called **quartzite**. Sandstone fractures or breaks around the quartz grains, whereas quartzite breaks across the grains, since individual rock particles and the cementing material are the same hardness. Quartzite is typically harder and more resistant to weathering and erosion than sandstone. Indeed, quartzite proves to be a very resistant rock in most climatic realms.

Since metamorphic rocks are derived from primary and secondary rock types, they are said to be **tertiary rocks.** As previously indicated, they may be broken down to form the parts of a new sedimentary rock, or they may be subducted to lower levels only to be remelted and issue forth as new igneous material.

Rock Cycle

As we have seen, rocks may be changed from their original form. Igneous rocks may become molten again if they are found in a volcanic area. Earth movements may modify the character of exposed rock. Over time, however, rocks are reduced by the agents of weathering and erosion to materials that become part of the **solum,** or soil. In turn, the soil may be reduced by agents of erosion and finally deposited in the sea where it again may become solidified to rock. The cycle of rock production and destruction is a never-ending aspect of structural movement, volcanic activity, and the forces of weathering, erosion, and deposition. Different rocks respond in different ways to the agents of weathering and erosion, and they have varying degrees of utility to man depending on their inherent physical characteristics and their situation. The varying hardness or resistance of different kinds of rocks subjected to various diastrophic forces and geomorphological processes is reflected in the kind of landforms produced.

The rock materials of the earth's crust are subjected to significant disturbances by convectional forces within the earth's mantle. These movements and their effects are explored in the section that follows.

Plate Tectonics

The crust of the earth is in constant motion, although usually that motion is undetectable by people, except for those scientists who may be using sophisticated devices to measure such motion. Of course, sudden catastrophic movements, such as those encountered in earthquakes are felt by lay people, and the destruction they create may cause property damage, injury, and loss of lives. Over time a number of factors have suggested to astute observers that crustal movement is a fact of life, but sophisticated theory and devices to test that theory are largely products of the twentieth century. The general "fit" of the continents into a kind of jigsaw-like pattern if they were moved across intervening oceans has long intrigued careful observers of global patterns. Even schoolchildren have noted this apparent fit as if complementary pieces were pulled apart by a giant hand. When the continental shelves are considered, the apparent fit is even more pronounced. Similarities in the fossil record, between plant and animal species in land areas widely separated by oceans, and evidences of past glaciation in areas now decidedly tropical, are all suggestive of past terrestrial connections, but such inferences are backed neither by compelling theory nor a persuasive array of observed facts.

A variety of data has come together in the past thirty years or so to cause all but a few earth scientists to accept the proposition of plate tectonics. Actually the present broad acceptance was long in coming, and a few visionaries, ahead of their time, were scoffed at by scientists and the scientific wisdom of their day. This has almost always been true. Those who have challenged the conventional wisdoms have been viewed as scientific heretics.

Eventually new truths become conventional wisdoms until the orthodox is again challenged by new insights, persuasive theory, and corroborative data. Scientific advances have always proceeded in this way, and it's reasonable to assume that this will remain the prevailing pattern. Scientists are people subject to the same human frailties that characterize all of *Homo sapiens*. No one likes to have long-cherished beliefs challenged, and many scientists resist changes that may make their knowledge or skills obsolete.

The Wegener Hypothesis

In the early part of this century, Alfred Wegener, a German scientist, hypothesized that all continents were joined together in one, or perhaps two supercontinent(s). The break up and movement of portions of the supercontinent(s) could explain the presence of tropical fossils in cold climates, the presence of coal seams in Antarctica, or glacial features in tropical environments. Although Wegener suggested that the lighter sialic crust did "float" on the plastic simatic rocks and that continents might drift from original locations, he was discredited by contemporaries, because he provided no theory as to the propelling mechanism for such movement. He was, in a sense, a scientific casualty of minimal information and imperfect understanding. There is no doubt that he was a visionary who articulated certain aspects of plate tectonics, and he deserves some scientific credit in spite of the shortcomings of undergirding theory and proof, which attended his early announcements.

It is true that Southern Hemisphere earth scientists gave the concept of continental drift somewhat more credence than their Northern Hemisphere counterparts, primarily because the cross-continent fossil record was difficult to explain in any other scientific context. Nevertheless, the skepticism of European and American scientists put the notion of such drift on the back burner for almost half a century. Advances in science usually come slowly and painfully as the result of the accumulation of large masses of data and inferences drawn from those data. The "Eureka, I have found it" moments are rare in science, and many unsung heroes pave the way for the subsequent Nobel laureates.

Evidence Supporting Plate Tectonics

Wegener's hypothesis remained in eclipse until after World War II, but bits of evidence from a variety of sources suggested that the earth's crust did, in fact, move significantly. The paleomagnetic record of different orientation of magnetic materials in ancient igneous rocks could be most readily explained by movements of continental proportions. Continuity of mountain ranges could be sustained by radioactive dating techniques. Close correlations between animal and plant species in areas now vastly different from a climatological standpoint provided further inferential evidence of movement. Curious apparent aberrations in paleoclimates could be explained by large-scale shifts in various segments of the earth's crust. Similarities of rock types and the fossil record in areas separated by thousands of miles of intervening ocean were noted by petrologists and paleontologists. Each of these aspects, singly, produced less than persuasive evidence of crustal shifts of continental proportions. Together, however, these data tipped the balance of thinking in new directions.

As these pieces of evidence suggesting continental or crustal movement began to coalesce, other pieces of data demonstrated conclusively that there had been major crustal movements. The accumulating evidence set minds to work to explain how such a thing might be. Subsequently, a theory deriving from this empirical evidence was developed to explain the movement. Oceanographic research revealed that a system of midocean ridges paralleled the continental form in remarkable fashion. Parallel bands of rocks with the same magnetic properties extend in identical fashion on both sides of midocean ridges in both the Atlantic and Pacific Oceans. Transverse faults extended from the ridges in comparable ways. Rocks in ocean basins are relatively young and on the open ocean floor older rocks are located nearest the continents. And, temperatures in rocks in midocean ridges are higher than those further away from those ridges.

These pieces of information led to the inescapable conclusion that hot, new earth crust is being formed at the midocean ridges, and cooler and older portions of the crust are disappearing near the continental margins. These warmer areas have been measured again and again, and various hot spots have been observed by manned and unmanned deep water submersibles. This movement is known as seafloor spreading. It seems apparent that large crustal blocks, or plates, are being moved over the plastic-like rocks of the asthenosphere. Presumably there are giant convection-like movements through the asthenosphere, not unlike those observed in convectional cells in the atmosphere, or in boiling cookpots on stoves. The exact nature of those convection-like cells and the driving mechanisms for them are still somewhat uncertain, but basic patterns are understood, and the principal scientific underpinnings of plate tectonics seem secure. The upward movement near the midocean ridges causes the plates to move away at a somewhat more rapid rate in the Pacific than in the Atlantic Ocean basins. Portions of the crustal plates are consumed at the continental margins. As these enormous plates grind and crunch together, one plate overrides the other, forcing the lower plate down to lower levels in a process known as **subduction** (figure 9.12).

Some rocks from near the earth's surface are carried to depths and are once again subjected to the heat of volcanic fire (figure 9.13). As previously indicated, these rocks may form the raw materials for new igneous rock intruded into or extruded onto the earth's crust. Some of

Figure 9.12

The convection-like activity driving plate tectonics.
From Carla W. Montgomery, Physical Geology. *Copyright © 1987
Wm. C. Brown Publishers, Dubuque, Iowa. All Rights Reserved.
Reprinted by permission.*

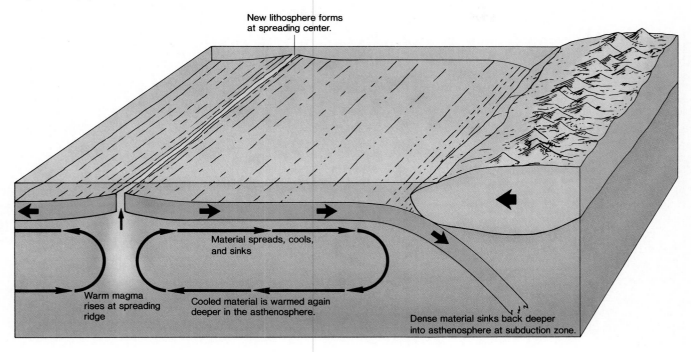

New lithosphere forms
at spreading center.

Material spreads, cools,
and sinks

Warm magma
rises at spreading
ridge

Cooled material is warmed again
deeper in the asthenosphere.

Dense material sinks back deeper
into asthenosphere at subduction zone.

Figure 9.13

*Subduction occurs along convergent plate boundaries. Such
convergence along an ocean–ocean boundary (A) may
produce island arcs, and along an ocean–continent boundary
(B), mountain building and volcanic activity are likely
results.*
From Carla W. Montgomery, Physical Geology. *Copyright © 1987
Wm. C. Brown Publishers, Dubuque, Iowa. All Rights Reserved.
Reprinted by permission.*

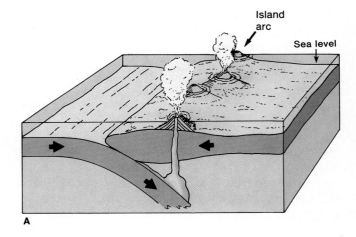

Island
arc

Sea level

A

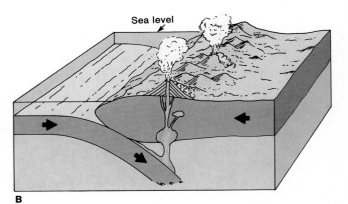

Sea level

B

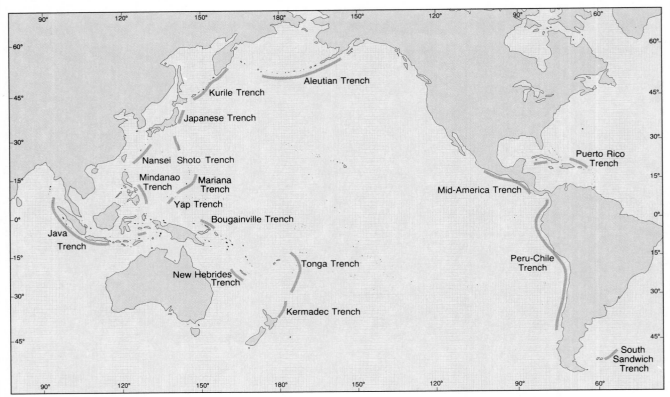

the great oceanic deeps are found in these areas of subduction. Places like the Mariana Trench, the Aleutian
Trench, and the Mindanao Trench contain the earth's
deepest ocean waters (figure 9.14). Movements along the
margins of the plates cause earthquakes and volcanic activity. These effects will be discussed in greater detail in
chapters 10 and 16 of this book.

Opinions vary as to the number of rigid plates—some
scholars place the number as high as twenty. At least seven
are major plates, and the movement of these plates towards, away, or tangent to others is responsible for the
building of mountains, the development of trenches, the
movement of the earth we observe as earthquakes, and

volcanic activity. The earth's restless activity is more
readily observable along these margins of crustal instability. **Plate tectonics** is the unifying theory that explains
the movement of continents and the development of primary landforms (figure 9.15). For the first time, geomorphologists have a valid and coherent explanation for
diastrophic movements and volcanic activity.

The details attaching to plate tectonics need to be
worked out, and a few problems exist in interpreting past
and present activity. There are apparent irregularities like
earthquake zones well away from the margins of known
plates, but geologists and geomorphologists are confident
that corollaries to the basic theorems of plate tectonics

Figure 9.15
*Earth's plates and apparent movement with relationship to
each other. Seafloor spreading occurs along midocean ridges.*
From Arthur Getis, Judith Getis, and Jerome Fellmann, Introduction
to Geography, *2d ed. Copyright © 1987 Wm. C. Brown Publishers,
Dubuque, Iowa. All Rights Reserved. Reprinted by permission.*

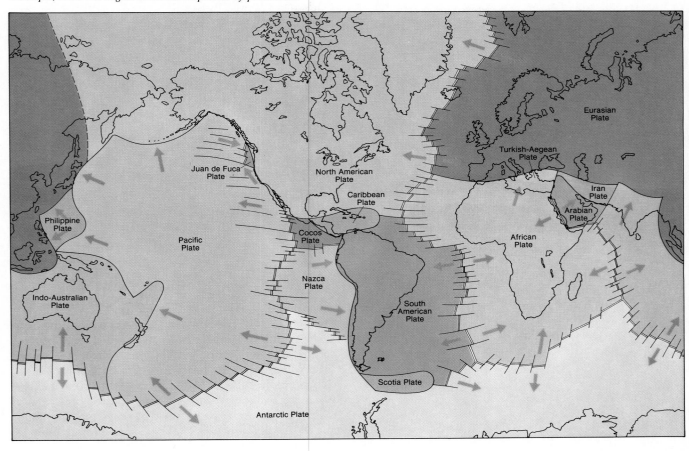

will eventually be developed to ensure coherence of the entire theory. The origin of continents is something of a problem too, although it may be assumed that central areas of very old crystalline rock serve as the locus of development. These areas, known as **shields,** have remained as large portions of plates for protracted periods, and continents have apparently spread out from those cores, presumably in two significant ways (figure 9.16). On the one hand, sediments composed of loose or weathered material from the shields have been deposited in marginal seas, and these sediments have been exposed as plates have converged. Similarly, as plates converged, activities around the margins led to new accretions of materials as a result of volcanic eruptions. The central core of the continent had and has additions at the margins by deposition, warping, folding, and vulcanism. The overall effect has been the expansion of the size of continents as they have grown outward from the shields.

Figure 9.16
Generalized shields and mountain areas of the world.
From Richard A. Davis, Jr., Oceanography: An Introduction to the
Marine Environment. *Copyright © 1987 Wm. C. Brown Publishers,
Dubuque, Iowa. All Rights Reserved. Reprinted by permission.*

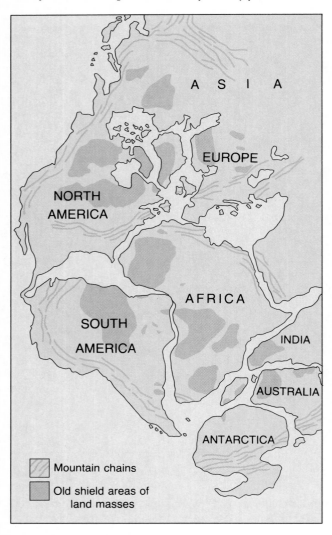

Mountain chains

Old shield areas of
land masses

Figure 9.17
Location of the San Andreas Fault.
From Charles C. Plummer and David McGeary, Physical Geology,
*4th ed. Copyright © 1988 Wm. C. Brown Publishers, Dubuque, Iowa.
All Rights Reserved. Reprinted by permission.*

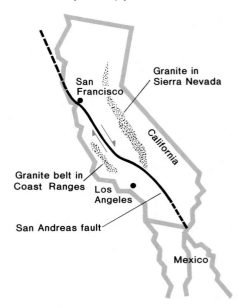

Earthquake Zones

When plates slide along laterally with reference to an adjacent plate, they may be responsible for significant earthquake activity. The famous (or infamous) San Andreas Fault (or Rift) of California is such a line of contact (figure 9.17). Movement along the fault is omnipresent, and the effect is to create an area of frequent earthquakes. Scientists are now predicting a high probability of a major destructive earthquake within the region prior to the dawn of the twenty-first century. If such a quake does occur, the density of population at or near the rift will probably create a catastrophe of major proportions. Past devastating

earthquakes, like the San Francisco quake near the turn of the twentieth century, are likely to pale in significance because a rapid increase in California's population seems certain to cause higher casualty figures. Although modern construction is superior to that that existed during the San Francisco quake, it's clear that many structures cannot withstand the force of a major quake. Frequent small quakes are but warnings of a major quake(s) to follow.

Not all of the major earthquake zones are so close to margins of the continents. For example, the great African rift zone, which is interconnected through the Red Sea, Dead Sea, and the Jordan River Valley into the Middle East, lies at the margins of the African and Arabian Plates, although the African portion is well in the interior of the African continent (figure 9.18). The entire zone shows evidence of past or present volcanic or earthquake activity.

Similarly, certain other earthquake-prone areas are located at some distance from the margins of existing plates (figure 9.19). The area near New Madrid, Missouri in the Missouri boot heel is the location of the most intense earthquake in North America in the early nineteenth century. The quake was so severe that it altered the course of the Mississippi River and produced a cutoff lake, Reelfoot Lake in Tennessee. Small quakes have been recorded since, and it is still regarded as an area where significant earthquake activity can be anticipated.

Figure 9.18
The Dead Sea occupies a portion of the East African and Middle East rift system.

Figure 9.19
Seismic risk map of U.S.
Source: National Oceanic and Atmospheric Administration.

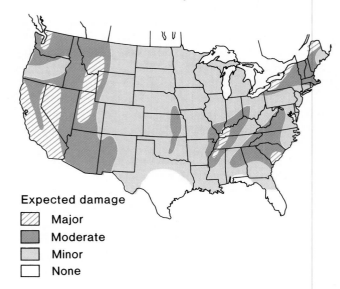

Expected damage

▨	Major
▤	Moderate
▥	Minor
☐	None

Another similar earthquake area exists in South Carolina, away from the coastal margin. Indeed, there are several areas in the eastern and western United States that have a history of earthquake activity. Because fault lines in earthquake-prone portions of the eastern United States are deeply buried, rational predictions of the sites and the timing of future quakes are much more difficult than in the western part of the country. It is also true that pronounced movement along fault lines has been much more prevalent in the western part of the country in the twentieth century.

The magnitude of earthquake destruction has been measured since 1935 with a scale devised by C. F. Richter. This is really a measurement of energy expressed as the wave motion radiating out in all directions from the earthquake's focus. The **epicenter** (the point on the earth's surface above the focus) is the area of maximum intensity, and the seismic waves gradually dissipate away from that center. The Richter Scale estimates the energy released by the ground motion. Richter scale numbers extend from

Figure 9.20
Epicenters of major earthquakes and location of young volcanoes.
Source: *Map plotted by the Environmental Data and Information Service of the National Oceanographic and Atmospheric Administration; earthquakes from the U.S. Coast and Geodetic Survey.*

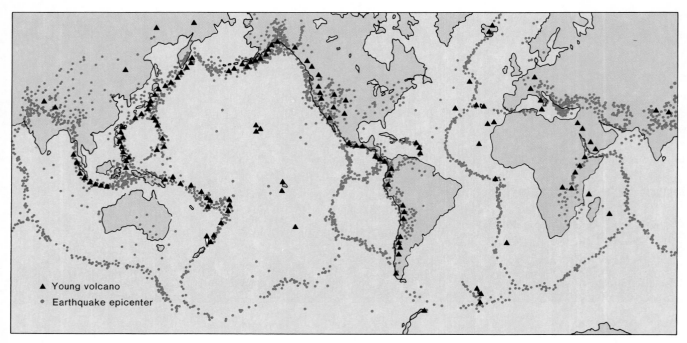

▲ Young volcano
• Earthquake epicenter

0–9, although there is presumably no upper limit to earthquake intensity. The most intense earthquake measured to date has reached 8.5–8.6 on the Richter scale. It should be recognized that an increase of one on the Richter scale represents an increase in magnitude of ten times. At the upper end of the scale total destruction could be expected.

The theory of plate tectonics seems to be founded on sound geophysical principles, and the basic tenets of the theory are known, but detailed explanations await the discoveries of scores of dedicated scientists. Certainly the movement of continental plates over a semiplastic interior seems to be responsible for the development of first-order landforms, and it sets the stage for an elaboration on landform orogeny and development in subsequent chapters.

Secondary and tertiary landform features add detail and definition to the gross aspect of primary features. From a human perspective, the global effects of plate tectonics may be difficult to comprehend. On the other hand, the catastrophic effects of earthquakes or volcanic eruptions have direct effects on human life and are certainly common experiences for millions of people (figure 9.20).

The topographic barrier represented by a mountain range may produce climatic opportunities or austerity. It may serve as a barrier to human intercourse between regions or it may serve as an opportunity for recreation or settlement. A river valley may represent an area of fertile, productive, agricultural soil, but the river may also prove to be a flood threat. The bedrock may weather into fertile or infertile soil. Unstable areas create problems for builders, and unwise construction or waste disposal may create environmental hazards with serious potential consequences.

This surface manifestation of the earth's crust, and its soil and vegetation cloak in association with climate represents our home. It will always be so. It is appropriate for us to have some appreciation for and understanding of this stage beneath our feet. Subsequent chapters describe the forces that are constantly changing the face of the world's landscapes.

We seek to develop an appreciation of and understanding for the lithosphere comparable to our knowledge about the atmosphere. Together, the atmosphere and lithosphere combine with the hydrosphere to produce a habitat for the life of the biosphere.

Study Questions

1. Draw a cross-sectional profile of the presumed physical structure of the earth.
2. Why is the melting point of rocks at great depths higher than at or near the surface of the earth?
3. What is Moho?
4. Explain the difference between simatic and sialic rocks.
5. What is the difference between a mineral and a rock?
6. Reproduce the rock cycle.
7. Explain the physical differences between extrusive and intrusive rocks.
8. Explain why some extrusive rock has a porous, sponge-like character.
9. What is the difference between clastic sediments and chemical precipitates?
10. Explain the concept of superposition.
11. What are the evaporites? Under what conditions do they develop?
12. Briefly explain the driving mechanism of the theory of plate tectonics.
13. What pieces of evidence have tended to validate the theory of plate tectonics?
14. Who was Alfred Wegener?
15. Explain subduction.
16. What are shields in a geomorphological sense?
17. What is the relationship of earthquakes to plate boundaries?
18. What were the effects of the New Madrid earthquake?
19. Explain how shields affect the development of continents.
20. Compare and contrast seafloor spreading and subduction.

Selected References

Birot, P. 1966. *General physical geography.* New York: John Wiley and Sons.

Butzer, K. W. 1976. *Geomorphology from the earth.* New York: Harper and Row, Publishers.

Dott, R. H., and Batten, R. L. 1971. *Evolution of the earth.* New York: McGraw-Hill Book Company.

Gersmehl, P.; Kammrath, W.; and Gross, H. 1980. *Physical geography.* Philadelphia: Saunders College Publishing.

Kolars, J. F., and Nystuen, J. D. 1975. *Physical geography: Environment and man.* New York: McGraw-Hill Book Company.

Scott, R. C. 1989. *Physical geography.* St. Paul: West Publishing Company.

Strahler, A. N., and Strahler, A. H. 1984. *Elements of physical geography.* 3d ed. New York: John Wiley and Sons.

Trewartha, G. T.; Robinson, A. H.; Hammond, E. H.; and Horn, L. H. 1977. *Fundamentals of physical geography.* 3d ed. New York: McGraw-Hill Book Company.

10

The Shaping of Landforms

Antithetical forces build up the surface of the land while others tend to reduce uplift towards sea level.
Department of Regional Industrial Expansion photo, Government of Canada.

"The hills,
Rock-ribbed, and ancient
as the sun."
William Cullen Bryant, *Thanatopsis*

*T*he earth's crust is in constant movement because of the action of internal forces, and, as a result, the earth is deformed and changed in a never-ending cycle. This general deformation of the crust is known as **diastrophism.** Diastrophic forces are of three main types: warping, folding, and fracturing. These forces are occasionally simple, but they are usually complex and interconnected, and frequently they are attended by volcanic fire, and the tremors of earthquakes.

Warping

As the earth's surface is slowly bent over large areas by forces such as the collision and movement of adjacent tectonic plates, the increase of load on seafloors induced by the addition of great quantities of sediments, or the decrease in load by the melt and retreat of continental glaciers, the crust is **warped.** Large upwarped and downwarped features may cover thousands of square miles. Indeed, these large upwarped and downwarped areas are major landform components, and they may bear on their

surfaces numerous examples of landforms covering smaller areas. Their surfaces are etched by the effects of weathering and erosion, producing the kaleidoscopic variety that is characteristic of the face of the earth. This warping continues as a part of the never-ending movement of this restless planet.

The Atlantic and Pacific coasts of the United States, for example, show antithetical effects of warping—the Atlantic coast is showing signs of submergence in the formation of embayments and drowned river mouths because of some downwarping, whereas the Pacific coast has been warped upwards. Those upwarping and downwarping forces are apparently continuing. Such downwarping on the Atlantic and Gulf margins may be partially responsible for the powerful erosion of beaches on the coastal strand. Certain areas of Canada, the northern United States, and portions of northern Europe are slowly rebounding upwards in response to the removal of glaciers of enormous size, which depressed land surfaces during the last great Ice Age. Although ice left the areas 10,000–20,000 years ago, the upward rebound of land surfaces depressed by the weight of the glaciers is expected to persist for an indefinite amount of time. The rate of rebound of such overloaded areas is greatest immediately after removal of the load, whereas the rate of rebound diminishes the longer the area has been relieved of its load. Nevertheless, the rebound continues for protracted periods. The exact length of time for measurable rebound is contingent upon the load removed and the nature of the underlying material.

Folding

The surface may be subjected to **folding** as a consequence of earth movements usually involving the horizontal collision of tectonic plates. These folds vary greatly in dimension and type, but a few common types will suffice to meet the descriptive and analytical needs of the physical geographer. Essentially, all folds are the result of lateral movements in the earth's crust. The intricacy of fold patterns reflects the asymmetries in direction and intensity of lateral movements. Both compressional and tensional movements can create folds of different form and dimensions, although compression appears to be dominant in the creation of most folds. The magnitude and intensity of the forces and the nature of the crustal material contribute to the variety and character of the folds. Some of the common types of uplifted or folded features are considered in the sections that follow.

Domes are characterized by essentially symmetrical slopes downward in all directions from a crest. They usually result from some sort of material intruding into almost horizontal sedimentary beds from below. In other instances, such domes may be created by compressional forces acting from several directions at the same time rather than from two opposite directions. This may occur as the result of volcanic activity (which will be discussed in a subsequent section of this chapter) or because of the intrusion of salt, as is the case along portions of the American Gulf Coast. In Louisiana and Texas, especially, salt plugs, a number with sulphur caps, have intruded through nearly flat-lying sedimentary beds and pushed up overlying beds into dome structures.

Frequently, the flanks, or margins, of these salt domes are excellent sites for oil and/or natural gas to be trapped. Indeed, significant quantities of these important minerals are produced on the coastal margin and offshore in Texas and Louisiana. The last best areas for oil exploration in the conterminous United States appear to be along the coastal margins and on the continental shelf where buried dome structures, especially, are sought. The coastal waters of Alabama have recently revealed a treasure trove of buried gas, whereas drilling in the coastal waters of west Florida has so far yielded disappointing results, although there were some hints in late 1988 that paying quantities of oil and/or natural gas may have been discovered in the gulf waters off the panhandle of West Florida.

When compressional forces operate, a series of different folds may be recognized. A **monocline** occurs as a one-sided fold marking the transition between essentially horizontal strata to those that are more rigorously folded. Monoclines apparently result from less intensive forces operating over short periods of time. A simple symmetrical upfold is called an **anticline,** and the downfold equivalent is called a **syncline** (figure 10.2). The amount of bending serves as an indirect gauge of the magnitude of the forces that created the fold. Exceedingly large folds may be referred to as **geoanticlines** and **geosynclines.** Geosynclines frequently are reservoirs that receive enormous amounts of sediment at, or near, the margins of continents. As tectonic plates grind together, these geosynclines or troughs may contain the sedimentary rock reservoirs that are folded into massive mountain ranges. Such features may have many anticlines or synclines as constituent elements (figure 10.3).

Figure 10.2
A small syncline.

Figure 10.3
Patterns of simple folds.
From Arthur Getis, Judith Getis, and Jerome Fellmann, Introduction
to Geography, *2d ed. Copyright © 1988 Wm. C. Brown Publishers,
Dubuque, Iowa. All Rights Reserved. Reprinted by permission.*

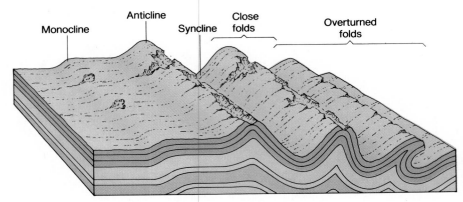

Figure 10.4
Reversed topography in the ridge and valley country of Pennsylvania.
From Arthur Getis, Judith Getis, and Jerome Fellmann, Introduction to Geography, *2d ed. Copyright © 1988 Wm. C. Brown Publishers, Dubuque, Iowa. All Rights Reserved. Reprinted by permission.*

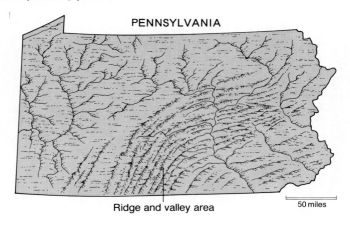

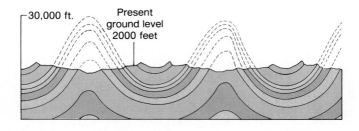

Figure 10.5
The relationship between dip and strike.
From Charles C. Plummer and David McGeary, Physical Geology, *4th ed. Copyright © 1988 Wm. C. Brown Publishers, Dubuque, Iowa. All Rights Reserved. Reprinted by permission.*

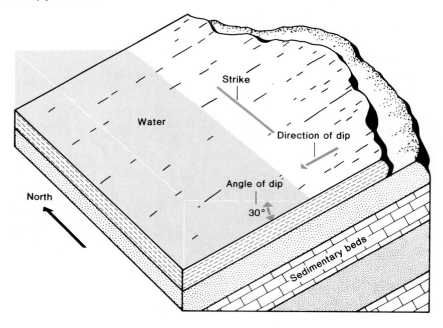

If compressional forces are asymmetrical, the fold may be an **overturned fold;** if the force exceeds the elastic limits of the affected rock, a break may occur, and the fold may become overthrust along the break. Such an overthrust fold is sometimes known as a **nappe.**

In certain circumstances, weathering and erosion may attack initial upfolds more effectively because of zones of weakness near the apex of such folds. If this occurs, initial upfolds may become erosional valleys and the former synclines may stand as ridges, since compression may have reduced their susceptibility to erosion (figure 10.4).

In all these instances of folding, the compressional or tensional forces result in the deformation of affected rocks. They are subjected to the stress of crustal movement and, as a result, the rocks respond in size, volume, or shape reflecting the strain on a particular rock. In **plastic strain** characteristic of folding, the rock does not return to its original shape. **Elastic strain** results when rock returns essentially to the form it had before the stress was applied. Rock rebounding from a reduction in load as in the case of the melting of the continental ice sheet usually exhibits elastic strain. Forces producing folds in rocks usually result in plastic strain of the affected rocks. When forces applied extend beyond the plastic limits of rocks, they fracture, or break. Clearly, some rocks are more brittle and subject to **fracturing** than others.

In the slow movements reflective of most crustal activity, rocks tend to be plastically deformed. In the more rapid movements along certain fault lines during earthquakes and at times of violent volcanic eruptions, a great deal of fracturing occurs as rock is pushed beyond its plastic limits. The enormous pressures that exist along zones of crustal movement cause rocks to grind and crunch together to produce a variety of landforms, and the very rocks may be metamorphosed by heat and pressure.

During the tilting of rock beds implicit in warping or folding, it is important to determine the tilt of these beds relative to a horizon plane and to ascertain the general trend of the tilt across the earth's surface. **Strike** is the compass direction of the intersection of the inclined plane represented by the rock tilt and a horizontal plane. The strike is then described in terms of the compass direction. The **dip direction** is at right angles to the strike, and the **dip angle** is measured within a vertical plane that is perpendicular to the bedding plane and the horizontal plane (figure 10.5). The dip angle may vary between 0° and 90°. Few rock outcrops in nature are characterized by angles of dip approaching vertical.

In the field, the strike and dip are typically measured with a Brunton compass, which combines a magnetic compass with a clinometer to measure slope angle. Strike and dip are depicted on geologic maps with appropriate symbols, and where they are depicted represents a surface exposure that was actually measured. The dip and strike are often key pieces of information in following the trend of a given structure or in analyzing the basic geomorphological history and character of a particular region.

The **limbs** of folds represent the tilting beds extending downward from a crest or upward from a trough. The axis is the hinge line from which the folds emanate.

Faulting

When rock exceeds its elastic limits, it may ultimately fracture, or break. Such breaks are called **joints** (figure 10.6). In addition, joints may develop as a natural consequence of rock formation. Basalt, for example, frequently develops a set of columnar joints as it cools. In certain areas these hexagonally jointed forms produce interesting landscape features. The Giant's Causeway in Ireland is such an exposure of jointed basalt. Joints provide avenues for the invasion of weathering elements, and movement may also occur along joint surfaces.

Figure 10.6
Joints in sandstone.

Figure 10.7
The relationships between common faults.
From Arthur Getis, Judith Getis, and Jerome Fellmann, Introduction
to Geography, *2d ed. Copyright © 1988 Wm. C. Brown Publishers,
Dubuque, Iowa. All Rights Reserved. Reprinted by permission.*

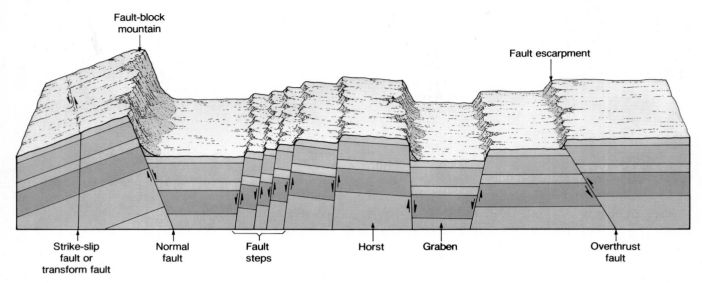

If there is movement along a joint in rock, **faulting** occurs (figure 10.7). Such movements occur along zones of weakness in the earth's crust known as **fault zones** or **fault planes.** Where such a zone intersects the earth's surface, it is known as a **fault line.** Movements along fault lines can be horizontal, vertical, or any of myriad combinations. Geologists may describe faults that occur up or down dips as **dip-slip faults.** Those with movements parallel to the strike are known as **strike-slip faults.** Those with both dip-slip and strike-slip components are known as **oblique-slip faults.**

There are several common kinds of faults: normal, reverse, transcurrent, and thrust (figure 10.8). Along an inclined fault plane, the area below the inclined plane of the fault line is called the footwall. The area above the footwall is called the hanging wall. In a **normal fault,** the hanging wall block has moved downward relative to the footwall. It is usually not possible to ascertain which block moved relative to the other. Such faults may develop from either tensional or compressional forces. In a **reverse fault,** strong compressional forces cause the hanging wall block to be pushed over the footwall block. If erosion did not occur, this would result in a dramatically oversteepened fault scarp. Such overhang, if it occurs, is a transient feature that quickly succumbs to forces of gravity, weathering, and erosion. A **transcurrent** or **strike-slip fault** occurs when horizontal movement occurs along the fault line as the result of shearing. **Thrust faults** occur when blocks of material subject to severe compressional forces along low-angle fault planes may completely override the underlying materials, sometimes for significant distances. **Transform faults,** at right angles to areas of ocean floor spreading, are characteristic of midocean ridges. The existence of such faults provided one piece of evidence for ocean floor spreading. In a sense, it provided some of the underpinning for the theory of plate tectonics.

When a block of material is downthrown between two essentially parallel faults, or if the area on either side of such faults moves upward relative to a central block, the resulting depression is known as a **graben.** If grabens persist over large areas, as in certain sections of east Africa and the Middle East, they are known as **rift valleys.** Lower portions of some rift valleys may be occupied by lakes, and the margins of active rifts may be subject to volcanic activity and earthquakes. The rift system of East Africa is marked by lakes such as Nyasa, Rudolf, and Victoria. Volcanic peaks like Kilimanjaro and the giant Ngorongoro Crater stand as a testimony to the instability of this major plate boundary. This area also experiences earthquakes, and areas of hot springs testify to crustal instability. The Red Sea, Dead Sea, and Jordan Valley occupy portions of this rift system as it extends into the Middle East (figure 10.9).

Figure 10.8
Effects of compressional forces (A) in a reverse fault. The same area (B) after erosion. In a thrust fault (C) movement along a low-angle fault line has placed older rocks above younger rocks.
From Charles C. Plummer and David McGeary, Physical Geology, 4th ed. Copyright © 1988 Wm. C. Brown Publishers, Dubuque, Iowa. All Rights Reserved. Reprinted by permission.

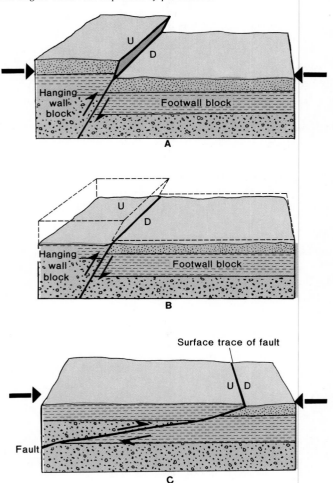

Figure 10.9
The interconnecting rift systems of East Africa and the Middle East.
From Arthur Getis, Judith Getis, and Jerome Fellmann, Introduction to Geography, 2d ed. Copyright © 1988 Wm. C. Brown Publishers, Dubuque, Iowa. All Rights Reserved. Reprinted by permission.

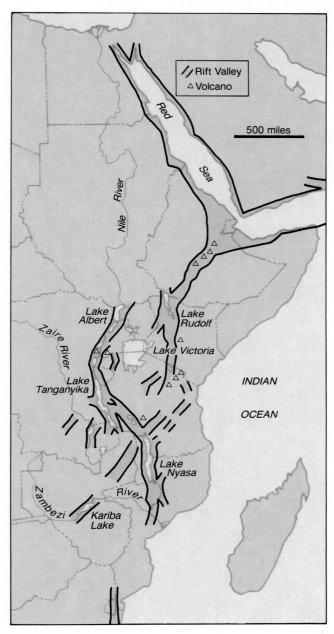

The far reaches of this rift valley extend into Lebanon and are marked by river valleys between the Lebanon and Anti-Lebanon Mountains.

The antithesis of a graben—an upthrown block between two essentially parallel faults—is known as a **horst. Fault-block mountains** may occur when pieces of the crust are elevated significantly above the surrounding terrain. A typical form is characterized by a steep slope along the fault line, whereas the slope leading away from the fault line is less severe. Many of the western mountains of the Great Basin and the Sierra Nevada Range are examples of fault-block mountains (figure 10.10). Many of these

Figure 10.10
Origin of some basin and range topography.
From Charles C. Plummer and David McGeary, Physical Geology,
4th ed. Copyright © 1988 Wm. C. Brown Publishers, Dubuque, Iowa.
All Rights Reserved. Reprinted by permission.

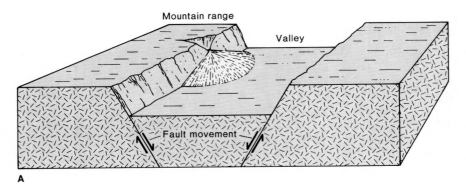

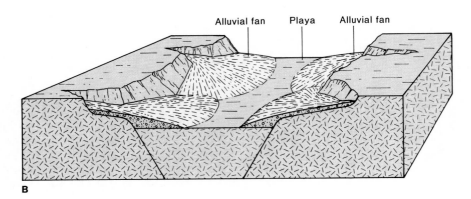

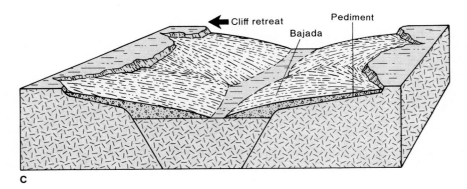

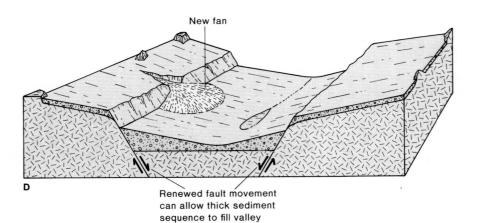

fault blocks have been tilted up asymmetrically and are currently characterized by one side with a fairly steep slope and the other side with a gentler slope. Indeed, such forms are quite characteristic of many arid and semiarid regions.

Some geomorphologists argue that this may be the most common form of mountain building, but in humid areas much of the structural evidence is buried under a thick mantle of weathered rock and debris. Authentication of such buried faults requires direct evidence from drilling or seismological surveying, although earth movements, or earthquakes, in such regions may provide inferential clues about the location and character of buried faults. It seems that fractures at some depth beneath the surface are quite common in mountain regions. As a matter of fact, most mountainous areas are quite complex composites of diastrophic forces, frequently containing volcanic features, which have been sculpted into present surface forms by an array of weathering, erosional, and gravity transfer mechanisms (figure 10.11).

Unconformities

An **unconformity** represents a disjunct in the geologic record. It is marked by the fact that rock immediately above a line or zone is considerably younger than the rock immediately below it. Frequently, this represents a surface where erosion removed large portions of the preexisting geologic section prior to subsequent deposition.

A **disconformity** occurs when several beds have been removed from the cross section, but the beds above and below the line are parallel to each other. Such lines are potential snares for the unwary who may be attempting to assign dates or geomorphological characteristics, unless a zone of weathering or a clear break in the fossil record reveals the missing beds. Perhaps erosion removed the intervening beds, or a period when sedimentation did not occur can account for the disjuncture. Certainly such lines of contact deserve careful study and analysis prior to pronouncements about causal mechanisms in their development.

An **angular unconformity** occurs when tilted beds representing a period of past folding are overlain by later horizontal sedimentaries. Clearly, such a line of demarcation records postfolding erosion and subsequent submergence and deposition. Such unconformities are striking features of exposed landscape. When buried, they sometimes suggest prospective drilling sites in the search for hydrocarbon fuels.

A **nonconformity** occurs when an erosion surface on igneous or metamorphic rocks is covered by later sedimentary or igneous rock. Usually a nonconformity represents a long period of erosion prior to subsequent sedimentary or igneous deposition. At such contacts, substantial portions of the geologic record may be missing, and great skill must be employed in detecting the dimensions and probable character of the missing record.

Diastrophic features of all kinds often precede or accompany volcanic activity. Volcanic pyrotechnics, or slow intrusions or extrusions of igneous material add form and variety to the earth's landscape. A consideration of such activity and forms follows. Since volcanic activity is often spectacular and sometimes dangerous, it catches the attention of people. The legacy of volcanoes may be death and destruction, but beautiful scenery or productive soil, or both, may be positive outcomes.

Figure 10.12
Diagram illustrating how an angular unconformity develops.
From Carla W. Montgomery, Physical Geology. *Copyright © 1987*
Wm. C. Brown Publishers, Dubuque, Iowa. All Rights Reserved.
Reprinted by permission.

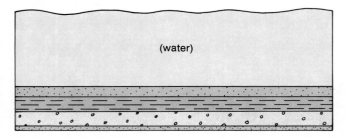

(water)

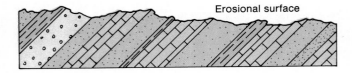

Erosional surface

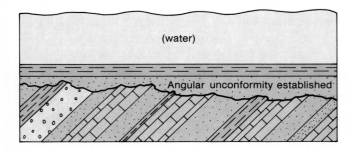

(water)

Angular unconformity established

Vulcanism

Vulcanism, named for the ancient Roman god Vulcan, is defined as the movement of molten rock materials from the interior of the earth towards the surface. Vulcanism may occur in either an explosive or quiescent manner. The island volcano of Vulcan in the Mediterranean has helped to characterize the volcanic mountain form. Such activity is called **extrusive vulcanism.**

Similarly, *magma* (molten rock) may never reach the surface and may solidify within the earth's crust. This type of volcanism is known as **intrusive vulcanism.** Evidence of both types of activity is characteristic of volcanic regions. Both provide rock types and landforms that react

in their own unique ways to agencies of weathering and erosion, which begin their inexorable attack as soon as the rock and initial landform are exposed.

Both forms of volcanic activity have significant impact upon the landscape including the creation of landforms, the exposure of new soil-forming material, and, occasionally, explosive eruptions that cause loss of life and property. Intrusive features are frequently exposed by subsequent erosion and, because of differential resistance to weathering and erosion, they may create interesting surface landscape features. The terror of the death-dealing explosive eruption in certain areas may be contrasted to intensively cultivated slopes on weathered basic volcanics in areas where volcanoes have long been inactive.

Volcanic rocks tend to be dominated by the silicates, although the amount of silica present may vary from about 45 percent to as much as 75 percent of the rock materials. Rocks that are relatively silica-poor are called **mafic rocks.** They typically contain relatively high contents of magnesium, calcium, and iron. Basalt is the most common of the mafic rocks, and it may produce interesting forms, including columnar jointing, as it hardens. Such mafic rocks typically weather down into soils that are favorable for agriculture both in terms of mineral constituents and physical properties. In such areas, cultivation on steep slopes through terracing may be extensively employed, as it is in Java (Jawa), to use these fertile soils even on very steep hillsides.

Felsic rocks, in contrast, are silica rich with significant admixtures of alumina and oxides of sodium and potassium. These rocks tend to disintegrate when weathered, often leading to physiological aridity. Coarse texture and the lack of the basic minerals essential for plant growth reduce the value of soils derived from these rocks for agricultural purpose. Rhyolite is a typical felsic rock.

Rocks, essentially midway between the felsic and mafic rocks are called **intermediate rocks.** The typical intermediate volcanic rock is andesite. Obviously, the weathered products of these rocks yield soils somewhere between the very fertile and the quite infertile volcanically derived soils.

It is useful to recognize that volcanic soils are *not* universally fertile and productive. Some are, however, and, as in the case of Java (Jawa), may be very densely settled. On the other hand, some of the outer islands of Indonesia are composed of felsic rocks that prove to be somewhat unproductive when weathered to soil. Their productivity is limited both because important minerals may be missing from the soils derived from felsic rocks, and coarse texture may produce **physiological aridity** (i.e., dry conditions created by rapid percolation of water through the permeable soils).

Figure 10.13
Schematic diagram of (A) *fissure eruption and* (B) *overall extent of flood basalts of the Columbia Plateau.*
From Carla W. Montgomery, Physical Geology. *Copyright © 1987 Wm. C. Brown Publishers, Dubuque, Iowa. All Rights Reserved. Reprinted by permission.*

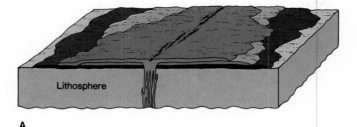

A

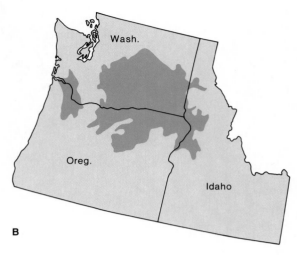

B

Extrusive Vulcanism

Molten magma that is extruded on the surface of the earth is known as **lava.** This lava may flow out in repeated sheets to cover vast areas, sometimes to depths of hundreds of feet. Examples of such large areas of lava flow include the Columbia Plateau in the Pacific Northwest of the United States; the Deccan Plateau of India; the Paraña Plateau in portions of Brazil, Uruguay, and Argentina; and large areas of East Africa. Where these rocks have weathered to soils of some depth, they are quite productive, although productivity may be curtailed somewhat by climatic conditions, notably in the Columbia and Deccan Plateaus. These areas receive light and uncertain amounts of rainfall, which restrict their agricultural utility somewhat and increase agricultural risks. Such flows may also issue from a single vent, as well as from fissures, building up certain types of volcanoes, such as Mauna Loa in Hawaii.

Depending on the constituent materials, the amount of trapped gases, and the rate of cooling, the lava may exhibit some interesting variations in form. In the Hawaiian region, for example, surfaces of molten lava have sometimes solidified while molten materials still flow beneath the surface in channels, which, when finally solidified, leave lava caves or tubes. In several areas, these caves are almost labyrinthine in character and scope. Some carry water after rains when openings are exposed at the surface, while others remain dry even during downpours of rain.

At the surface, lava that has cooled in large rope-like or taffy-like strands is called **pahoehoe.** Pahoehoe lava extruded beneath the surface of the sea often produces **pillow lava,** a form that resembles a pillow, which has formed as the external surface is cooled rapidly by the sea while the interior remains viscous and continues to push against the outer hardened margins. There may be continuing breaks through a hardened pillow skin producing repeated examples of new pillows. In fact, the typical situation is to discover a broad field of such pillows representing a persistent flow and repetitive rapid cooling of the outbreaks. Pillow lava is often seen at the areas where seafloor spreading is occurring. Although several areas of pillow lava are now observed on land, it's clear that such lava initially flowed out and solidified beneath the sea. Angular, blocky surfaces are called **aa** by the Hawaiians.

Magma (called lava when it reaches the surface) has its source somewhere within the upper 60 miles (about 100 kilometers) of the earth's crust. The partial melting of rock material within the asthenosphere may be due to decreased pressure of overlying material or it may occur when hot plumes of rock from the mantle increase the temperature of crustal rocks. Apparently, these conditions operating separately or in tandem result in rock melting. Reductions in pressure, with associated lowering of rock melting temperatures, may occur as plates move, creating cracks that descend to considerable depth.

Figure 10.14
Magma may make its way to the surface along a spreading center of a midoceanic ridge. Lava deposited beneath the sea in such circumstances usually produces pillow lava.
From Charles C. Plummer and David McGeary, *Physical Geology*, 4th ed. Copyright © 1988 Wm. C. Brown Publishers, Dubuque, Iowa. All Rights Reserved. Reprinted by permission.

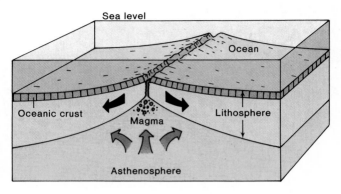

Figure 10.15
Magma may make its way to the surface along subduction zones.
From Charles C. Plummer and David McGeary, *Physical Geology*, 4th ed. Copyright © 1988 Wm. C. Brown Publishers, Dubuque, Iowa. All Rights Reserved. Reprinted by permission.

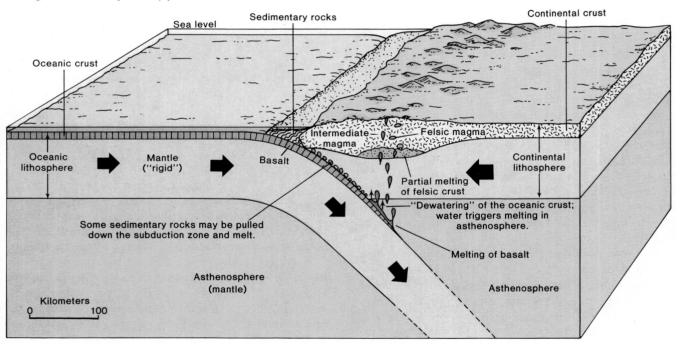

Figure 10.16
Blocky lava surface in Oregon.

Figure 10.17
Sequence of events during eruption of Mount Saint Helens on May 18, 1980. The profile (A) before eruption; (B) after the landslide, which reduced pressure on magma; and (C) magma blasts outward.
From Charles C. Plummer and David McGeary, Physical Geology, 4th ed. Copyright © 1988 Wm. C. Brown Publishers, Dubuque, Iowa. All Rights Reserved. Reprinted by permission.

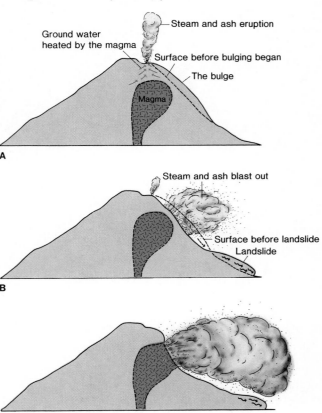

Heat may be generated in part from the radioactive decay of rocks as well as from pressure generated by an overlying load and through earth movements associated with plate movements and subduction. Most geologists and geomorphologists have assumed that this heating has come largely from the mantle. A body of evidence is developing, however, that suggests that a hotter than heretofore postulated earth core may be contributing significant quantities of heat up through the mantle into the crust. As indicated earlier, it appears that there are irregular protuberances of the core into the surrounding mantle. These protuberances may contribute to heating of the mantle as well as to the wobble in earth rotation. Whatever the details may be, molten rock is formed and intrudes itself into the earth's crust and makes its way through fissures and volcanic vents to the earth's surface.

Volcanoes may erupt explosively, throwing up quantities of lava, ash, steam, and gases. These **pyroclastic materials** or **tephra** tend to build the volcano, although large explosions may markedly change the preexistent mountain profile sometimes resulting in a lowering of the preexistent volcanic form.

The profile of Mount Saint Helens in the American Cascades has changed from a typical volcanic mountain shape to an irregular form after eruptions in the early 1980s. The profile after the May 1980 eruption reduced the symmetry of the preexisting cone. Renewed activity at Mount Saint Helens was something of a surprise, since

it had long been considered a dormant volcano. After the explosive, destructive eruption, the volcano has continued to show signs of activity with plumes of smoke, occasional modest ash eruptions, and frequent earth tremors. The major recent eruption in 1980 caused significant damage to forests and wildlife, but the loss of human life was limited to sixty-three primarily because precursor activity at the site allowed geologists to warn that an eruption was imminent. If all who were in danger had heeded the advanced warnings, there need not have been any loss of human life at all.

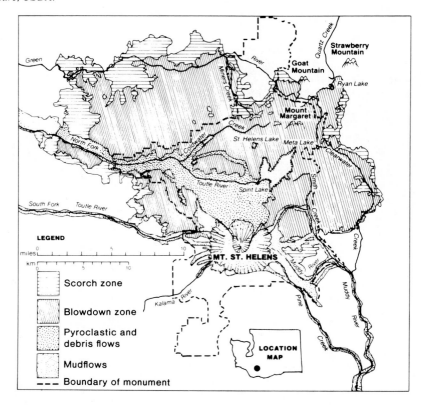

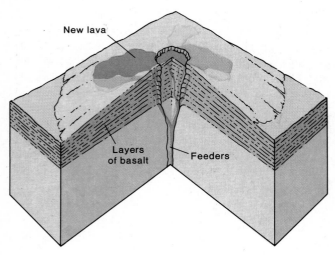

Types of Volcanoes

Volcanoes vary significantly in type from place to place
depending on how they are formed. For example, volcanoes that are made up of repeated quiescent flows of lava,
and are characterized by relatively gentle slopes, are
termed **shield volcanoes.** Many of the Hawaiian volcanoes,
like Kilauea, Mauna Loa, and Mauna Kea are of this type.
Those that are made up of alternate layers of ash and lava,
and possess a more nearly conical or prototypical volcanic
form, are called **composite volcanoes.** Composite volcanoes that have a solidified intrusive plug at their core are
termed **plugdome volcanoes.** Those that are built almost
entirely of ash (which usually stands in a high angle of
repose) are known as **cinder cones.** Parícutin, in Mexico,
typifies the cinder cone type.

Some of the most symmetrical volcanic forms are
composite types. This extraordinary symmetry is exhibited by volcanoes like Mount Fuji in Japan and Mount
Mayon in the Philippines. It is said of Mount Mayon that
it is the most perfect volcanic cone in the world. If symmetry of slope is considered, that may be the case, but
Fuji with a snow-capped summit has a kind of ethereal
beauty about it. Mayon is not high enough to be snow clad
and may not be considered quite as picturesque as Fuji.

Figure 10.20
Mt. Mayon—a composite volcano.

Figure 10.22
Black Butte in Oregon is a very symmetrical cinder cone.
Daniel Ehrlich.

Figure 10.23
Mt. Fuji.
Japan National Tourist Organization.

Figure 10.21
Parícutin is an excellent example of a cinder cone.
Daniel Ehrlich.

The Shaping of Landforms   171

Figure 10.24
Crater Lake and Wizard Island in Oregon.
Daniel Ehrlich.

Figure 10.25
Caldera formation resulting from collapse. Magma (A) helps to support overlying rock. The magma chamber (B) has been emptied through eruption. The roof (C) has collapsed producing a caldera.
From Carla W. Montgomery, Physical Geology. Copyright © 1987 Wm. C. Brown Publishers, Dubuque, Iowa. All Rights Reserved. Reprinted by permission.

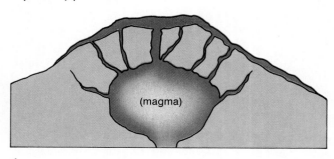

A

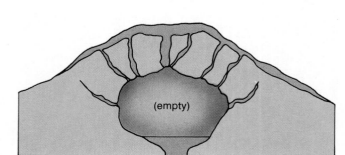

B

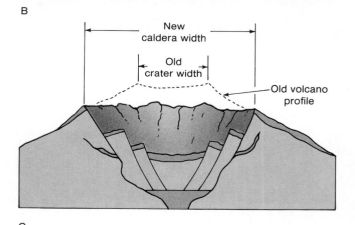

C

Calderas develop when volcanoes literally "blow their tops" in a vast explosive eruption that destroys much of the mountain top. The vast remnant circular area may then fill with water from rainfall and surface runoff. Of course, subsequent cones may be built above the surface of the water, as is true of Wizard Island in Crater Lake. Crater Lake in Oregon is the caldera left after Mount Mazama erupted explosively about 7,000 years ago. Lake Taal and associated Tagaytay Ridge south of Manila in the Philippines are, in effect, a similar caldera form. The array of islands in Lake Taal are a series of secondary volcanic cones. Explosions of the kind that produce calderas may frequently result from steam pressure developed when groundwater or surface supplies have come in contact with the magma source. The explosive ejection of large quantities of magma may leave the magma chamber depleted. Subsequent collapse of the crater into the depleted chamber may produce a caldera.

Current agricultural activities within the caldera of Taal and on adjacent slopes show that people are willing to take the risk of subsequent eruptions as a trade-off for the agricultural productivity of the volcanic soils. Calderas, especially those that have subsequently filled with water, often become outstanding tourist attractions. Much of the charm of Crater Lake derives from the water of the lake and the presence of the parasitic cone of Wizard Island.

Volcanoes tend to be transient forms, and most are active for only a relatively short period of earth history. They occur along zones of crustal weakness, and the vast majority develop along the margins of tectonic plates, although some develop in areas where the earth's crust is quite thin at considerable distances from plate boundaries. As plate boundaries become more stable or when a subsurface "hot spot" cools, volcanic activity recedes and volcanoes become dormant. Occasionally, dormant volcanoes, as in the case of Mount Saint Helens, are revivified by crustal movement, and a new cycle of vulcanism begins. Obviously, as soon as the magma is exposed, it is subjected to forces of weathering and erosion, which will eventually reduce it to base level.

Some Famous Volcanic Eruptions

Explosive eruptions usually occur when the gas volume in a lava plume exceeds about 75 percent. Below that level the lava usually pours from the vent in an effusive rather than explosive fashion. Perhaps the most explosive volcanic eruption of all time occurred at Santorini, an island about 70 miles (112 kilometers) from Crete in the eastern Mediterranean. This occurred about 1450 B.C., and the 5,000 foot high (about 1,500 meters) mountain destroyed about 50 square miles (130 square kilometers) of land.

Figure 10.26
Lake Taal caldera in the Philippines.

Figure 10.27
Areas of greatest volcanic activity.
From Charles C. Plummer and David McGeary, Physical Geology,
4th ed. Copyright © 1988 Wm. C. Brown Publishers, Dubuque, Iowa.
All Rights Reserved. Reprinted by permission.

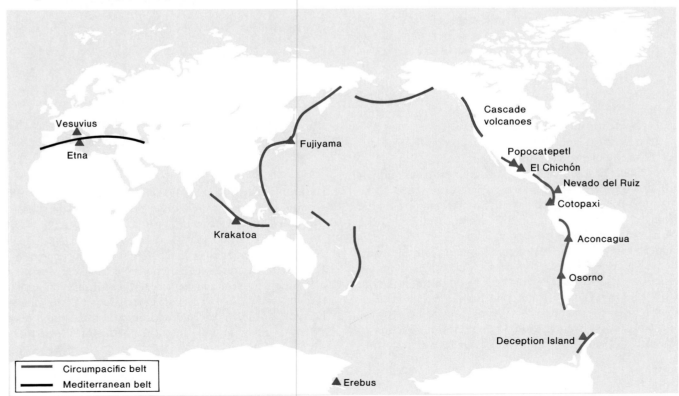

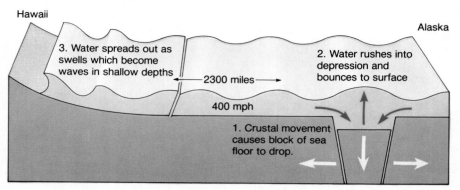

Ash 100–200 feet (30–60 meters) thick covered the rest of the island. Ash extended over at least 80,000 square miles (208,000 square kilometers), and the tsunami generated by the blast may have been 150 feet (45 meters) high by the time it reached Crete. Geologists and archaeologists speculate that the eruption may have been a major factor in the decline of the Minoan civilization on Crete.

Explosive eruptions within historic times have had profound impact on the environment. It should be recognized that the energy released in a violent eruption is greater than the explosive force of a medium-sized hydrogen bomb (i.e., about 10^{24} or 10^{25} ergs compared to such a bomb's release of about 10^{21} ergs of energy). Krakatau (Krakatoa), a volcanic island between Java (Jawa) and Sumatra (Sumatera) in Indonesia, erupted so violently in 1883 that the entire island was essentially blown away. The explosion was heard as far as 1,500 miles away, and the **tsunami** (tidal wave) generated by the eruption killed more than 30,000 people. The dust and debris thrown into the upper atmosphere and carried around the globe created brilliant sunsets for more than a year. The pattern of the island and environs was significantly different in the pre- and posteruption situation.

Since large quantities of volcanic dust may inhibit the amount of solar insolation received at the surface, many paleoclimatologists have attributed major climatic changes to periods of significant volcanic activity. Other climatologists have attributed short-term effects such as colder winters or cooler summers to major volcanic eruptions. The exact relationship of such eruptions to global climatic patterns has been difficult to establish, although it appears that the particulate burden thrown into the atmosphere by the eruption of Krakatau was principally responsible for an average global temperature reduction of about 2° C (about 3.6° F) for almost two years following

Figure 10.29
Before and after the eruption of Krakatau in Indonesia.

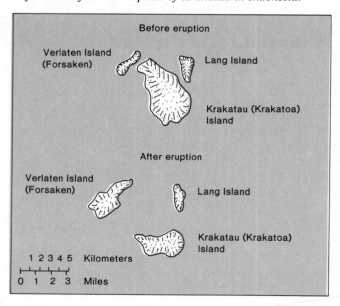

the eruption. Effects on global climatic circulation are uncertain, since the lack of a close global network of stations precluded effective correlative studies.

Mount Tambora erupted violently in the Dutch East Indies (now Indonesia) in 1815, and 90,000 people lost their lives. In addition, the enormous cloud of dust and ash thrown into the atmosphere reduced insolation so that the winter of 1815–16 was particularly severe in the middle latitudes of the Northern Hemisphere. In that same year, there was snow in June in the United States, and there were frosts throughout the summer in the middle latitudes.

Another historic explosion occurred in 1912 when Mount Katmai erupted in Alaska. The explosion was heard 750 miles away, and ash was carried for hundreds of miles from the site of the eruption. Loss of life was negligible, however, since the eruption occurred in a remote, sparsely populated area.

In 1983 El Chichón, a volcano in southern Mexico, erupted in a series of violent explosions sending ash and gases into the atmosphere and killing thousands of people. The event received remarkably little press coverage apparently because of its remote location and relatively difficult access. Climatologists believe that the vast quantities of ejected materials also had serious negative climatological effects. Studies are still under way in an attempt to ascertain the dimensions of those climatological effects.

Mount Kilauea in Hawaii has been the most active volcano in the last decade. Its quieter eruptions, usually not explosive, may add land to Hawaii as the lava descends to the sea. Kilauea is a common tourist site, and its ready accessibility means that viewers of the evening news often get a view of lava fountains or see homes consumed as tongues of lava invade settled areas. Tongues of lava may reach the sea where they are quenched by the water in hissing clouds of steam.

Nevada del Ruiz erupted in Colombia in late 1985, and several thousand people were buried in viscous mud and ash, which choked surrounding streams and forced them over their banks. Villages were inundated with mud, water, and debris, and many buildings were buried. Both people and things will remain entombed in a kind of modern day Pompeii.

Until the time of the eruption of Mount Saint Helens, only minor evidence of activity was detected among volcanoes in the conterminous forty-eight states. After the eruption of Mount Saint Helens, it seems clear that mountains like Hood, Baker, Rainier, Jefferson, Adams, the Three Sisters, and others standing above the Cascade–Sierra Nevada system also have potential for becoming active. In Alaska and the Aleutians, several other volcanoes continue to be active.

During volcanic eruptions, the friction developed between the adjacent ash particles and the water released from subsurface rocks may produce a pyrotechnical display of lightning, and convectional currents developed by heating may produce towering cumulus clouds and rain, which turns ash to steaming mud. Scientific observers were not at Nevada del Ruiz during the eruption, but it appears that the great loss of life came from mudflows, which were created by a mixture of storm water and ash as well as from streams, which were completely overloaded by ash. A long period of quiescence had led unsuspecting citizens to extend their settlements up the slopes of the mountain to take advantage of the fertility of the volcanic soils. Such settlement is always a kind of Russian roulette in which people must weigh the attributes of agricultural productivity against the dangers of a catastrophic eruption. The citizens of Colombia in the vicinity of Nevada del Ruiz lost that deadly game in 1985.

Volcanoes continue to develop in a number of places around the world. Within the recent historical past, for example, Parícutin began abruptly out of a Mexican cornfield in 1943 and grew to a height of more than 1,500 feet (450 meters) before becoming quiet in 1952. Imagine the surprise and concern of a Mexican farmer when the earth cracked and steam began issuing from a field he had been plowing. Fortunately, the village priest recognized the beginning of the eruption for what it was. An alerted scientific community was able to study Parícutin intensively almost from the beginning, and a great deal was added to the science of volcanology.

Surtsey began as a series of spectacular eruptions out of the sea off Iceland in 1963. Doubtless, more volcanoes will be born and some will fall silent. Volcanic activity will continue to be awe-inspiring evidence of the restlessness in the earth's crust.

A few have given rise to specific geomorphological terminology, as in the case of Mount Pelée in Martinique, which had a particularly vigorous eruption in the early 1900s. The choking clouds of incandescent gas and dust, which moved down the side of the mountain and caused such severe loss of life in the city of Saint Pierre, provided the nomenclature for all such subsequent eruptions—**Peléean eruption.** After the major eruption, subsurface magma forced a solidified portion of a volcanic plug to project above the crest. This curiosity was known as **Pelée's spine.** The destruction was so complete at Saint Pierre, then the dominant city on the island, that it never recovered its former preeminent position. Fort de France became the leading city on Martinique and continues to carry that distinction to the present time.

Bizarre examples of volcanic destruction continue to emerge. In the late summer of 1986, a curious gas eruption in Cameroon in Africa killed as many as 2,000 people. The exact nature of the eruption is unknown, but the volcanic lake literally hiccoughed and heavy gas (apparently a combination of carbon dioxide and hydrogen sulfide) bubbled up through the lake and down the outside of the crater. Virtually all those in the path of the gas were asphyxiated. In addition, large numbers of cattle were killed, adding to the posteruption hardships of the people. Apparently, very special circumstances must exist for such eruptions to occur, but less destructive similar eruptions have occurred in Cameroon in the past. This is but another example of bizarre happenings that may occur in conjunction with volcanic activity.

Certainly volcanoes continue to be a major hazard to life and property. More than 60,000 people have died in volcanic eruptions in the twentieth century.

Figure 10.30
The relationship of Cascade volcanoes to subduction zone.
From Carla W. Montgomery, Physical Geology. *Copyright © 1987*
Wm. C. Brown Publishers, Dubuque, Iowa. All Rights Reserved.
Reprinted by permission.

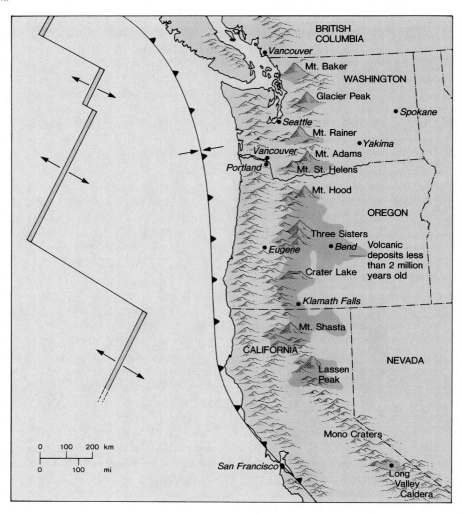

Major Volcanic Regions

Volcanoes are clustered at or near the margins of the great tectonic plates, and the area around the margins of the Pacific Ocean is particularly active. This vast area is sometimes described as the **Pacific Ring of Fire.** It extends through the Andes, Central America and Mexico, Sierra–Cascades, into Alaska, through the Aleutian Island chain, through the Kamchatka Peninsula of the U.S.S.R., through Japan and the Philippines, into New Zealand and even to the margins of Antarctica. Other major volcanic regions include a zone through the Mediterranean, areas along the rift zone in East Africa, and in the Lesser Antilles at the margins of the Caribbean. The young circum-Pacific zone and the Mediterranean area are marked by volcanoes with andesitic extrusives.

The new ejecta adds to the profiles of preexisting surfaces. Lava and ash exposures are immediately subjected to the ravages of weathering and erosion. The profile of the volcanic mountain and the exposure of new rock materials at the surface are only a part of the story, however. What is happening beneath the surface may create surface manifestations, and the subsequent exhumation of subsurface forms by erosion may have geomorphological and economic significance.

In the United States quite a number of volcanoes have erupted in the last 200 to 300 years. Several in the Aleutians have erupted in the 1980s, as have Hawaiian volcanoes and Mount Saint Helens. Several volcanoes of the Sierra–Cascade system (i.e., Lassen, Shasta, Hood, Baker, and Rainier) have erupted within this 200- to 300-year time span, as have Hualalai and Haleakala in Hawaii.

Figure 10.32
The volcanic area near Parícutin has resulted in the destruction of property.

In the Sierra–Cascade chain, Adams, Crater Lake, Jefferson, and Three Sisters have all erupted within the past 1,000 years. Those that last erupted 10,000 or more years ago include San Francisco Peak near Flagstaff and the Yellowstone Caldera. All told, there are about thirty-three volcanoes in the United States (including Alaska and Hawaii) that have the potential in terms of magma source and instability to produce a destructive eruption in the near future.

The elements of extrusive vulcanism are matched by an intricate array of intrusive features that are described and analyzed in the section that follows.

Intrusive Vulcanism

Intrusive volcanic forms, collectively known as **plutons,** come in myriad forms, shapes, and sizes. They may affect surface landforms so as to produce recognizable terrain features, or they may remain undetected. Subsequent erosion may reveal their presence and, because of different resistance to erosion, significant landform relics may remain.

Some of the common intrusive forms include the **batholith,** the largest and most amorphous type. These features often extend over thousands of square miles, and frequently they lie at the core of major mountain ranges. This is the case in several areas of the Sierra Nevada and Rocky Mountains of the American West. These exceptionally large forms may also include, as adjuncts, substantial numbers of other intrusive features. Technically, a batholith must involve an area of at least 60 square miles (100 square kilometers), but most are much larger than that. Obviously, the choice of such an areal figure is arbitrary, but it provides a reasonable differentiation between the batholith and other large plutonic forms.

Laccoliths are toadstool-shaped igneous intrusions that form when slow-moving magma is forced up through a tube-like connection between horizontal layers of rock already in place. Because the magma is so viscous, its flow is slowed and it pushes up in a blister-like form, which may cause a surface bulge. While many laccoliths are small, some are large enough to have pushed up significant terrain features. The Black Hills of South Dakota, for example, have a laccolithic core.

Stocks, or **bosses,** are similar to batholiths in being generally amorphous in shape, but they are much smaller in size. Stocks may occur as peripheral appendages of batholiths or laccoliths, or they may develop from a larger magma source. They are less likely to create surface landforms although they may be exposed, long after their formation, by erosion.

Dikes are intruded into vertical, or almost vertical cracks, or they may simply force their way upwards in weaker areas of overlying rock. They may extend for many miles. Frequently, they exist in a radial pattern around a volcanic vent. If subsequently exposed by erosion, dikes will frequently be more resistant than the rocks into which they were intruded. In such cases, they may stand like irregular walls above the surface of the surrounding terrain. From above, they may appear like radiating spokes of a wheel from a central hub. Shiprock, New Mexico has such a set of radiating dikes out from the center of an old volcanic neck.

Sills are nearly horizontal, having been intruded in sheet-like form between sedimentary strata or between layers of volcanic ash or lava. They rarely have significant surface expression, even when exposed, except in certain arid areas where they may serve as the cap rock for mesas or buttes. Sometimes, they are recognizable along the margins of valley or canyon walls where they have been exposed by erosion.

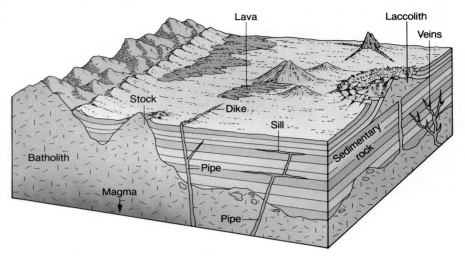

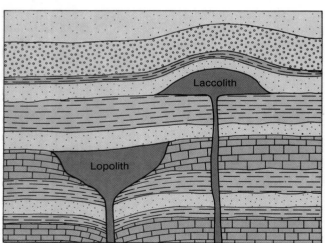

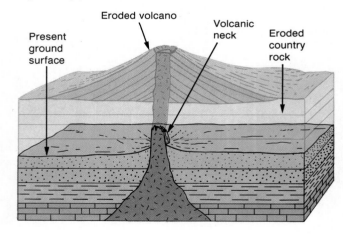

Figure 10.36
A dike in western Colorado exposed by erosion. Material in the dike proved to be more resistant to erosion, so it stands above the surface into which it was intruded.

Figure 10.37
Columnar jointing in basalt surrounded by tuff in Yellowstone National Park.

Veins are small magma-filled cracks that may exist in profusion. The stresses attached to earth movements usually produce large numbers of such cracks, which in an area of active vulcanism may be filled with magma.

A somewhat less common igneous intrusive feature is the **lopolith.** It is somewhat like the laccolith except that it has a flattened top.

Plugs or **necks** are not intrusive features in the usual sense, but they represent the solidified remains of magma contained in a volcanic vent. Subsequent erosion may expose them as prominent landform features. Shiprock, New Mexico is an example of such a volcanic plug.

Intrusive features that cut across existing sediments or layers of ash and lava are termed **discordant features** by geologists. Those, like sills, that are essentially parallel to preexisting bedding planes are termed **concordant.**

Some volcanic areas and adjacent contact metamorphics are sites for metallic mineral exploitation. Certain carbon materials subjected to great heat and pressure were crystallized into diamonds. Long quiescent volcanic areas within southern Africa contain the "pipes" that have yielded significant diamond deposits. In other areas, superheated water in contact with magma may have dissolved certain minerals that remain as veins or deposits to be exploited subsequently. These veins may prove to be sources of metallics, including the precious metals.

Terrain Building

The grinding together of great tectonic plates, usually at continental margins and often involving giant geosynclinal troughs filled with thousands of feet of sediment, helps to build mountains of different types (i.e., folded, faulted, or volcanic). More often than not the diastrophic forces are comingled in ways to produce quite heterogeneous combinations of folding, faulting, and vulcanism. In a world of complex nature, all of these forces interact in an exceedingly varied array of land uplift and mountain formation. Those diastrophic forces are in an inexorable and never-ending battle with an antithetical set of gradational forces that are working to reduce all slopes to grade. The pattern of landforms and the nature of terrain ultimately relate to the temporal associations between the forces that are pushing segments of the earth's crust upwards versus those that are wearing it down. It is a never-ending cycle that has been going on since the earth had an atmosphere. The inexorability and characteristics of agents of weathering and erosion are described in the subsequent chapter. The gross aspects of diastrophic forces and attendant landforms are etched in intricate detail at both the micro and macro level to produce a landscape of variegated and constantly changing form—a panorama at once fascinating and ever-changing. The same forces acting in the same way and increasingly augmented by technological man are continuing to produce and modify the face of the physical world in which we live.

Figure 10.38
Volcanoes sometimes become tourist attractions.

Figure 10.39
Volcanic materials may weather into productive agricultural soils.

Study Questions

1. What are the principal diastrophic forces?
2. Compare and contrast monoclines, anticlines, and synclines.
3. Explain the difference between elastic and plastic strain.
4. Define strike and dip.
5. What is the axis of a fold? The limbs?
6. Explain the relationship between joints and faults.
7. List and characterize the principal types of faults.
8. Define disconformity, unconformity, and nonconformity.
9. Characterize the principal types of volcanoes.
10. Differentiate between pahoehoe and aa.
11. In what kinds of areas are volcanoes found?
12. List some particularly destructive volcanic eruptions that have occurred in historic times.
13. What are plutons? List and briefly describe the principal plutonic forms.
14. Contrast concordant and discordant plutons.
15. Explain how pillow lava forms.

Selected References

American Geological Institute. 1984. *Dictionary of geological terms.* 3d ed. Garden City: Doubleday Publishing Company.

Dennis, J. G. 1987. *Structural geology: An introduction.* Dubuque, IA: Wm. C. Brown Publishers.

Ernst, W. G. 1969. *Earth materials.* Englewood Cliffs: Prentice-Hall.

MacDonald, G. A. 1972. *Volcanoes.* Englewood Cliffs: Prentice-Hall.

McKnight, T. L. 1984. *Physical geography: A landscape appreciation.* Englewood Cliffs: Prentice-Hall.

Montgomery, C. W. 1987. *Physical geology.* Dubuque, IA: Wm. C. Brown Publishers.

Plummer, C. C., and McGeary, D. 1985. *Physical geology.* 3d ed. Dubuque, IA: Wm. C. Brown Publishers.

Ragen, D. M. 1984. *Structural geology: An introduction to geometrical techniques.* 3d ed. New York: John Wiley and Sons.

11

Weathering and Erosion

Weathering and erosion may produce bizarre
forms.

*All streams run to the
sea, but the sea is not
full;*

Ecclesiastes 1:7, *Bible*

*T*he very diastrophic forces
that build up the surface of the land sow the seeds for the
agents of weathering and erosion, which reduce it. The
forces that fold and fault rock produce breaks, cracks,
and joints where agents of weathering, mass wasting, and erosion can intrude. The elevations that are produced magnify the effects of gravity and provide the slopes that facilitate the activities of the several degradational agents.

Over time, all landform profiles experience some
combination of weathering, gravity, and erosion. Elevation above grade and steepness of slope accelerate rates of
erosion. Conversely, gentle slopes reduce the impact of
erosional agents. Landform profiles constantly change in
response to incessant attacks by the various agents of
weathering and erosion. The rate of reduction may be rapid

or slow, depending on the nature of materials being reduced, the amount of time they have been exposed, the characteristics of the existing climate, the type of vegetative mantle, and the average slope and elevation of exposed surfaces. In some regions, weak rock on steep slopes in an area of copious precipitation may be quickly reduced; whereas, in other regions, hard rocks with gentle slopes and minimal precipitation may resist the ravages of erosion for long periods.

The earth's surface is **degraded** or **denuded** (i.e., reduced towards some base level, commonly sea level). Three types of activities are responsible for this denudation: weathering, mass wasting, and erosion. This is a never-ending, ever-changing weathering and erosion cycle that is constantly reducing prominences to grade. Antithetical forces initiated by diastrophism and vulcanism are constantly increasing elevations and modifying slopes, especially at the margins of the great tectonic plates. Just as inexorably, these new land surfaces are attacked by gradational agents.

Weathering involves the breakdown of rock materials in place by some combination of physical, chemical, or biological activity. **Mass wasting** includes the downslope movement of rock materials primarily by the force of gravity, although water may lubricate the interstices between rock particles and speed downslope movement. **Erosion** is the movement of rock materials by active transporting agents such as running water, moving ice, or blowing wind. Any surface may be exposed to more than one erosional agent at the same or at different times. Basically, weathering prepares material for transport, whereas mass wasting and erosion move such materials down grade. Weathering and erosion are inexorable in all environments; only the agents and time required for rock reduction vary from place to place.

Weathering

Massive rock materials must be reduced in size in order to be transported by agents of movement such as mass wasting or erosion. As soon as a landform is produced and rock is exposed as part of the earth's crust, weathering begins. It continues until the exposed surface has been reduced and transported away. No rock escapes this destruction. Just as people experience the effects of aging with time, so rocks break down under the constant assaults of weathering. Time is the ally of weathering and the enemy of exposed rock surfaces. In human terms, long periods of time are required to break down rock; but in geological time, even the most resistant rock succumbs to breakdown fairly rapidly.

Weathering is of two broad types: disintegration and decomposition. **Disintegration** involves the mechanical reduction of pieces of rock that maintain their essential chemical integrity. **Decomposition** involves the chemical modification of rock materials to produce softer, or less resistant, materials, which are more readily transportable by agents of erosion. Chemical bonds remain intact in the former; whereas in the latter, chemical bonds are broken and new compounds are formed, which are usually less resistant to further destruction and removal than the preweathered material.

Physical Weathering

Larger surfaces of rock are exposed to weathering by breaks, which have developed from a variety of causes. Original folding, faulting, or volcanic activity frequently produces cracks in rocks from large size to the microscopic. **Vesicles** (cavities) may form in lava rock when gases escape as the lava is hardening. **Solution cavities** exist in rocks where water dissolves soluble rocks, such as limestone or dolomite. These exposed areas are subject to the ravages of physical and chemical weathering. Every exposed surface is vulnerable to attack; every additional crack facilitates the work of reduction. The greater the surface exposed, the more pronounced the effects of weathering.

Frost wedging is one example of physical or mechanical weathering (figure 11.1). Water seeping into cracks expands as it freezes, and this expansion, over time, eventually splits rocks into smaller and smaller pieces. The wedging effect is made more pronounced and effective by the fact that the surface water freezes first and expands. Gradually, this wedging effect widens the crack more near the surface. In fact, the crack assumes a wedge-like cross section over time, which tends to persist until the rock is finally split. Although a number of scientists challenge the efficacy of this wedging, it seems certain that such wedging in jointed rock is significant, especially over protracted periods of time. Effects during a few seasons may be imperceptible, and the compass of a human lifetime may be too limited to ascertain significant changes.

A similar localized effect occurs in certain arid areas. This **salt wedging** develops when, through capillary action, water may be drawn upward through rock interstices. Such water almost invariably carries dissolved salts. As the water evaporates, the salts, such as gypsum or halite, crystallize. Gradually these crystals grow, exerting a wedging effect. This action is not nearly so widespread in areal extent, nor is the effect as potent a force as the effects of frost wedging, since salt wedging is restricted principally to arid or semiarid regions. Frost wedging, on the other hand, exists to some degree wherever temperatures drop below freezing.

Freezing and thawing temperatures, not involving frozen water, occur millions of times and has the effect of weakening the coherence of mineral grains causing the rock to begin to break apart. This process can be accelerated by fire. Just as any elastic material ultimately

Figure 11.1
Frost wedging breaks up rock materials.
From Carla W. Montgomery, Physical Geology. *Copyright © 1987*
Wm. C. Brown Publishers, Dubuque, Iowa. All Rights Reserved.
Reprinted by permission.

Water freezing in crack
expands, widening crack.

Eventually,
rock is broken free
by progressive
fracturing.

Figure 11.2
Tree roots have exploited the joint pattern in this sandstone
in eastern Oklahoma to wedge rocks apart.

Figure 11.3
Pounding waves along shorelines may exert significant
hydraulic and pneumatic pressure on rocks.
Daniel Ehrlich.

breaks when stressed millions of times, so rock ultimately has grains broken apart by repetitive heating and cooling. Many earth scientists are skeptical about the effects of heating and cooling, since a variety of rocks that were heated and cooled artificially thousands of times in the laboratory have shown almost no changes. Nevertheless, it seems reasonable that heating and cooling millions of times over eons ultimately has an impact. The granular disintegration that characterizes certain rocks testifies to the apparent efficacy of repeated heating and cooling in breaking down the rock materials.

Organic agents can and do invade cracks, and split rocks apart. Plant roots expand as they grow, creating a splitting effect. In fact, a tree or shrub root can exert enormous pressure as the plant grows (figure 11.2). Burrowing animals expand preexistent cracks, and, obviously, human beings may break rocks in quarrying, mining, or similar activities.

Rainfall may, in effect, pry grains of rock apart over protracted periods. Each raindrop has a significant hydraulic effect upon impact. Water driven into cracks by waves along coastlines may pry rocks apart. The pneumatic and hydraulic effects of waves repeatedly pounding against coastal shorelines pry rocks apart and break larger ones into smaller ones (figure 11.3). Air trapped between the waves and the rock face is compressed, producing enormous pneumatic pressures at the rock face as the waves break. Rocks may be shattered by volcanic eruptions, too. In short, there are a vast array of physical forces

Weathering and Erosion   183

Figure 11.4
Angular fragments of rock tend to become rounded as the result of weathering.
From Carla W. Montgomery, Physical Geology. *Copyright © 1987 Wm. C. Brown Publishers, Dubuque, Iowa. All Rights Reserved. Reprinted by permission.*

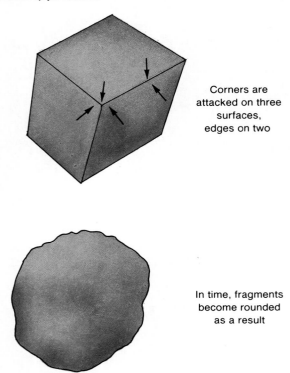

Corners are attacked on three surfaces, edges on two

In time, fragments become rounded as a result

that have the effect of breaking rock into small particles of the same material without moving it to another place. These forces are always present and persistent, and have been acting over most of geologic time. The smaller pieces produced by such activities are readily transportable by agents of erosion. Further, there is a tendency for angular rock to be rounded by the effects of weathering, since corners are attacked from three sides, whereas other portions of the rock can be attacked on only two sides (figure 11.4).

Physical weathering is accompanied almost always by chemical weathering. Indeed, in essentially all cases, the two work together, accelerating the effects of weathering. **Chemical weathering** usually involves water in some form and chemical reactions are typically speeded up by heat; hence, chemical weathering is normally more rapid in a hot, humid environment.

Chemical Weathering

The type of chemical weathering depends on the kind of rock, the amount of available water, and the environment to which the rock is exposed. Among the common types of chemical weathering are oxidation, carbonation, and hydrolysis. There are several examples of **oxidation.** Silicates or carbonates of iron or manganese, especially, combine with oxygen and water to produce less resistant materials, which may be more successfully removed by agents of erosion. Frequently, these new compounds carry

Figure 11.5
The rounding of these granite boulders is obvious.

a reddish, yellowish, or brown stain. When iron is involved, such combinations are described as **rusting**. Indeed, the influence of iron oxide is observable in the reddish stain so characteristic of many sedimentary rocks like the Permian Redbeds of the American Southwest. The reddish or yellowish color of many soils in hot or warm humid environments is attributable to oxides or hydroxides of iron. Interestingly, the chemical union of water with a substance is most often a precursor to subsequent oxidation.

Carbonation occurs where carbon dioxide and water combine to produce a weak solution of carbonic acid. Carbonic acid is especially effective in combining with limestone or dolomite to produce calcium bicarbonate, which is readily soluble in water. Water becomes a weak solution of carbonic acid through absorption of carbon dioxide released by the decay of organic material, or through absorption of carbon dioxide in the air. The vegetative mantle in most parts of the world contributes to a lower pH (higher acidity) of percolating rainwater. In some environments, especially those cloaked by a coniferous forest, the leaf litter may be strongly acidic. Limestone and dolomite, especially, are strongly affected by such acidified water and tend to break down quite readily.

Hydrolysis represents the chemical union of water with another substance. Almost inevitably, these materials are weaker than the original material and are, therefore, more amenable to reduction and subsequent transport. Such combinations of water with silicate minerals, found in abundance in many igneous rocks, leads to a volumetric change, which may induce physical as well as chemical changes in the rock.

As granite weathers chemically, for example, the feldspar breaks down into clay minerals. These clay minerals exhibit an internal crystal structure much like mica.

Orthoclase feldspar, a constituent of granite, for example, may undergo the following chemical transformation:

$$KAl\ Si_3\ O_8 + HOH \rightarrow HAl\ Si_3\ O_8 + KOH.$$

The aluminosilicic acid, which is produced, is an unstable compound that undergoes further changes to produce colloidal silica and a colloidal complex, which may ultimately become clay.

There are some interesting examples of weathering activities that involve physical and chemical modification concomitantly. Lichens will frequently attach themselves to bare rock surfaces. The rootlets have the effect of widening and deepening minuscule cracks, and there is some absorption of mineral materials from the rock by the plant. As portions of the lichens die, they contribute organic substances that, when dissolved in water, yield weak acids that continue an attack on and slightly below the rock surface. Although the weathering effects are slow, the combination of lichen attachment and life processes exhibit the effects of physical and chemical weathering occurring together.

Figure 11.7
Limestone weathers rapidly in humid environments with ample organic components added to rainfall. This monument is substantially less than one hundred years old.

Figure 11.6
A modernistic gypsum sculpture, which resisted weathering in subhumid conditions of central Oklahoma, is literally dissolving away in humid Florida.

Figure 11.8
Exfoliation resulting from unloading of a buried pluton.
From Carla W. Montgomery, Physical Geology. *Copyright © 1987*
Wm. C. Brown Publishers, Dubuque, Iowa. All Rights Reserved.
Reprinted by permission.

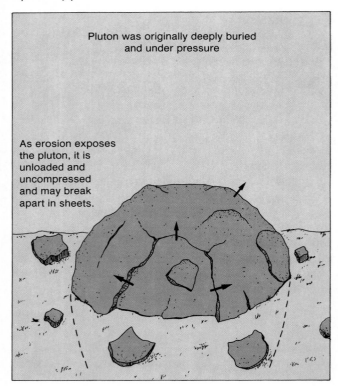

Figure 11.9
Material in an exposed road cut is spalling off.

Figure 11.10
Weathering wears away the less resistant materials first. The somewhat more resistant veins of material in this rock have produced bizarre shapes.

An extraordinary phenomenon occurs principally in granitic rocks. Under certain conditions, rock segments, like layers of an onion, spall or peel off in a concentric fashion. This process is called **exfoliation**—leafing off of rock materials (figure 11.8). The exact mechanism causing such a reduction of the rock is not well understood, but it seems to involve both physical and chemical actions. The rounding of blocky surfaces occurs because the corners and edges tend to be weathered more rapidly than the broad surface areas.

The process may be initiated by **unloading.** When the rocks have enormous quantities of overburden removed, the outer margins may crack as the rock slowly rebounds from the removal of such materials. It is also possible, indeed likely, that there is some weakening from the expansion and contraction of surface layers because of diurnal heating and cooling. The process may be accelerated by **hydration**—the attachment of water molecules to clay minerals. This wetting action may cause an expansion and physical separation of different kinds of rock materials in the rock matrix. Although the precise weathering mechanisms remain something of an enigma, the characteristic pattern created as rock layers spall off is known to all who have observed exposed granite surfaces.

In many humid tropical regions, rock, especially granite, exhibits this onion-like pattern long after it has become rotten (i.e., so soft that a person can literally reach in with bare hands and remove parts of the crumbly rock). Where feldspars have been reduced to clays, the rock has lost its character as granite. Granitic weathering is especially rapid in the hot humid conditions that characterize the rainy tropics. Exposed granite rock in such regions often weathers into deep residual soils.

Weathered rock products exposed at the surface over time become soil. In the chapter on soils, that process will be described in considerable detail.

Weathering mechanisms, acting singly or in combination, all have a similar effect (i.e., the reduction of rock size or coherence that makes it more available for transport). The agents of erosion begin the removal of this material in an endless cycle of erosion and deposition. These gradational agencies tend to smooth the contour of the land, reduce elevations, and generally diminish local relief. A description and analysis of these transport mechanisms follow.

Figure 11.11
Lichens growing on the surface of a conglomerate boulder in northwest Florida.

Mass Wasting

A typical intermediate step in the transport of weathered rock material is the movement of such material downslope in response to the forces of gravity. Such movements are known as **mass wasting** or **gravity transfer.**

The inexorable force of gravity, which draws everything towards the center of the earth, operates most effectively in areas of high relief and steep slopes. Some movements are abrupt, precipitous, and rapid, whereas others are gentle, slow, and protracted.

Different materials will remain essentially unmoved at different angles depending upon the nature of the material. For example, large, angular boulders are more likely to maintain their relative equilibrium along a given slope than a fine-grained material such as sand. Each aggregate of material tends to hold a particular **angle of repose** peculiar to the rock type (figure 11.12). At the angle of repose, cohesion and friction between the rock materials equal or exceed the downward pull of gravity. If anything such as an earth movement or the addition of water into the spaces between the rocks disturbs the materials, there is likely to be a downward movement. Ultimately, all unconsolidated materials must heed gravity's pull and move downslope (figure 11.13).

Several other factors may facilitate gravity's pull. In addition to water, clay—especially water-soaked clay—may lubricate and speed downslope movement. In areas where freezing and thawing may be frequent and persistent elements of the environment, the heaving created by the expansion of ice may cause downslope movements. And physical laws hold that materials set in motion tend to continue that motion until a counteracting force such as reduced grade slows and ultimately stops the process.

Figure 11.12
Angles of repose for different types of rock material.
From Carla W. Montgomery, *Physical Geology. Copyright © 1987 Wm. C. Brown Publishers, Dubuque, Iowa. All Rights Reserved. Reprinted by permission.*

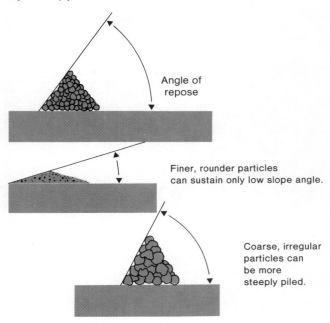

Angle of repose

Finer, rounder particles can sustain only low slope angle.

Coarse, irregular particles can be more steeply piled.

Figure 11.13
This rock segment has slumped away from the valley wall, and the force of gravity will ultimately cause it to collapse.

Rocks that have been broken by elements of weathering may simply fall down a steep slope. Along very steep slopes, there may be frequent falls of such rock, which is called **talus.** Often, these rock falls accumulate in enormous piles or cones. If there are a number of adjacent and coalescing **talus cones** along the margins of such steep slopes, these features may be called **talus aprons** (figure 11.14). Such talus cones or aprons are frequently observed at the base of very steep mountain slopes. These talus slopes are noted especially in arid mountainous regions because they are usually not masked by a mantle of vegetation.

Rock slides may also occur with stunning rapidity (figure 11.15). If talus cones are perched at an intermediate elevation, for example, earth tremors may dislodge materials that have been resting in place for protracted periods. Heavy snow cover may accumulate and dislodge in a catastrophic landslide, carrying snow, ice, rock, and debris along with it. Sometimes, in areas of heavy snow accumulation, people set explosive charges, or fire cannons or rockets to initiate a minor avalanche and dislodge the snow before it assumes the proportions of a major disaster.

There are, of course, less spectacular movements of materials from higher to lower elevations. **Mudflows** may move down established drainage channels when rains have saturated fine-grained soil, especially in areas that are sparsely vegetated. **Earthflows** are similar to mudflows except that there tends to be a slump of saturated earth downslope without reference to existing drainage channels. Such mudslides and earthflows have become particularly dangerous in closely settled areas on steep, devegetated slopes (figure 11.16). California and Japan have become particularly susceptible to such movements and the destruction that accompanies them. In California, especially, mudflows typically occur in the winter wet season on slopes devegetated by fire in the dry summers.

The prospects for damaging fires and subsequent mud slides increase year by year as more and more people build homes in vulnerable areas. Resinous vegetation and dry summers in a closely settled region are a fire hazard. When the fire occurs, the devegetated slope is a prospective mudslide or earthflow in the subsequent wet season. In fact, the devegetation of slopes, by accident or design, usually occurs as population densities increase.

Soil creep represents an almost imperceptible movement of soil materials downslope. Freezing and thawing, and wet and dry conditions facilitate such movements by the heaving of materials, along with the expanding, contracting, and lubricating effects of moisture. The effects of soil creep may be observed in a variety of ways: vertical slabs, such as tombstones or monuments, tilt downslope; building foundations crack; fences and power lines tilt; and tree trunks curve.

Solifluction is a special form of soil creep essentially restricted to areas underlain by permafrost (permanently frozen ground). During periods of thaw, the upper portion of the soil becomes saturated by water, since the water cannot percolate to a level below the frozen subsoil (figure 11.17). The hypersaturated soil then slides downslope in an irregular pattern over the upper surface of the permafrost. These actions recur every year, and the margins of the movement are typically somewhat higher than the centers. These tiny ridges frequently overlap, presenting a variegated hummocky surface.

Basically, the weathering of materials in place is followed by the movement of such materials downslope, principally in response to gravity (figure 11.18). Elements of erosion then move these materials long distances until a capacity load, a diminished velocity, or both cause the materials to be deposited in another place—usually a remote location.

Figure 11.16
Classification of various mass movement processes.
From Charles C. Plummer and David McGeary, Physical Geology,
4th ed. Copyright © 1988 Wm. C. Brown Publishers, Dubuque, Iowa.
All Rights Reserved. Reprinted by permission.

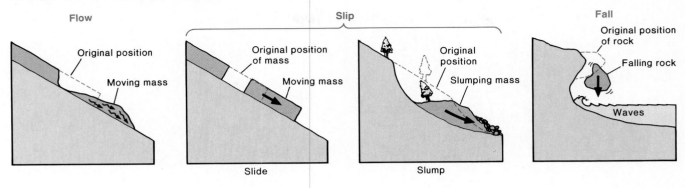

Figure 11.17
Thawing above permafrost zone resulting in solifluction.
From Charles C. Plummer and David McGeary, Physical Geology,
4th ed. Copyright © 1988 Wm. C. Brown Publishers, Dubuque, Iowa.
All Rights Reserved. Reprinted by permission.

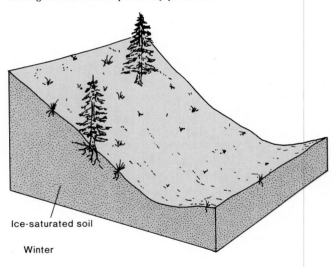

Figure 11.18
Schematic diagram of slumping in soil.
From Carla W. Montgomery, Physical Geology. Copyright © 1987
Wm. C. Brown Publishers, Dubuque, Iowa. All Rights Reserved.
Reprinted by permission.

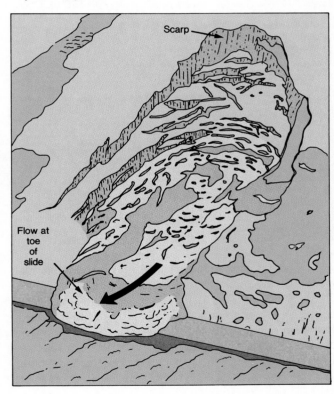

Figure 11.19
Joint patterns in rock may facilitate weathering and mass wasting.

Figure 11.20
Weathering in this 2,500-year-old figure has been slight in the dry environment of Iran.

The constantly shifting load may have subtle, occasionally dramatic, impact on the preexisting surface. These changes ensure that no piece of landscape is precisely the same on two succeeding days. In the space of a human lifetime, landform change—barring a dramatic event like a landslide, earthquake, or volcanic eruption—is almost imperceptible, although in the larger context of geologic time, vast changes occur in relatively short spans of time. Various estimates have been made as to the rate of landscape reduction, but the variables are so great that the estimates are of little use.

In subsequent chapters, erosion and deposition will be discussed. The surface of the earth is etched and changed inexorably, persistently, and pervasively, and changes occur at both the microlevels and macrolevels. The most potent and pervasive of erosional agents, running water, is discussed in the following chapter.

Study Questions

1. Carefully distinguish between weathering and erosion.
2. List and characterize the various types of physical and chemical weathering.
3. Discuss the process of exfoliation. In what types of rock does it characteristically occur?
4. Explain angle of repose. Why do these angles vary from one kind of earth material to another?
5. Where and under what conditions does solifluction occur?
6. What is permafrost?
7. Explain why the number and severity of mudslides may increase along with population growth.
8. Where are talus cones and talus aprons most likely to develop?

Selected References

Birot, P. 1966. *General physical geography.* New York: John Wiley and Sons.
Butzer, K. W. 1976. *Geomorphology from the earth.* New York: Harper and Row, Publishers.
Gabler, R. E.; Sager, R. J.; Brazier, S. M.; and Wise, D. L. 1987. *Essentials of physical geography.* 3d ed. Philadelphia: Saunders College Publishing.
Gersmehl, P.; Kammrath, W.; and Gross, H. 1980. *Physical geography.* Philadelphia: Saunders College Publishing.
Ritter, D. F. 1978. *Process geomorphology.* Dubuque, IA: Wm. C. Brown Publishers.
Strahler, A. N., and Strahler, A. H. 1984. *Elements of physical geography.* 3d ed. New York: John Wiley and Sons.

12

The Work of Running Water

Running water is the most pervasive of all agents of erosion.

I chatter, chatter, as I flow
To join the brimming river,
For men may come and men may go,
But I go on forever.
Alfred Lord Tennyson, *The Brook*

*A*fter materials are weathered and gravity has assisted in moving them to grade, various agents of erosion transport materials down towards base level in a never-ending process. Inexorably, various diastrophic movements modify terrain features, usually increasing elevations and slopes, and agents of erosion attack those modified surfaces. Aggradational and degradational forces alternately build up and wear down in a kind of rhythm that never ends. Diastrophism, vulcanism, and deposition act to build up land surfaces, whereas an array of erosional forces tend to lower them.

The most efficient and effective erosive agent by far is running water. Running water accomplishes this erosion both when contained within channels and when moving off surfaces in sheets. Normally, running water is more efficient as an eroding agent when contained within a stream channel, although in certain circumstances, especially on denuded slopes, sheet flow may remove enormous quantities of surface material.

Figure 12.1
Headward erosion in an Oklahoma gully.
Jerome Coling.

Figure 12.2
The Yellowstone River is a very effective agent of erosion because of a steep gradient and significant volume of water.

Figure 12.3
The land between adjacent streams are the interfluves. This is the drainage basin of Hellroaring Creek, and the subdrainage basin of North Fork, Silver Creek, and Clover Creek in Gallatin National Forest.
From Carla W. Montgomery, Physical Geology. Copyright © 1987 Wm. C. Brown Publishers, Dubuque, Iowa. All Rights Reserved. Reprinted by permission.

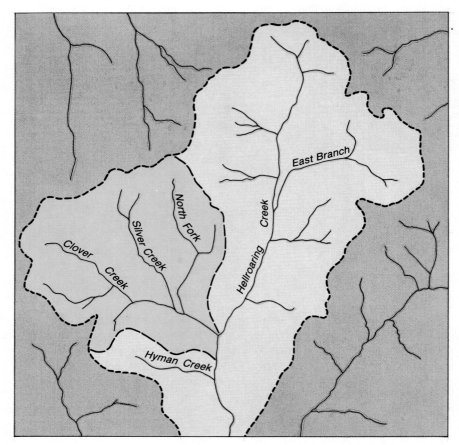

Running water is an important agent of erosion in virtually all climatological environments, even regions of considerable aridity, because periodic precipitation affects such regions. Areas covered by ice, which experience above freezing temperatures, have the subsurface and peripheral environments etched by the force of running water. The effects of running water are diminished when a close cover of vegetation is present. Erosional effectiveness is greatly enhanced when surfaces are denuded of vegetative cover. For example, steep selva-covered slopes in tropical rainy environments greatly reduce soil wash, whereas adjacent cleared slopes are quickly cut into steep-sided valleys with associated knife-edged interfluves, or areas between adjacent streams.

Running water usually produces small channels almost immediately in natural lowlands in response to slope on an exposed surface, although until such channels are developed, sheetflow is a significant erosive agent. These first small channels are known as **rills.** Subsequently, rills are enlarged into **gullies,** and the gullies develop into stream valleys. Upland areas between adjacent valleys are called **interfluves** (figure 12.3). For all intents and purposes, virtually all of the earth's surface is either a valley or an interfluve. Some interfluves are relatively narrow and others quite broad, reflecting the density of the existing drainage net. Some interfluves are heavily dissected, exhibiting the effects of numerous tributaries; whereas others are little affected, perhaps reflecting recent exposure to the ravages of erosion, the resistance of rock materials, or the lack of moisture.

Streams erode their beds by abrading the bottoms and sides with materials carried by the running water, which strike the confining walls of the stream bed, breaking up materials contained therein. A **stream gradient** is defined in terms of the vertical fall occurring in a specific horizontal distance. It is usually expressed in feet/mile in the United States, whereas in much of the world it is expressed in terms of meters per kilometer. Obviously, a stream with a 10 feet/mile gradient would flow at a much greater velocity than one with a fall of one foot/mile (figure 12.4). A swift and turbulent stream carrying a small load is more effective as an erosive agent than a slow-moving one with a heavy burden of sediment if the streams are moving over areas of comparable erodibility with similar volumes of water. Indeed, a stream's erosive capacity increases almost exponentially with the increase in velocity if other factors are essentially comparable. Early geomorphologists believed that stream erosive capacity might increase at a rate equal to the sixth power of the increase in velocity. Clearly, a number of factors in addition to velocity and turbulence affect the erosive power of streams (e.g., the nature of the rock materials through which the stream is flowing and the volume of water moving in a given stream channel). It seems certain that the sixth

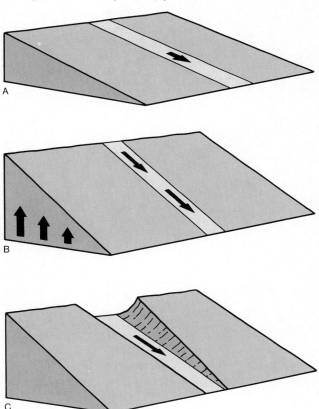

power increase in carrying capacity does not, in fact, occur because of frictional drag and asymmetrical stream channels. Most modern geomorphologists accept that the carrying capacity is proportional to the third power of mean current velocity. This latter figure is based on thousands of empirical measurements. The useful fact to remember is that equal volumes of rapidly moving water are much more effective erosive agents than comparable quantities of water moving slowly.

Stream discharge refers to the volume of water that flows past a particular place in a given unit of time. It is determined by multiplying the cross sectional area of a stream by its velocity (i.e., width × depth × velocity). It may be expressed in cubic meters [(m³)/second] or cubic feet [(ft³)/second]. The rate of discharge in most streams increases downstream because of additional volumes of water added by tributaries. The discharge may vary dramatically from wet to dry season. Exotic streams like the Nile may show such marked contrasts in discharge. The discharge of a river in flood may be fifty to one hundred

times greater than normal. The erosive effects of a stream may be multiplied many times by a significant increase in stream discharge.

Most streams increase their rate of discharge downstream, but a few do not. Intermittent streams in dry areas may have a high rate of discharge after a heavy downpour of rain, whereas a few miles away from an isolated storm, the stream might sink into the sand and disappear. Similarly, a stream fed by mountain snow might have a high rate of discharge, whereas a few miles away in a desert basin stream, flow might be reduced to a trickle because water volume has been reduced by infiltration and evaporation.

Streams move materials in several ways (figure 12.6). **Traction** is the actual rolling of materials along in the bed of a stream. Modest velocities allow the movement of only fine particles, whereas great velocities moving over a smooth rock bed might allow quite large materials to be moved along by traction. **Saltation** occurs when

smaller particles are literally bounced along a stream bed by turbulent water flow. Other very fine particles, like clay, are held in **suspension** and moved along; whereas certain other materials may dissolve and be transported in **solution.** Virtually all streams move materials in all these ways at all times.

When the velocity of a stream diminishes, either because the stream gradient has been reduced or because the stream has entered a quiet body of water such as a lake or a sea, materials are deposited. Similarly, as the load of a stream increases, often because of the addition of materials over a lengthy course or because of the addition of material from tributaries, it may be forced to drop a part of the suspended load or the load in solution.

Any stream varies in its erosion-deposition characteristics from place to place and from time to time. A stream that rises in rough, mountainous terrain may be a vigorous erosive agent near its headwaters where velocity and turbulent flow are at a maximum and where the suspended load is small; whereas that same stream may lose much of that erosive power in adjacent plains country as the velocity and turbulence are markedly reduced and the load has increased. Eventually, of course, as the stream empties at base level, it will deposit the load it has been carrying. **Base level** is the theoretical limit for erosion of the earth's surface. Base level slopes gently upward from the surface of the body of water into which the stream empties (figure 12.7). Theoretically, sea level should be the ultimate base level, but, in point of fact, it rarely is. Usually there are different base levels in response to rock materials as well as elevation.

Further, streams that may be able to carry an enormous load after heavy rains have greatly diminished capacities in periods of drought. Streams that exist in areas of highly seasonal rainfall patterns may be significantly different in erosive capacity from season to season or year to year. Obviously, knowledge about rainfall amounts, seasonality, and characteristics, derived from a study of meteorology and climatology, is important to an understanding of the relative significance of running water as an agent of erosion and deposition in a particular area. A growing body of evidence suggests that streams quickly gain an equilibrium condition in which the erosive power and load is matched to earth materials in the drainage basin, significant aspects of climate, and prevailing slope. That equilibrium may be changed if any significant aspect in the original equation, such as rainfall, for example, differs markedly from initial equilibrium conditions.

Figure 12.6
Material is carried in solution and suspension, is bounced along (saltation) and moves by traction along the streambed.
From Charles C. Plummer and David McGeary, Physical Geology, 4th ed. Copyright © 1988 Wm. C. Brown Publishers, Dubuque, Iowa. All Rights Reserved. Reprinted by permission.

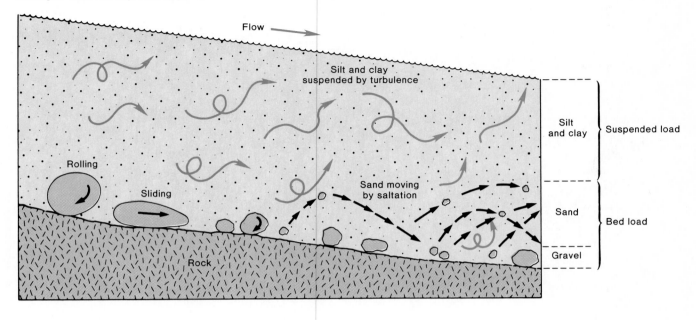

Figure 12.7
Base level may vary depending upon the lowest possible level of downcutting.
From Charles C. Plummer and David McGeary, Physical Geology, 4th ed. Copyright © 1988 Wm. C. Brown Publishers, Dubuque, Iowa. All Rights Reserved. Reprinted by permission.

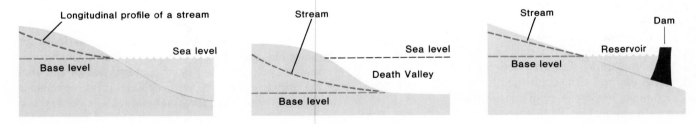

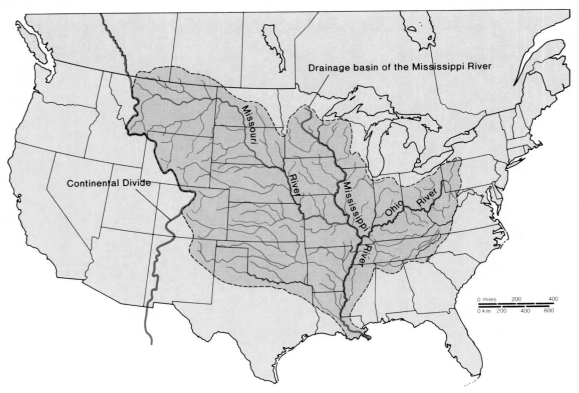

Stream Systems

As streams develop, an intricate array of tributaries form in response to the amount of precipitation received, the nature of the material through which the streams flow, and the characteristics and size of the **drainage basin,** or **watershed,** that serves the stream and associated tributaries (figure 12.8). The area between adjacent drainage basins is known as the drainage divide. North America, for example, has drainage divides that cause drainage to flow either to the Gulf of Mexico and Atlantic Ocean, the Pacific Ocean, or the Arctic Ocean.

The smallest stream in a drainage basin is known as a **first-order stream.** When two first order streams unite, a **second-order stream** is formed, and so on up a hierarchical ladder until the master stream in a given drainage basin is reached (figure 12.9). There are certain predictive elements attached to these associations: (1) there will be more first-order streams in a drainage basin than all other stream orders combined; (2) the stream length increases with stream order; (3) the size of watershed drained increases as the order of the stream; (4) the average gradient decreases with increasing order; and (5) the volume of water carried increases with increase in stream order.

Clearly, the network of streams varies greatly depending on precipitation amount, intensity, distribution, surface material, gradient, and similar factors. The **drainage density** may be determined by dividing the area of the watershed by the length of all streams contained within the same watershed. Another useful index of drainage characteristics is **stream frequency,** which is determined by dividing the number of streams by the area of the drainage basin. Humid regions with compact soils will obviously have a much higher drainage density and drainage frequency than desert regions with permeable soils, since abundant moisture falling on compact soils facilitates runoff, whereas minimum precipitation and permeable soils increases water infiltration. All sorts of variations between these two extreme examples are created by the nature of surface materials and the presence or absence of steep slopes. An increased drainage density and stream frequency will produce a finer grain to a particular piece of terrain as opposed to a rather coarser grain in an area where the density is much lower. Extremely fine-grained terrain is found in certain compact soil areas where hard showers produce an intricate erosional network. Extremely fine networks of hills and gullies deeply cut into compact surfaces produce **badlands** topography.

Figure 12.9
Stream relationships in a fourth-order watershed.
From Carla W. Montgomery, Physical Geology. Copyright © 1987
Wm. C. Brown Publishers, Dubuque, Iowa. All Rights Reserved.
Reprinted by permission.

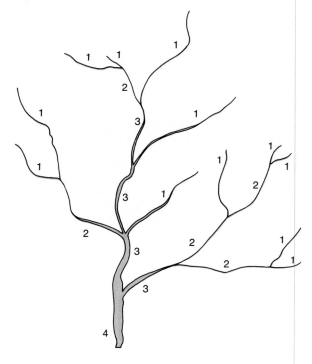

Figure 12.10
Badlands topography at Hell's Half Acre, Wyoming.

Streams and Structure

There are enormous differences in the character of streams, which have developed in response to the staggering array of structural forms, earth materials, and climatological regimes over the earth's surface. Nevertheless, streams have relationships to certain structural forms, which are repeated often enough to produce recognizable patterns, which are a part of the natural order in the landscape. These relationships are briefly explored in the paragraphs that follow.

Streams that follow the original slope of the land, often down dip, are known as **consequent streams** (figure 12.11). **Subsequent streams,** on the other hand, usually follow a zone or area of crustal weakness, frequently along the strike of rock outcrops and may enter consequent streams at a high angle. **Obsequent streams** are tributaries of subsequent streams and generally flow in a direction opposite to the initial consequent stream. Typically they are short, or at least shorter than most consequent or subsequent streams. **Resequent streams** flow in the same direction as the consequent stream, but they have developed on a slope at a different level from that originally responded to by the consequent stream. **Insequent streams** occur in a kind of aimless pattern of limited slope in areas such as those covered by glacial drift or in areas underlain

Figure 12.11
The relationships between consequent, subsequent, and obsequent streams in dipping beds of varying resistance to weathering and erosion.

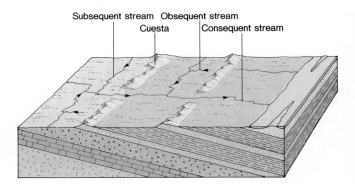

by permafrost, or they may be simply the initial stages of a consequent stream, which is developing in an area of limited slope.

An **antecedent stream** is one that has maintained its essential pattern in spite of the fact that land may have been uplifted across its direction of flow (figure 12.12). It has, in other words, been able to cut the land down as rapidly as such land has been lifted across its course. In certain instances, very convoluted courses have been maintained across uplift (figure 12.13). Deeply incised meanders may be indicative of such stream rejuvenation. Of course, such uplift must be at a rate not exceeding the ability of the stream to erode, or the initial pattern and stream course would be destroyed.

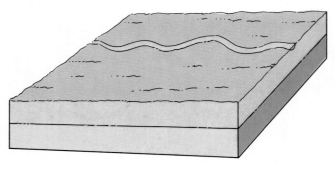

Superimposed streams began their erosive efforts on one kind of rock material, structure, and surface, and, over time, the stream and tributaries now rest on a stratigraphic environment very different from the one on which they formed initially (figure 12.14). The new surface is, in short, at variance to the one where the stream formed. It has, in cutting through one environment, superimposed itself on a different geologic structure. Streams respond to structure and slope, and principal streams and tributaries develop recognizable geometrical patterns. A description and analysis of common stream patterns follows. Stream geometry reveals a great deal about geologic structure.

Stream Patterns

Streams and their tributaries produce recognizable patterns when viewed from above, or when air photographs or maps are examined (figure 12.15). Probably the most common drainage type is the **dendritic.** Such a pattern develops in essentially similar earth materials with no extraordinary structural control. Tributary streams enter principal streams and main tributaries at an acute angle, and the overall pattern of streams and tributaries is much like the trunk and the branches of a tree. The pattern tends to become more intricate over time as tributaries develop to the maximum possible extent.

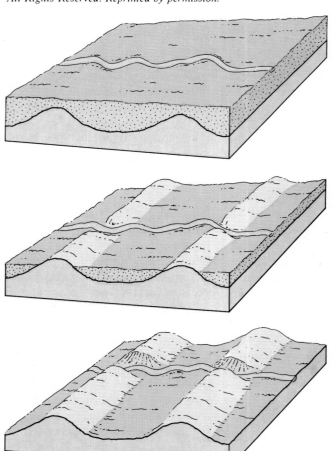

Figure 12.15
The geometry of some common stream patterns.

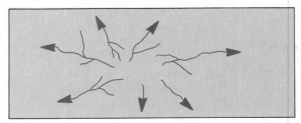

Radial Pattern—streams moving out from the center of a dome or volcano with fairly symmetrical slopes in all directions from center.

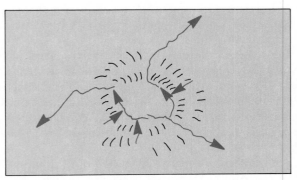

Annular Pattern—A pattern developing in an eroded dome. Original consequent streams eroded away the top of the dome and they have been joined by subsequent streams flowing inside the top of the eroded dome. In turn, the subsequent streams have been joined by obsequent streams flowing down the inward-facing escarpment of the eroded dome.

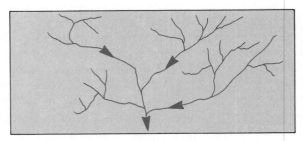

Rectangular Pattern—streams follow zones of weakness created by joint patterns in rock.

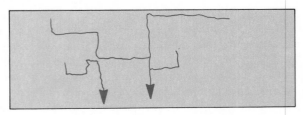

Dendritic Pattern—develops in areas of similar rock with no extraordinary structural control.

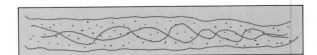

Braided Pattern—develops in a stream with a heavy sand burden. Stream deposits bars within its banks causing new channels to be cut.

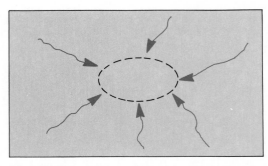

Centripetal Pattern—streams moving into a basin of interior drainage, which may be occupied by a playa lake.

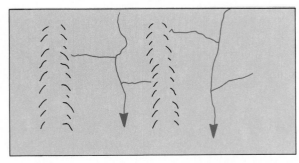

Trellis Pattern—often develops in areas of symmetrical folds or in alternating zones of soft and resistant rock.

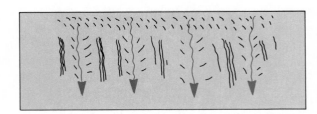

Parallel Pattern—streams flowing down a newly created slope with very symmetrical gradient.

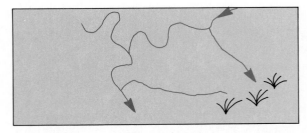

Deranged Pattern—Original stream pattern has been modified by blocking of original courses or interruption of initial gradient by unusual deposition (e.g., glacial deposition).

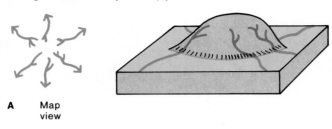

A   Map
        view

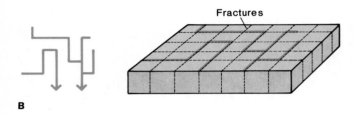

Fractures

B

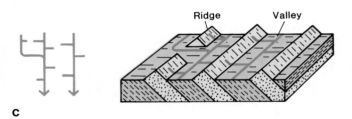

Ridge   Valley

C

Figure 12.17
*Braided stream pattern on the Canterbury Plain of New
Zealand.*
New Zealand Tourist and Publicity Office.

Where a variety of structural controls exist, drainage patterns develop, which give a major clue as to the type of earth materials and/or geomorphological features that exist. For example, a **trellis drainage pattern** often develops where there are differential zones of rock resistance or in an area with repetitive anticlinal and synclinal folds. Consequent streams flowing downdip are met by subsequent streams flowing along the strike at almost right angles, and obsequent and resequent streams enter subsequent tributaries at similar high angles.

The **rectangular drainage pattern** appears similar to a trellis pattern when examined in plain view except that streams flow along fault zones or joints that intersect at almost right angles. Rock strike and dip are of less significance, and joint or fault patterns in rock exercise drainage control.

An **annular drainage pattern** develops, usually on a dome where consequent streams have removed more resistant rocks near the crest of the dome, allowing subsequent streams to follow a curving pattern along weaker rock exposed below. Obsequent streams flow to subsequent streams along the exposed face of the eroded inner margin of the dome.

A **radial drainage pattern** occurs when streams flow down the slopes of a dome of essentially homogeneous material, or out from the center of a volcanic cone in a radiating pattern from a high central point (figure 12.16). Streams flowing down into a basin from a number of directions produce a **centripetal drainage pattern.** Centripetal patterns occur most often in basins of interior drainage.

In areas of similar slope over long distances, a series of small streams may develop essentially parallel to each other. Such a pattern is called a **parallel drainage pattern.** A parallel pattern may be the first to evolve on a newly exposed steep slope. In areas of very low relief or in areas with a heavy sediment burden, stream patterns, if they exist, may be so distorted or unrecognizable as to be termed **deranged,** or **aimless.**

In subhumid or semiarid areas, streams may deposit materials in their own beds, and break up into two or more channels, perhaps rejoining downstream within the confines of a main river channel. Such a pattern is said to be **braided** (figure 12.17). It obviously differs from the other patterns described, since the braiding occurs within the channel of the master stream. Braiding usually develops when a high percentage of the stream's burden is sandy in character. Braiding is less common when the stream's burden is principally clay or silt. Streams burdened with heavy loads of silt and clay tend to meander in low gradient segments. Braided streams usually have a dendritic tributary pattern.

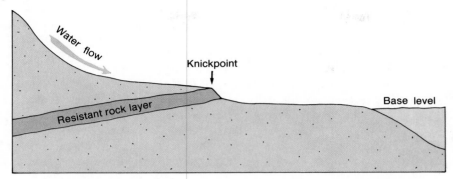

Near their mouths, streams may break up into a series of distributaries. These distributaries frequently splay out from the main stream like the frayed ends of a rope. In a sense, the pattern is dendritic in reverse. Streams branch out from a main stream, rather than coalescing to form a main stream.

The patterns of streams and tributaries or distributaries affect the erosion of interfluves or the patterns of deposition in deltas or alluvial fans. Like the wrinkles in a human face, the pattern of stream valleys adds character to the landscape.

Streams and Their Valleys

Streams exhibit several characteristics at different stages of development and in different segments of their courses. A stream tends to lengthen itself headward, because weathered materials slump into the valley. The effect of flow is to cut a valley downward and, primarily because of turbulence and the Coriolis effect, laterally as well. In its early stages, a stream tends to be **ungraded** (i.e., it is interrupted by a series of rapids or falls, and an irregular gradient). Over time, as those irregularities are smoothed out, the stream profile assumes a slightly concave shape from the mouth upwards towards the headwaters. This is by no means a symmetrical curve, and the concavity may in a real stream have a number of irregularities. In fact, these irregularities are normal occurrences.

For example, even in a stream that is mainly at grade, small rapids or waterfalls may intrude at exposures of very resistant rock. These interruptions, known as **knickpunkte** by the Germans, or **nickpoints** in English, have represented points of argument about a real uplift or stream rejuvenation. Although it is difficult to generalize, it seems likely that such areas are most often exposed edges of resistant rock, which have petrologic rather than uplift significance (figure 12.18).

A graded stream maintains a delicate balance between stream flow and the load of sediment available to it. In a graded stream, there tends to be a great deal of lateral cutting and deposition, whereas the downcutting is diminished. As a stream seeks to reduce grade to base level (a very gentle upward sloping line extending from sea level inland, which would just maintain enough slope for a reasonable flow), it cuts down, and the stream and its tributaries plane laterally. A stream also ultimately lengthens itself near the mouth as well, as it flows across its lengthening deltaic deposits, usually through a system of distributaries.

The Geomorphic Cycle

The history of the development of streams has been described in a variety of ways, but a continuing useful concept is one developed by an American scientist, William Morris Davis, in the last part of the nineteenth and first part of the twentieth century. What he termed the geographical cycle has been, over time, termed the **geomorphic cycle,** or geologic cycle (figure 12.19). Concepts that Davis developed dominated thinking about stream erosion and deposition until about three decades ago. Certain conceptual aspects of his theory remain valid interpretive tools, although the theory has been attacked for its general lack of quantification and development as well as for the difficulties in observing certain terrain features that he postulated, like peneplains.

The basic premise of the Davis theory was that newly uplifted land was immediately attacked by agents of weathering and erosion, and over a protracted period of time, the land was degraded and planed off to yield a lower surface almost at grade. According to the geomorphic cycle he proposed, a surface exposed in an initial uplift might be well above sea level, but because there had been very

Figure 12.19
The Davis geographic (geomorphic) cycle showing sequence from (A) initial, (B) youthful, (C) mature, and (D) old age stages.
From Charles C. Plummer and David McGeary, Physical Geology, 4th ed. Copyright © 1988 Wm. C. Brown Publishers, Dubuque, Iowa. All Rights Reserved. Reprinted by permission.

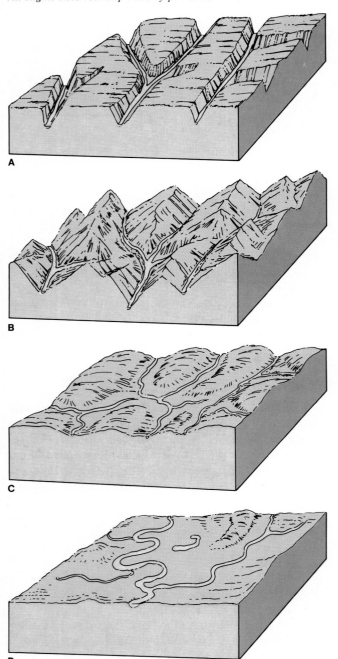

little time to operate, the surface would be little affected by erosion. The **initial stage** had not yet developed streams of significance. It was, in effect, a new landscape waiting for erosion to begin. Of course, all surfaces are subject to the effects of weathering and erosion as soon as they are exposed, but in the initial stage, the land, according to Davis, had been minimally influenced by weathering and erosion. Whatever such a surface is called, it's obvious that diastrophic forces are constantly changing the landscape and, just as such forces effect change, they are precursors of further change.

The **youthful stage,** or youth, saw the development of consequent streams and the beginnings of subsequent streams. The drainage net was not yet well established, and the streams were characterized by swift currents and a period of vigorous downcutting with broad, relatively flat, minimally dissected interfluves separating adjacent streams. Falls and rapids marked the streams' courses, and the valley profile was V-shaped, since lateral planation had had little opportunity to produce a floodplain, and slope wash, although it existed especially along valley sides, was not yet developed to its maximum potential extent. The streams, at this stage, were ungraded. A great deal of downcutting remained to be done before such streams would be graded.

As these streams exhibited characteristics of **maturity,** the principal streams continued to cut downward, but a great deal of lateral movement of the stream along with significant valley slope wash resulted in the development of a broad valley and floodplain. Falls and rapids had long since disappeared, and the streams meandered back and forth across the floodplain. Tributaries developed to their maximum extent, and the interfluves were heavily dissected. Lateral planation had produced a significant floodplain, and the valley profile was a modified U-shape. The streams had become graded.

In **old age,** the landscape was reduced almost to base level. Tributaries had consolidated and, although less in number than in maturity, were more numerous than in youth. A few major streams meandered across the countryside in almost featureless valleys. Valley profiles as the result of significant slope wash were broad with gentle slopes extending up to the interfluves. The streams were graded.

The end product of the geomorphic cycle, according to Davis, was an almost featureless plain near base level, which he called a **peneplain.** He recognized that there might be certain knobs of very resistant material that would break the surface of the peneplain. He called these features **monadnocks.** Early geomorphologists spent an inordinate amount of time in search of these peneplains, and certain attempts to identify peneplain surfaces bordered on the ludicrous. In mountainous areas, especially, geomorphologists were fond of identifying numerous peneplain surfaces, which they believed indicated cycles of erosion and renewed uplift. In a number of areas almost any broad upland surface was in danger of being identified as a portion of a peneplain.

In the Davis scheme, the whole process could be renewed if the land was uplifted at the completion of the cycle or at some time within the cycle. This process, called **rejuvenation,** would renew the cycle; hence, the justification for finding a number of peneplain remnants, especially in mountain or hilly terrain.

Streams certainly exhibit many of the characteristics described by the geomorphic cycle. Indeed, a given stream may, over the course of several hundred miles, be "youthful" near its headquarter, "mature" in mid-course, and "old" near the sea. A vigorous mountain stream, which has all the characteristics of youth in most of its course, as described by Davis, can meander and show many characteristics of maturity in a high mountain meadow of low relief. A sluggish stream may recapture periods of renewed vigor during periods of flooding or upon exposure of resistant rock at nickpoints. In short, the terms perhaps best describe characteristics rather than temporal sequential developments. Perhaps the selection of terms like youth, maturity, and old age was unfortunate, especially in light of the fact that no fixed time period accounted for differences in stream characteristics.

Because peneplains are not found in nature and because of other shortcomings of the Davis concept, a variety of challenges to his theory have been issued at different times. In fact, the geomorphic cycle cannot be demonstrated. Nevertheless, the characteristics applied to streams in various stages by Davis are useful descriptions, especially for the student of physical geography. As a matter of fact, at one place or another, streams exhibit all of the characteristics that Davis attributed to developmental stages. William Morris Davis has remained a geomorphological giant in spite of attacks by his contemporaries and present-day critics. His ability to compass the scope of change and his efforts to bring coherence of landscape analysis are legacies of inestimable value to us. Indeed, his breadth of vision has been rarely matched by his contemporaries or his successors.

The Equilibrium Theory

It is fair to say that the **equilibrium theory** dominates geomorphological thinking in the last part of the twentieth century. Basically, this theory holds that slope forms and characteristics adjust to geomorphic processes so that there is a balance of energy. That is, the energy provided is just enough for the work to be done. In general, it can be stated that softer rock develops gentler relief and slopes, and

Figure 12.20
Data seem to show that valley profiles remain nearly the same with a parallel retreat over time. (Note that this runs counter to the Davis theory.)
From Charles C. Plummer and David McGeary, Physical Geology, *4th ed. Copyright © 1988 Wm. C. Brown Publishers, Dubuque, Iowa. All Rights Reserved. Reprinted by permission.*

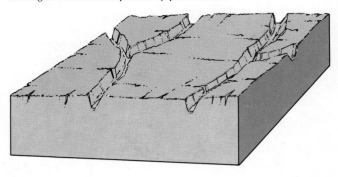
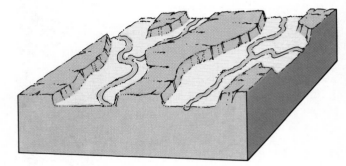
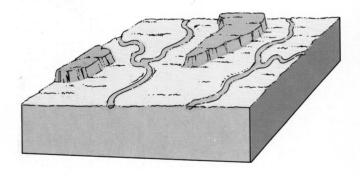

harder rock develops greater relief and steeper slopes. The theory suggests that valley profiles tend to continue in about the same form over time, unlike the Davis postulation that valley slopes become gentler over time (figure 12.20). The equilibrium theory suggests that there is a kind of parallel retreat of existing slopes, which tend to persist in about the same form. There really is no adequate way to test either hypothesis in nature, since no one has been able to observe valley profiles for a long enough period of time. Because of the enormous span of geologic time, which may be involved in landscape development, and because of the short time compassed by even generations of geomorphologists, the theoretical models produced, the computer simulations developed, and the mathematical expressions used cannot provide the ultimate empirical testing, which only the forces of nature operating over protracted periods can provide. The preponderance of opinion tends to support the hypothesis that valley profiles tend to remain the same over protracted periods.

Rough terrain in an area where uplift is occurring, for example, maintains similar slopes and characteristics as erosion counterbalances the effects of uplift. Equilibrium conditions are maintained unless there is some diastrophic interruption or some major modification in stream flow induced by climatological change. Obviously, if the rate of uplift or the rate of erosion changes, there is a disjuncture until a new state of equilibrium is reached.

Such disjunctures could occur relatively quickly in a specific environment. For example, if an area is denuded of its vegetation quickly, as in the case of a fire or logging activities, erosion will be rapidly accelerated, and equilibrium will be disturbed. Similarly, if there is rapidly accelerated uplift, as in the case of a volcanic eruption, disequilibrium exists. In such circumstances, a new equilibrium must be established. That new equilibrium fits the new slopes and materials developed by the volcanic eruption, or it adjusts to the additional sediment burden that exists on denuded slopes. An event like the eruption of Mount Saint Helens (described in an earlier chapter of this book) clearly interrupts one pattern of equilibrium and sets into motion the establishment of a new one. The equilibrium theory has the persuasiveness of logic about it. It seems to make sense, and localized and laboratory experiments tend to corroborate at least parts of it.

The equilibrium theory is by no means a panacea, however, for it deals less effectively with tectonically stable areas or with regions with limited stream flow. The patterns of streams do seem to change, and slope contours and profiles are modified in areas of tectonic stability and climatological constancy. Further, it fails to deal effectively with landform evolution. In short, a unified theory that seems to compass all situations effectively has yet to be developed.

The nature of slopes, whether convex, concave, or some combination of the two, is quite significant to equilibrium theory geomorphologists. Convex slopes predominate, although concavity can be dominant in regions of earth slides, near the headward margins of mountain glaciers, and along slopes of composite volcanoes. As we have already seen, the profile of a stream from mouth to head typically produces a curve that approaches concavity. Some mesas and buttes have planar top surfaces, and their marginal slopes are typically concave. Geometric forms help to explain evolutionary changes in landscape, but slope characteristics are inadequate to explain the development and evolution of all landforms.

The dynamism of a constantly pulsing earth with enormous plates moving over the surface of viscous rock, the ebb and flow of climatological cycles, the certainty of catastrophic events, and the increasingly potent force of human activities place never-ending challenges before the scholar seeking to bring order out of conditions that are often chaotic. The search to establish order and to gain predictability is at once frustrating, fascinating, and fruitful. The interconnectedness of various elements of the landscape are demonstrated repeatedly.

These interconnections become especially noteworthy at the micro levels of interpretation, although they are certainly prevalent at all levels of geomorphological investigation. Landforms at the micro level are especially interesting, since a scholar can compass the entire landform personally both in terms of seeing the entire feature from a few feet or inches away, and observing the construction and destruction of features in a few hours or days. Sand tables in laboratory settings provide simulations of certain geomorphological events. Conversely, in landforms of ordinary size, one is always observing a part of the whole. Although remotely sensed images, maps, and extensive field work tend to mitigate the problem, the reduced scale masks details that might prove to be significant. Remotely sensed images, especially, have provided numerous opportunities to make important regional connections and associations, which often escape the interpretive eye of scientists using more conventional investigative techniques.

Certain elements affecting aspects of the physical environment appear to be chaotic. Turbulent flow in atmosphere, water, or mantle has placed enormous challenges before earth and atmospheric scientists. The chaos engendered in these and other circumstances has resisted past attempts to develop order in process or form; however, a new generation of scientists who are studying chaos are beginning to find order in some processes and events formerly considered to be disordered. Their sophisticated mathematics and high-speed computer programs hold the prospect for developing a greater understanding of earth and atmospheric processes and forms. In the meantime, there is much that we do know, and it is appropriate to get on with the business of analyzing and synthesizing those elements of the environment that we do understand.

Characteristics of Stream Erosion and Deposition

We have seen how streams erode and carry material, lengthen themselves, and broaden their valleys, while tributaries and slope wash attack interfluves. Certain features developed in the erosion (degradation) and deposition (aggradation) processes are worthy of closer scrutiny. Indeed, these forms dominate in most present expressions of surface landscapes. Landscape surfaces are most often the product of protracted periods of erosion and deposition.

Figure 12.21
Stream piracy has occurred here. The stream flowing to the right has eroded more rapidly than the one to the left, resulting in its capture and flow reversal.
From Charles C. Plummer and David McGeary, Physical Geology, 4th ed. Copyright © 1988 Wm. C. Brown Publishers, Dubuque, Iowa. All Rights Reserved. Reprinted by permission.

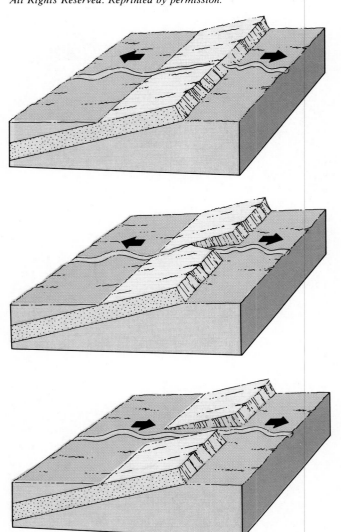

Figure 12.22
Stream piracy also occurs when some of the drainage on the left side of the divide has been diverted to the right.
From Carla W. Montgomery, Physical Geology. Copyright © 1987 Wm. C. Brown Publishers, Dubuque, Iowa. All Rights Reserved. Reprinted by permission.

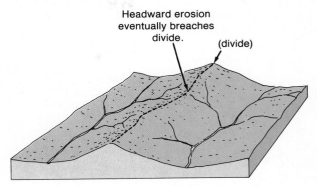

Headward erosion eventually breaches divide.

(divide)

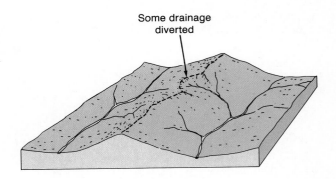

Some drainage diverted

When tributaries are actively developed, one may lengthen much more rapidly than an adjacent stream and actively cut into the adjacent valley. The actively eroding stream may capture the headwaters of the other in an act of **stream piracy** (figure 12.21). The stream that has lost its headwaters has been **beheaded.** The remnant lower portion of that stream now occupies a valley disproportionate in size to the reduced volume of the beheaded stream (figure 12.22). Such a stream is termed **misfit** because it does not appear to be appropriate for the valley it occupies (i.e., it obviously doesn't have the present capacity to have produced the valley it occupies). A stream too small for the valley it occupies is an **underfit** stream.

Streams that have cut through topographic barriers produce **water gaps.** These valleys have often provided routes for pioneering, exploration, and surface transportation. Where such water gaps are no longer occupied by a stream because of stream capture or tectonic shifts, the notch, or valley that remains is termed a **wind gap.** Both water and wind gaps are quite common in the Appalachians, and both features served the westward migration of Americans well since they provided easier access routes across mountain barriers. Indeed, famous routes of westward migration through the Appalachians (e.g., the Cumberland Gap) were produced by the erosive power of streams.

Streams may have well-developed floodplains and an uplift(s) may occur. In such a case, a stream that was involved principally in lateral planing and deposition may suddenly be rejuvenated (i.e., begin to downcut rapidly again). When this happens, remnants of the floodplain level are left as relatively flat benches known as **terraces.** Many streams will exhibit several terrace levels illustrating, graphically, several periods of rejuvenation. Terraces may be of several types (i.e., they may be benches cut in rock, sometimes sediment veneered and sometimes

Figure 12.23
Sequence in the development of paired stream terraces: The stream (A) has deposited sediments in the floodplain. The river (B) has cut to a lower level leaving former floodplain surface as terraces. Lateral erosion (C) develops a new floodplain below terrace levels.
From Charles C. Plummer and David McGeary, Physical Geology, *4th ed. Copyright © 1988 Wm. C. Brown Publishers, Dubuque, Iowa. All Rights Reserved. Reprinted by permission.*

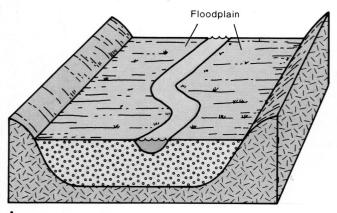

Floodplain

A

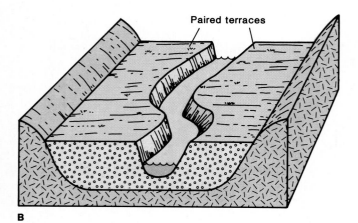

Paired terraces

B

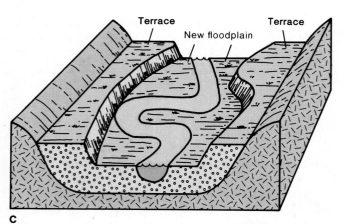

Terrace   New floodplain   Terrace

C

Figure 12.24
Unpaired terraces may result when downcutting and lateral erosion occur simultaneously.
From Charles C. Plummer and David McGeary, Physical Geology, *4th ed. Copyright © 1988 Wm. C. Brown Publishers, Dubuque, Iowa. All Rights Reserved. Reprinted by permission.*

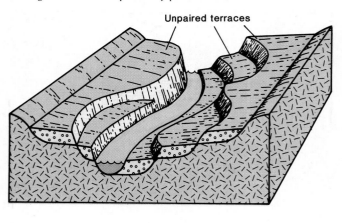

Unpaired terraces

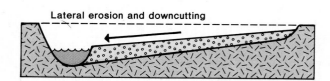

Lateral erosion and downcutting

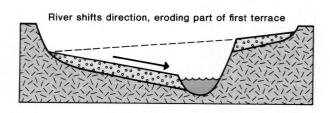

River shifts direction, eroding part of first terrace

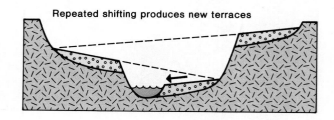

Repeated shifting produces new terraces

Figure 12.25
Terraces along the Snake River in Wyoming.

Figure 12.26
Portion of map showing meandering, ox-bow lakes, and meander scars.
Portion of Menan, Buttes, Idaho, 7 1/2 Quadrangle, USGS.

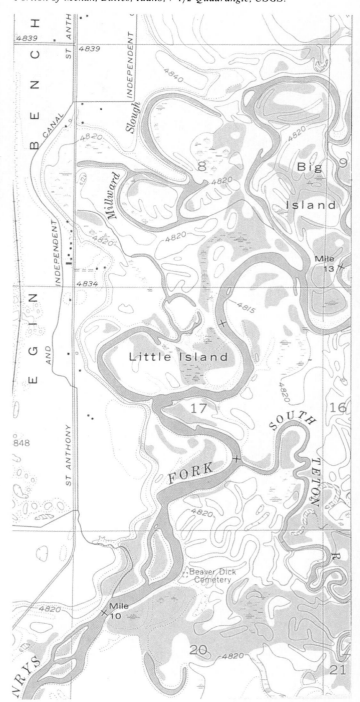

not, or they may be alluvial benches left as a floodplain remnant above the existing floodplain as the stream resumed downcutting). The terraces may be paired (at the same level) or they may be unpaired (figures 12.23 and 12.24). Because of lateral planation and the constant reworking of flood-plain sediments, they are usually unpaired.

In addition to regional uplift, which increases stream gradients and erosive capacity, other mechanisms may account for a shift from lateral planation and deposition to vigorous downcutting again. A change in climate to more pluvial (wetter) conditions can increase a stream's erosive capacity. Those issuing as meltwater in glaciated regions may have their erosive capacity enhanced by a warming period, which increases glacial melt and water volume in a stream.

Removal of vegetation through purposeful clearing, or as the result of biological destruction or fire may greatly alter the sediment burdens of streams. This disequilibrium will, of course, set new responses in motion, working towards new equilibrium conditions. Streams receiving a significant new sediment burden may have that erosive capacity drastically altered.

Streams with heavy loads of sediments that meander across broad floodplains often produce a variety of land or water forms, which have resulted largely from deposition. The twisting of a river in a series of intricate meanders, or loops, may result in such a meander being cut off by the effects of lateral erosion and deposition. Eventually these meanders may be separated from the main river entirely, producing **oxbow** or **horseshoe** lakes (figure 12.26).

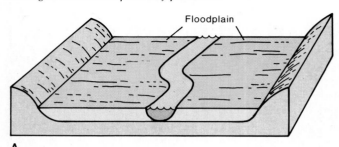

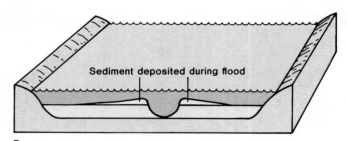

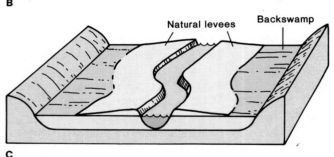

As indicated previously, streams carrying loads of fine
sediments like clays and silts tend to meander, whereas
those carrying coarse sediments like sand tend to exhibit
a braided pattern. Since stream velocities are greatest on
the outside bend of a meander, erosion is characteristic
along those banks (figure 12.27). At the same time, the
inside portion of the bend will have a reduced velocity, and
deposits will often build in the inside bends. Meanders migrate back and forth in intricate whorls and loops across
the floodplain in response to shifting periods of erosion and
deposition. Aerial photographs, especially, reveal the scars
left by long-filled oxbow lakes. Vegetation may vary in response to different sediment and drainage conditions. The
patterns of the moment are inevitably altered by the constant working and reworking of floodplain sediments.

Heavily laden rivers, when they are forced out of
their banks during floods, drop significant loads of sediment along the stream bank to produce small ridges known
as **natural levees** (figure 12.28). When water levels drop,
areas a few hundred feet away from the river may be lower
than the crests of the natural levee. Many such areas
remain swampy because there are no good drainage connections to the river, and very fine silts or clays and a high
water table may inhibit downward percolation of water.
Such swampy areas are called **back swamps.**

In some areas where subsequent streams developed
at or adjacent to a river floodplain, the natural levees may
be formidable barriers to the entry of the tributary into
the main stream. In these cases the tributary stream may
parallel the master stream in the floodplain for considerable distances before the tributary ultimately enters the
master stream. Such a stream is known as a **Yazoo tributary.** Since they usually flow in areas of low gradients,
Yazoo tributaries tend to have a tortuous course similar
to the master stream to which they are tending. As in so
many instances, the geomorphological feature takes its

name from the prototypical example—in this case, the
Yazoo River of Mississippi, which enters the Mississippi
River near Vicksburg.

In a few areas where man has augmented natural
levees with artificial levees, rivers may be substantially
above the level of the floodplain. Streams in such regions
with low gradients are typically heavily laden and must
deposit some of their load within their own river bed. The
Hwang Ho of north China is such a river. When the dikes
are breached, floodwaters may inundate enormous areas.
Areas as large as Texas have been flooded in certain catastrophic outpourings of water from the Hwang Ho, which
flows with a considerable hydrostatic head of pressure.

Streams with high natural levees, or artificial levees,
or both, may be slow to return to the confines of their banks
after a flood. The problems of resettlement of displaced
people and the difficulties of cleanup are magnified in such
circumstances. The undesirable effects on communities,
transportation links, and agricultural pursuits are obvious.

Table 12.1
Property Loss in Floods (Billions of Dollars)

1974	1975	1976	1977	1978	1979	1980	1981	1982	1983	1984	1985	1986
0.6	1.1	1.0	1.4	1.0	4.0	1.5	1.0	2.5	4.5	3.3	3.1	2.0

Source: NOAA

Human interference with river courses through deepening or straightening, although well intended, often add to the frequency and severity of flooding. Harmful effects on the water table may also occur.

Erosional Forms that Have Developed in Response to Structural Control

Large areas of the earth's surface are underlain by sedimentaries that have dips, which vary from the gentle to the quite steep. These structures, called **homoclines,** can influence the development of certain interesting erosional landforms, especially if they are composed of alternate layers of weak and resistant rock.

For example, in areas such as the Gulf Coastal Plain, where alternate weak and strong rocks dip towards the sea, forms called **cuestas** may develop. A cuesta is marked by a relatively sharp escarpment where the edge of resistant rock is exposed and stands above the adjacent belt of weak rock, which has been eroded to a lower level by a subsequent stream. The steep escarpment, which faces inland in the case of the Gulf Coastal Plain, is contrasted with the gentler slope downdip. A number of escarpments at cuesta fronts, essentially parallel to the coast, produce a series of undulations with intervening lowlands as a part of a so-called **Belted Coastal Plain.**

Cuestas are not restricted to coastal plains, and are frequently found far in the interior. Niagara Falls, for example, falls over a cuesta escarpment. The margin of the Allegheny Front marks a cuesta edge.

When rocks have a very steep dip, generally in excess of 45°, erosional activity may produce an almost knife-edge ridge. Such **hogback** ridges are frequently found at the margins of major mountain uplifts (figure 12.29). Good examples may be found along the flanks of the Front Range in Colorado.

Domes that have had their crests eroded, revealing alternating strata of weak and resistant rock may produce cuesta-like escarpments with steep edges facing in towards the center of the dome and gentler slopes away from the center. The Nashville Basin in Tennessee has such an array of in-facing escarpments.

An antithetical situation occurs in a structural basin when alternating weak and strong strata may present steeper outward-facing escarpments with gentler slopes downdip towards the basin center (figure 12.30). The Paris Basin exemplifies such a geomorphological area.

Figure 12.29
Hogbacks in Garden of the Gods, Colorado.

Figure 12.30
Structural dome (A) *and structural basin* (B).
From Charles C. Plummer and David McGeary, Physical Geology, *4th ed. Copyright © 1988 Wm. C. Brown Publishers, Dubuque, Iowa. All Rights Reserved. Reprinted by permission.*

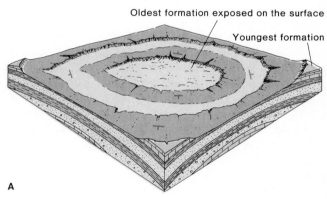

Oldest formation exposed on the surface

Youngest formation

A

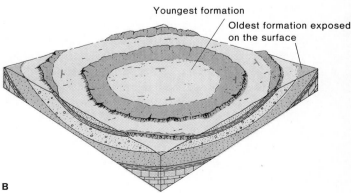

Youngest formation

Oldest formation exposed on the surface

B

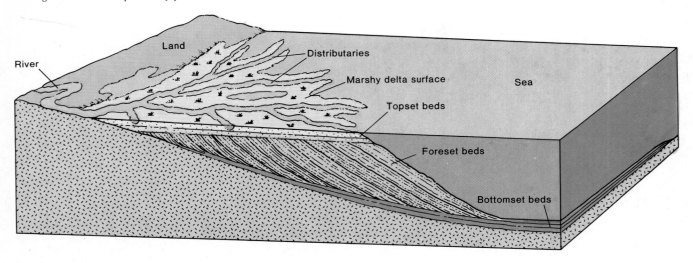

Other erosional forms, some prosaic and others bizarre, combine to give character to the face of the land. The erosional products that are subsequently deposited add another dimension to the modification of the landform surface of the earth. Some major depositional forms will be considered in the section that follows.

Deltas

As streams enter the quiet waters of a sea or lake and leave the confining channel(s) of the stream, the load of sediment is dropped. The stream may extend some measurable current and sediment load well out into the quiet body of water—most do. The Amazon, for example, is clearly observable in terms of current and load scores of miles out to sea. The turbid waters are readily seen on aerial photographs or satellite images taken at great heights.

The materials dropped to the bottom of the lake or sea almost immediately produce a layer of sediment, which is called the **bottomset bed** (figure 12.31). As the stream pushes material in a steady extension of deposits on the outer slopes of deposits, the **foreset beds** are deposited. The very fine materials, including the smallest particles of silt and clay, which settle last, produce the **topset beds** of sediment. The nature of sediment deposited and the attitude of strata deposition can be observed, especially, when cross-sectional trenches are cut through them. The alluvial deposits built into the sea in this way produce a landform, usually composed of fine-grained materials, known as a **delta.** The name comes from the Greek letter, Delta, and was logical nomenclature for Greeks to use, since the largest delta within their ken was that of the Nile, which happened to have a form generally like the Greek letter Delta (Δ).

Delta Forms

Several major types of delta have been produced and described (figure 12.32). The **arcuate,** or fan-shaped, delta is the type represented by the Nile. The river breaks into two main channels near Cairo, and several lesser distributaries spread out fan-like from the main river. Each stream aggrades its bed and deposits material between adjacent distributaries in times of flood, creating a vast low-lying area of land that is biologically highly productive. Distributaries extend themselves across the expanding delta as new deposits add to the surface area (figure 12.33). Like many deltas, the Nile is a fertile agricultural region where deposits of silt renew fertility and subsurface movements of water through soil and/or surface irrigation, when necessary, combine to produce abundant crops.

Figure 12.32
The geometry of delta form depends upon the load and type of sediment carried, wave and current action, and other factors. The Nile is an arcuate delta, the Mississippi is a bird's foot delta, the Niger is a modified cuspate delta, the Ganges-Brahmaputra is the compound form, and the Escambia is the estuarine form.
A, B, and C are from Charles C. Plummer and David McGeary, Physical Geology, 4th ed. Copyright © 1988 Wm. C. Brown Publishers, Dubuque, Iowa. All Rights Reserved. Reprinted by permission.

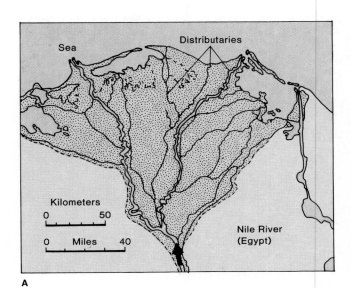

A

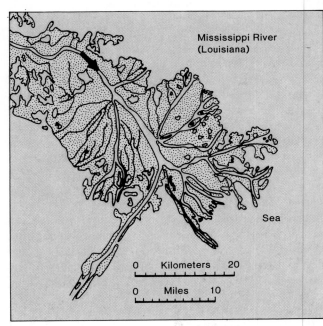

B

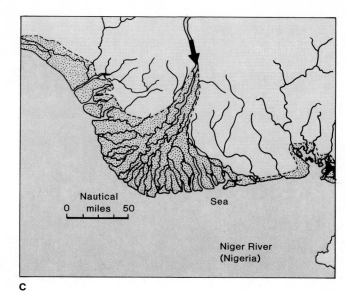

C

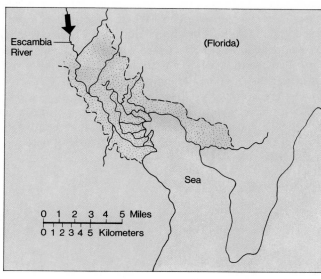

D

E

The **bird's foot,** or digitate, delta develops when major distributaries build deposits adjacent to them, but open water interdigitates between these finger- or toe-like projections. This form is facilitated by dredging of principal stream channels. The Mississippi River has produced such a delta form in the Gulf. The enormous load of sediment has actually caused some subsidence of the Gulf floor, so that only the natural levees adjacent to the principal distributaries stand above sea level. The sediment may eventually fill the interstices creating a new delta form, although the antithetical factor of seafloor subsidence caused by deposited load and continued dredging militates against filling above sea level between major distributaries.

A **compound delta** exists where two major streams have co-mingled distributaries and deposits as they've entered the sea. Such an enormous delta has been developed by the Ganges and Brahmaputra Rivers in India and Bangladesh. The seaward margin, known as the Sundarbans, is an intricate array of distributaries and tidal creeks marked by the steady buildup of new land assisted by the intricate network of mangrove roots, which support one of the largest mangrove areas in the world.

Enormous areas of wetland characteristic of deltas are very important in the web of life around the world. Such wetlands are among the most productive of habitats, supporting, as they do, many vital elements in aquatic and terrestrial food chains. Many are threatened as human developments in such regions encroach on the nurturing wetlands.

Estuarine deltas develop where former drowned river mouths are steadily filled by a stream and distributaries depositing materials and filling the drowned area. The Escambia River of northwest Florida and the Mobile River of Southern Alabama, like a number of streams entering the northern Gulf of Mexico, exhibit this type of delta form. Such estuarine wetlands are precious resources in providing habitat for a host of aquatic and terrestrial creatures. They should be protected at all costs since their impact as nurseries for all kinds of creatures inhabiting the sea is very significant. Without them, life in the sea and many creatures dependent on such life would be impoverished.

Cuspate deltas may develop as the result of wave action modifying the form along its seaward margin. The Niger River of west Africa, the Ebro River of Spain, and the Rhone River of France exhibit the cuspate form to varying degrees. In a sense, several deltas exhibit characteristics of both arcuate and cuspate form. This is true of the Niger, which has had its seaward margins only moderately influenced by longshore currents.

Deltas, as low-lying, relatively flat features, are often areas where agriculture is significant. This is not surprising, since the fine-grained materials are usually fertile topsoil eroded from upriver sources. Obviously, drainage and flooding may create problems for the farmer. Some delta regions are very heavily settled because of their fertility, which is constantly being renewed. For example, the Nile Delta of Egypt, the Ganges–Brahmaputra Delta of India–Bangladesh, and the Hwang Ho Delta of China are very thickly populated. Some delta regions, notably the Nile and the combined delta of the Tigris-Euphrates, have nurtured major civilizations for thousands of years. As population expands, especially in urbanizing Western societies, there is a constant push of people towards the coastal margins to make use of coastal attributes. The temptation to drain deltaic wetlands for housing areas or commercial development is high. The risk in disturbing the wetland web of life is equally high. The author has observed steady encroachment on the coastal margins of northwest Florida in the last two decades. The negative effects on habitat have become obvious with reports of fish kills, with increases in coliform bacteria above acceptable levels, and with increasing turbidity of coastal and estuarine waters.

Running water offers opportunities and creates problems for people. The landforms created give character to the face of the land. Other agents of erosion and deposition will be discussed in subsequent chapters.

Study Questions

1. Explain the sequence of stream development on a newly exposed slope.
2. Explain how streams transport materials and contrast the various methods of transport.
3. Explain the concept of base level.
4. Draw a pattern of streams that might exist in a fourth-order stream system.
5. Define the following terms:
   a. consequent
   b. subsequent
   c. obsequent
   d. resequent
   e. insequent
6. Compare and contrast antecedent and superimposed streams.
7. Draw the principal types of drainage patterns.
8. Describe the geographical (geomorphic) cycle of stream development as developed by William Morris Davis.
9. Explain the concepts of the peneplain. What are the principal weaknesses of the concept?
10. How may rejuvenation occur?
11. What are the basic tenets of the equilibrium theory?
12. Explain the concept of stream piracy.
13. What is a Yazoo tributary?
14. What is the difference between cuestas and hogbacks?
15. Explain the different circumstances in which beds of materials are deposited in deltas.
16. List the several types of deltas and give examples of each type.
17. Explain why human settlement in coastal and delta regions may prove to be a threat to the environment.
18. What is a monadnock?

Selected References

Bloom, A. L. 1978. *Geomorphology: A systematic analysis of late cenozoic landforms.* Englewood Cliffs: Prentice-Hall.

Gabler, R. E.; Sager, R. J.; Brazier, S. M.; and Wise, D. L. 1987. *Essentials of physical geography.* 3d ed. Philadelphia: Saunders College Publishing.

Kolars, J. F., and Nystuen, J. D. 1975. *Physical geography: Environment and man.* New York: McGraw-Hill Book Company.

Ritter, D. F. 1986. *Process geomorphology.* 2d ed. Dubuque: Wm. C. Brown Publishers.

Thornbury, W. D. 1969. *Principles of geomorphology.* 2d ed. New York: John Wiley and Sons.

Tuttle, S. D. 1980. *Landforms and landscapes.* 3d ed. Dubuque: Wm. C. Brown Publishers.

13

Glaciation

*The Tasman Glacier in the Mt. Cook region
of New Zealand.
New Zealand Tourist and Publicity Office.*

> *As chaste as unsunn'd
> snow.*
> William Shakespeare, *Cymbeline*

For reasons still undetermined, the earth from time to time undergoes climatic changes that may produce colder conditions than normal, allowing for the accumulation of ice and snow from season-to-season and year-to-year. If such conditions are more than transient, an ice age may develop. Apparently, the earth has been subjected to a number of such ice ages in its long history, but the vestigial remnants of glacial effects are small indeed, except for those of the most recent such ice age. Those of ice ages, prior to the most recent, have been so affected by subsequent diastrophism and/or erosion that they exhibit few present landscape characteristics. In specific locales this generalization may not apply. For example, the tillites of peninsular India lithified from Permian glacial deposits are important parts of the stratigraphic sequence and, in some instances, significantly affect landform evolution. In general, however, only the most recent ice age is significant in the existing geomorphological landscape.

That glacial epoch, which began about 1,500,000 years ago and persisted until about 10,000 years ago, is known as the **Pleistocene.** The landform effects of that period of glacial erosion and deposition can be observed over vast areas of Eurasia and North America, and in restricted sections of the other continents, especially at higher elevations. In certain areas, glacial landforms, resulting either from ice scour or ice deposition, continue to be the dominant aspect of the landform environment. Effects of unloading of vast ice covered areas, as seen in the rebound of that land, are still being observed. Preglacial drainage patterns in many regions have been permanently altered. The present configuration of some of the world's great water bodies, notably the Great Lakes of the United States and Canada, is primarily due to the effects of glaciation. The rise in sea level attendant to ice melt inundated many former coastal areas.

Glacial ice, notably in Antarctica and Greenland, as well as in certain mountainous regions, continues to affect the terrestrial landscape. Advances and retreats of glaciers in some mountainous regions continue to create unanticipated hazards. The current overall pattern is for a slow shrinkage of glaciers, although Antarctica and Greenland show no significant diminution in area or thickness of ice cover and some reports suggest that Greenland glaciers may be growing. Occasionally, a mountain glacier will advance to lower elevations, apparently in response to heavier snowfall or because of dynamic changes within the ice cover or in the slope on which it rests.

Although it's not possible to ascertain with any degree of certainty what caused periods of glaciation, it is useful to consider some of the possible causes. The following section will elaborate on some of the possible mechanisms that produced periods of glacial activity.

Possible Causes of Glaciation

The mechanisms contributing to glaciation are uncertain, but worthy of consideration because, in addition to Pleistocene glaciation, glaciers of continental scope affected the earth about 2,300,000,000; 700,000,000; 460,000,000; and 250,000,000 years ago. It seems logical to assume that glaciation is likely to be a phenomenon of the future as well as the past. One such possible contributor to a modification of the global climates, which might result in glaciation, is the variability of solar radiation. Data are not available for really long periods, but it appears that periods of minimum sunspot activity are times when cold weather exists on earth. As we have seen earlier, periods of lesser amplitude of swing between sunspot maxima and minima apparently have precipitation ramifications. Do these periods of sunspot minima occur in some cyclical response to changes in thermonuclear reactions on the sun? Conversely, do they somehow affect the rate and intensity of reactions? Is there some tidal response to cyclical alignments of the planets? The answers to those questions are highly speculative, but variations in solar radiation over protracted periods *may* contribute to continental glaciations. Short-term variability is quite probably responsible for climatic fluctuations of more limited scope and impact.

The reflectivity of the earth and its atmosphere may also be a factor. Certainly there is greater reflection from a snow or ice clad surface than from a surface of rock or earth clad in vegetation. Clouds are also quite reflective, and increased cloudiness may have prefaced increasing amounts of snow. What caused such periods of increased cloudiness if they did, in fact, occur? These are classic examples of Which came first, the chicken or the egg? Was increased cloudiness a precursor to greater reflectivity and cooling, or were such clouds producers of snowfall that was not completely melted in the warm season? Perhaps a period of more active vulcanism created dust clouds that reduced the effectiveness of solar radiation, although other data do not seem to corroborate such large scale and protracted periods of volcanic activity.

More plausible explanations of glaciation may exist within the framework of the earth's various motions about the sun. Sedimentary cores in ocean deposits have revealed quite regular oscillations of climate in the last 300,000 years with periods of about 100,000, 42,000, and 23,000 years.

An astronomer, John Herschel, suggested such periodicities in 1830. Later, his ideas were expanded and elucidated by Milutin Milantovich in 1941. The theory points to the periodicities associated with certain eccentricities of earth's orbit caused by astrophysical interrelationships. Gravitational interactions within the solar system modify the nature of the ellipse made by the earth in its revolution around the sun at about 105,000-year intervals. This modification of average distance from the sun could, in fact, have the effect of diminishing or increasing the effectiveness of insolation, although, as we have seen, the angle at which the sun's rays strike the earth's surface is apparently much more significant than distance from the sun from one season to the next. Have those variations been of major moment in periodically reducing average global temperatures? Data are inconclusive.

The tilt of the axis of rotation in relationship to the plane of the earth's orbit, which affects summer and winter contrasts, follows a 41,000-year cycle. Finally, the precession of the earth's axis moves through an imaginary circle changing the orientation of the axis by about 1° in 21,000 years. These orbital idiosyncrasies and wobbles show a close relationship to the climatological inferences drawn from sedimentary cores. Such climatological periodicities do not appear to be borne out by glacial records in rocks. Why not? If the theory had validity, the same pattern should have existed in past geological time.

But all is not lost. Quite probably, continental drift over time took the protocontinent, Pangea, as well as large continental successors, out of high latitudes where continental glaciation might have occurred. Further, the eons of time that have elapsed since glaciation may have occurred have allowed certain land masses to be buried beneath seas, while others have had glacial forms erased by subsequent diastrophic movements or erosion. Clearly, these very old records of past history have been obscured by time and circumstance. As more facts are acquired, it seems logical to expect that the theory's validity may be corroborated or negated.

The key to any glaciation, of course, are summers that are cooler than normal. This would allow for some snow to remain unmelted and eventually would produce accumulations that could change to ice. Milantovich suggested 100,000 year intervals between periods of cooling, with approximately 25,000 years of spring, summer, fall, and winter. All of these are attributed to the various rotational and orbital eccentricities.

A 2° to 3° C decline in average summer temperatures could trigger the accumulation of snow and ice. Assuming that the theory of varying insolation is correct, and that land and water bodies remain essentially as they now are, a period of glaciation in about 100,000 years may be anticipated. It would be imprudent to make such a prediction, but the possibility of future glaciation is certainly something that must be considered.

Unfortunately, an understanding of the causal mechanisms for glaciation remains elusive. Perhaps glacial epochs are initiated by complex interactions of solar variability, orbital eccentricities, and rotational wobble. There are, however, enough examples of past repetitive glaciations to strongly suggest that such episodes may well occur again.

Evidences of Permian glaciation, for example, are quite pronounced in several segments of India. At the time of glaciation, however, India was not situated in a tropical latitude. Continental drift probably moved India from a high-latitude position in Permian times to its present location.

Long- and short-range climatic changes are evident from the fossil record and more recently from the written record. Changes of only slightly greater amplitude than recently recorded could trigger another period of glacial activity.

Pleistocene Glaciation

Although events in distant-past glaciations are unknown, occurrences during the Pleistocene are unmistakable. The evidence of the effects of glaciation is clear and persuasive. During at least five periods, ice accumulated and moved out from centers of accumulation over vast expanses of North America and Eurasia. The five periods of active growth and movement were divided by at least four interglacial periods when most of the ice disappeared. The advance and retreat of glacial ice left an indelible imprint on the middle and higher latitudes of North America and Eurasia. In many such regions, the effects of ice erosion and deposition are of greatest significance in interpreting the present face of the landscape. The effects are less pronounced in the Southern Hemisphere because of the general lack of large landmasses in high latitudes, except for Antarctica, which is still ice covered. Here, too, however, the effects of Pleistocene glaciation can be seen in areas not now covered by ice and snow, especially in higher elevations and adjacent piedmont slopes.

During the maximum extent of ice advance, virtually all of Canada and about half of Alaska were covered with glacial ice, and the conterminous forty-eight United States were covered to a line from about Long Island to the Ohio River to the Missouri River (figure 13.1). Few areas south of this general line were affected directly by glaciation, although indirect effects included a change in floral and faunal associations, and significant variations of river flow attendant to ice accretion and diminution. Certain sections of northwestern Canada and Alaska were not covered by ice, as well as a "driftless" area in southwestern Wisconsin and small adjacent segments of Illinois, Iowa, and Minnesota, which were surrounded by glacial ice, but not covered by it.

These driftless regions exhibit characteristics in sharp contrast to adjacent glaciated areas. Their soils are different, since they derive from preexisting parent materials rather than glacial drift. Further, they have typically developed deeper profiles because of longer exposure to normal soil forming processes. Their surfaces show more of the effects of stream erosion, since they were not subsequently masked by glacial debris. Valleys have been notched deeply into the plain, and drainage systems are somewhat better developed. Streams and their tributaries produce a more intricate erosional network than in adjacent glaciated regions.

In Eurasia, the Scandinavian countries, the Baltic States, much of Poland, northern Germany, the United Kingdom, and about the northern 30–40 percent of the Soviet Union were glaciated. In addition, much of the Alpine region of Switzerland, Austria, and northern Italy was a detached island of glacial ice. All of these glaciated areas exhibit marked contrasts to those areas not directly affected by the ice cover.

Figure 13.1
Extent of Pleistocene glaciation in North America.
Source: After C. S. Denney, U.S. Geological Survey, National Atlas of the United States.

In the other continents, mountain glaciers were substantially larger than at present, and alpine forms of many of the higher elevations were created by glacial activity during the Pleistocene. The advancing ice had climatological and biological effects that matched their geomorphological influences. Plant and animal species were driven south, and certain segments of the Deep South were colonized by subarctic plant and animal species. Small relict stands of such boreal forests still exist in secluded areas in the South as remainders of stands that predominated when ice cloaked the northern one-third of the country. As these relict stands are disturbed to accommodate agricultural expansion or urban accouterments, they tend to be succeeded by different species more characteristic of the current humid subtropical climate of the region. Similar evidence of migration of boreal plant and animal species further south can be obtained for Europe.

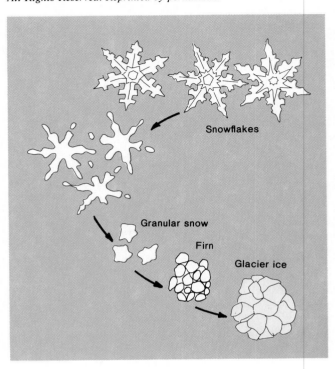

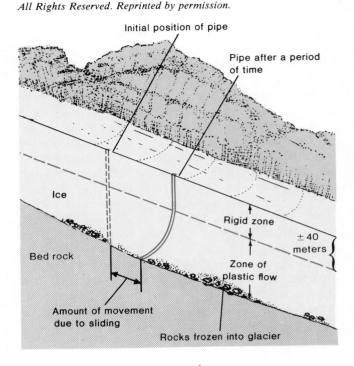

Formation and Movement of Glaciers

When snow accumulation exceeds melt over substantial periods of time, the snow crystallizes into **névé.** Snowflakes progress from a hexagonal flake form to granular characteristics and finally to ice. The intermediate stage between granular snow and ice is known as **firn** (figure 13.2). This ice continues to undergo compression as a result of the overlying weight of additional material. Compaction occurs in significant part because of the expulsion of air trapped in snow and granular ice. Apparently, at least 150 feet (about 45 meters) of granular snow must accumulate before compaction to ice occurs.

By the time the snow and ice has accumulated to depths of several feet, it begins to move outward in all directions in the case of continental glaciers and downslope in mountain glaciers. The characteristics of movement are complex and fall properly within the purview of the trained glaciologist. Nevertheless, basic descriptive aspects of movement are important to an understanding of and an appreciation for the character of landforms produced by glacial action. Crystalline ice undergoes very slow plastic deformation and flow. This **laminar flow** is not greatly dissimilar to the movement that would occur if weight were applied to an inverted bowl of gelatin. The basal portion of the ice moves, sometimes, over a lubricating film of meltwater created by friction and pressure as the enormous load of ice moves over bedrock. There is a kind of inexorability to this ice movement, and the sheer volume and weight of the ice produce an enormous erosive engine. The mechanisms of movement are still not entirely understood, but the effects of gravity are quite significant in the downslope movement of mountain and valley glaciers (figure 13.3). Glaciers are nourished by accumulations of snow that exceed wastage. These accumulations may represent increased precipitation during the cold season or decreased mean temperatures, especially in the summer season. Rapid nourishment occurs when both situations occur simultaneously.

In North America, three main focal points for ice accumulation in western, central, and eastern Canada coalesced to form an enormous expanse of glacial ice. At centers of accumulation, the ice may have been as much as a mile (1.6 kilometers) in thickness. At its outer margins, it must have been scores of feet thick. Glacial ice in Greenland currently exceeds 8,000 feet (2,500 meters) in the interior, and the weight of material is so great that the land beneath the ice in the interior has been depressed below sea level. Since ice moved out in all directions from centers of accumulation, the advance of ice into a particular area might be from almost any quadrant. In the United States, continental ice came principally from a northerly quadrant, although in any given locale, the ice movement may have been dominantly from a westerly or easterly quadrant.

Ice retreat during the interglacial periods occurred at a time when melt exceeded accumulation. Movement still occurred towards the periphery of the ice, but **ablation** (melting and sublimation at the periphery) exceeded accumulation and rate of advance. The net effect was a retreat, or meltback, (called **backwasting**) of the ice front, as well as a thinning of ice called **downwasting.** When backwasting and downwasting proceeded far enough, essentially dead ice was left, and most glacial influences at such times are generated by glacial meltwater. During movement of the ice, freezing and thawing of various portions occurs repetitively creating numerous strains and changes in the crystalline form of the ice. Some isolated segments of dead ice may still have existed in Scandinavia as recently as 5,000 years ago.

Movements of a material like ice, in spite of the fact that it may assume semiplastic characteristics, results in innumerable cracks being formed on and in the ice (figure 13.4). Many of these cracks fill with debris and add to the complexity of glacial forms. Eroded earth materials are carried on, in, and beneath the surface of the ice. Debris is heterogeneous in type, form, and size. Some of the materials frozen in the ice serve as more effective etching tools than the ice matrix that carries them. This heterogeneity is carried over in the array of glacial deposits. Some are composed of unassorted materials deposited directly by the ice, and others have sorted characteristics by size reflecting their deposition by glacial meltwater.

Glacial Erosion and Deposition

The volume of ice and the rate of flow have significant impact on the rate and effectiveness of erosion. As a general rule, the greater the volume of ice and the faster the movement, the greater the amount of erosion, although very resistant bedrock might resist erosion quite effectively even from a large volume of ice that was moving rapidly. This relationship of volume and flow is, of course, analogous in water erosion. In ice sheets, the central area is not much affected by either erosion or deposition. In an

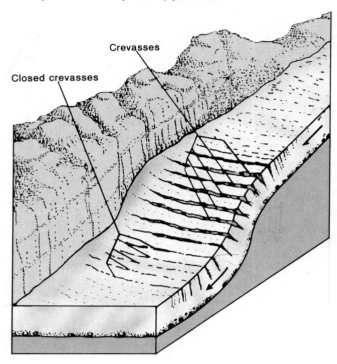

intermediate area, the effects of ice scour are pronounced, whereas near the ice periphery, depositional forms tend to predominate.

Glaciers erode principally through two processes: plucking or **quarrying,** and **abrasion. Plucking** develops principally as friction and pressure cause some of the basal portions of ice to melt, and as this water refreezes in cracks and interstices in rock, the movement of flowing ice plucks or quarries out pieces of rock to be transported in the direction of ice movements (figure 13.5). Since pieces of rock are carried along in the frozen ice, the movement of such materials over the surface is like an enormous sander, or grinder, under huge amounts of pressure. The effect is to abrade and to produce a landscape that is likely to be rougher than an original smooth surface or smoother than an original rough surface.

In continental glaciation, the common forms developed from this scouring and plucking of the underlying surface create an interesting array of features. Some are subtle modifications of the preexisting landscape, whereas others are features unmistakably attributable to glacial action.

Figure 13.5
The effects of plucking in glacial erosion.
From Carla W. Montgomery, Physical Geology. *Copyright © 1987*
Wm. C. Brown Publishers, Dubuque, Iowa. All Rights Reserved.
Reprinted by permission.

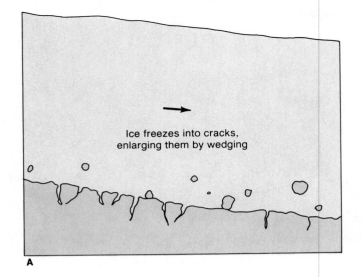

Ice freezes into cracks,
enlarging them by wedging

A

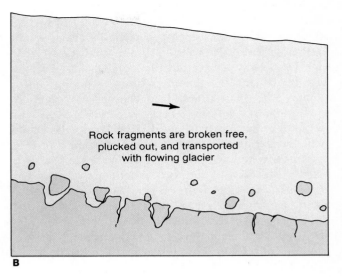

Rock fragments are broken free,
plucked out, and transported
with flowing glacier

B

As glaciers move over the land, hills are rounded, valleys are modified, and surfaces are gouged. Ice moving along valley bottoms or depressions may deepen such valleys in a general U-shape. The Finger Lakes of New York represent a series of parallel stream valleys that were deepened significantly by the ice scour and subsequently filled by water.

The profile, depth, and character of several of the Great Lakes were affected by the scouring and eroding effect of the continental glaciation in preexisting lowland areas. Indeed, numerous arms, bays, and inlets owe their existence to the glacial scouring, which took place during the Pleistocene. Similar effects can be observed in several of the lakes of Scandinavia.

Smaller, but still conspicuous features of the ice scour landscape are **rôche moutonnées.** These asymmetrically rounded hillocks have their **stoss** sides (direction from which the ice came) abraded in a fairly smooth and symmetrical fashion. On the other hand, the **lee** side (direction away from which the ice came) has been plucked and quarried to produce a steepening of the profile.

Any exposed bedrock in glaciated terrain is almost certain to have **glacial grooves** cut in the surface along with smaller scratches known as **striae.** The orientation of these grooves or striae, cut by rock debris frozen in the basal portion of the ice, is sometimes helpful in determining the direction of ice movement in a particular locale. The grooves or striae are parallel to the direction of ice movement in a given place. Sometimes this scouring effect is so pronounced that it produces a kind of lineation to major terrain features.

The **glacial flour**—very fine particles of rock produced by the grinding action of the ice—sometimes polishes exposed rock to exceptionally smooth surfaces. Meltwater streams often contain such enormous quantities of glacial flour that they are milky in appearance. Far from the pristine stream of television commercial lore, glacial meltwater streams are discolored and full of grit and sandy materials. After the period of glaciation when glacial flour is not being replenished, the streams may become clear.

Some spectacular ice scour features have been produced by glaciers in mountainous terrain, and a description of their development and characteristics will be set forth in a subsequent section of this chapter. In fact, the alpine forms, which we have come to associate with high mountain scenery, have generally been produced by ice scour. In continental glaciation, the effect of ice scour has largely been to round and smooth preexisting rock outcrops, whereas in mountainous terrain, ice scour has tended to add sharp edges and angularities to the preglaciated terrain. Vast areas, especially on the Laurentian Shield of Canada, have been modified by ice scour producing **ice scour plains.** These ice scour plains are frequently marked by interspersed glacial lakes representing declivities etched or modified by plucking and abrasion. They may also be marked by localized deposits of materials even in zones where dominant features owe their origins to ice scour.

Materials scoured from one area were moved along only to be deposited elsewhere. A consideration of those depositional forms follows.

Deposition by Glaciers and Depositional Landforms

The enormous amount of debris frozen in the bottom of the ice, around the ice edges, frozen within the ice, and carried on the surface of the ice is a heterogeneous mixture of unsorted material, which when deposited without reworking by water is known as glacial **till.** Where such materials are dropped beneath the ice usually in an irregular, hummocky surface of little local relief, it is known as a **till plain,** or sometimes as a **ground moraine.** Clayey till may produce an almost flat surface or one with many small swales and very low hillocks. Such gentle slopes with low local relief may mask quite irregular preglacial terrain. Extraordinarily large boulders, dissimilar to their present surroundings, left on the surface of an ice scour plain or a till plain on certain occasions, are called glacial **erratics** (figure 13.6). Till is frequently different in character, having come from a different source region, than the preglacial surface, which has been buried. Till may vary from very fine materials like clays through a whole gamut of silts, sands, gravels, cobbles, and boulders.

Moraines

Other conspicuous landform features of ice deposition are an array of **moraines**—hummocky hills or hill-like features formed at the margins of advancing ice. This debris is transported by the laminar flow of the ice, almost like a conveyor belt. At a zone where ablation is equal to or exceeds this forward component of force, the debris is dropped to produce the moraine. Small amounts of such material may be moved towards the margin as the ice scrapes up surface material. Three principal moraine forms are associated with continental glaciation: terminal moraines, recessional moraines, and interlobate moraines. The principal distinction among them is their relative position with respect to the position of the ice front. Obviously, **terminal moraines** have formed at the margins of the maximum extent of the ice advance. As the moving ice continues to carry materials conveyorlike towards the ice front augmented by debris pushed along, larger and larger quantities of material are deposited. Some of these terminal moraines, especially those of the Wisconsin (last) stage of glaciation, are quite hilly zones marching across the otherwise flat lands of central Illinois or Indiana.

Recessional moraines develop at a position in back of the ice front where a readvance or period of protracted stability leads to the formation of a moraine parallel to the terminal moraine, but at some distance closer to the source of the ice movement. **Interlobate moraines** are coalescing terminal or recessional moraines developing between lobes or tongues of the advancing ice front. The development of lobes of ice is principally responsible for the curving ground plan of most moraines.

Figure 13.6
Glacial erratics have been bulldozed to the side of a field to permit cultivation.

Other Ice Depositional Forms

Blocks of ice were often buried in the debris of moraines. Subsequent melting and slumping of material into the space left by the melting ice produced depressions called **kettles** (figure 13.8). Moraines are frequently pitted by the large numbers of kettles that have formed along the zone of stagnant ice. Some of these depressions have subsequently filled with water, producing small ponds or lakes.

The other principal ice deposition feature of vast ice sheets is the drumlin. **Drumlins** are hills of glacial till of uncertain etiology, which typically have a steep stoss end and a more gentle slope on the lee side. Occasionally, they contain lens-shaped deposits of glaciofluvial material surrounded by till. Drumlins are oriented essentially parallel to the direction from which the ice came and may exist in large groups of hills. They have the approximate profile of a hard-boiled egg cut in two longitudinally, or lengthwise. The exact cause of the asymmetry is uncertain, but it may derive from a temporary readvance of the ice after initial deposition. They almost always exist in swarms and may exist in compound combinations. They may vary in height from 20 feet to as much as 200 feet (about 6 to 60 meters), and they may be only a fraction of a mile in length or as long as a few miles. In certain locales, drumlins may have bedrock cores. Such features are called **rocdrumlins.** Drumlins are fairly common features in upstate New York, portions of Massachusetts, and eastern Wisconsin (figure 13.10). Indeed, certain famous hills like Breed's Hill and Bunker Hill of Revolutionary War fame are drumlins. They also exist in several areas of Canada, notably in Ontario and Nova Scotia. In Europe, they are encountered in Ireland, England, and Germany, especially.

Figure 13.8
The development of kettles.
From Carla W. Montgomery, Physical Geology. *Copyright © 1987 Wm. C. Brown Publishers, Dubuque, Iowa. All Rights Reserved. Reprinted by permission.*

Ice block
stranded
in outwash

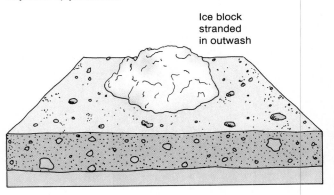

Hole left
in outwash
after ice
melted

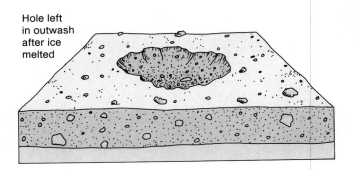

Figure 13.9
Partially filled kettle (swampy area) and adjacent kame in Wisconsin.

Figure 13.10
Portion of a drumlin field as depicted on a contour map.
Portion of Fair Haven, New York Quadrangle, 7 1/2 minute series,
USGS.

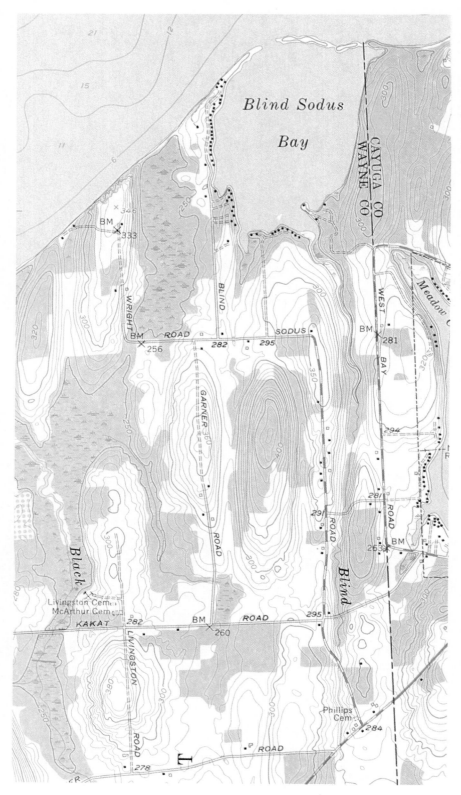

Figure 13.11
*Kettle hole that has been almost filled by glacial erratics
bulldozed into it.*

Ice deposition produces irregularities in the preglacial landscape to such an extent that preexisting
drainage is often completely disrupted. The period since
the departure of the ice has been sufficiently short that
certain such regions continue to have streams wandering
in an almost aimless pattern, and small lakes, ponds, and
swamps may be frequent occurrences. Not only is terrain
affected in such areas, but soils and vegetation reflect the
geomorphological and drainage conditions. A variety of
circumstances may create lakes and ponds, including kettles, moraine damming, or simply the irregular, hummocky deposits created by uneven debris loads.

As indicated earlier, for much of the northern one-
third of the conterminous forty-eight states, the influence
of glaciation is the salient aspect of the present geomorphological landscape. In some areas, people's activities
have been enhanced; in others, they have been inhibited.

Glaciofluvial Features

Glacial ice ablates because of frictional drag over underlying rock material, because of the pressure of the overlying ice, in response to the summer season, and when the
ice retreats during a period of climatological warming.
Since glaciers contain materials on the ice, in the ice, beneath the ice, and along the ice margins, there is an abundant source of weathered material available to be moved
by meltwater. Meltwater streams typically carry an enormous load of material in suspension and solution, and, as
has been mentioned previously, these glaciofluvial forms
tend to predominate near the margins of the ice.

These glaciofluvial forms are composed of stratified
glacial drift as opposed to the heterogeneous size and mix
of materials deposited directly by the ice. **Outwash plains**
occur as broad alluvial aprons, which form beyond the

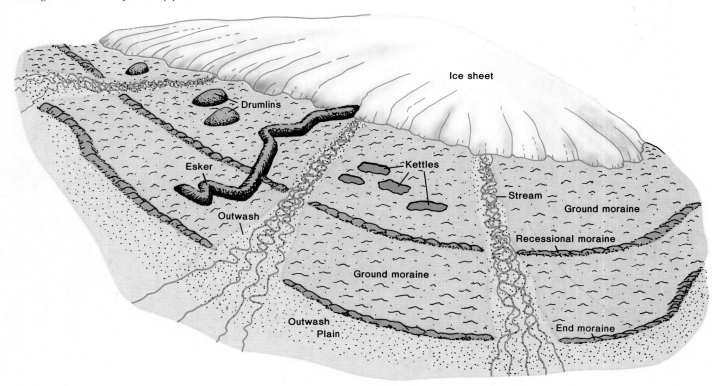

margins of terminal or recessional moraines. **Pitted outwash plains** are frequently pitted with depressions left when blocks of stagnant ice—some of them transported by meltwater streams—melted.

Eskers are long sinuous ridges that develop when overburdened streams within or beneath the surface of the ice deposited heavy loads in their own channels during long periods of ice stagnation. They stand up like long sinuous railroad embankments after the ice has melted. Some may extend for scores of miles, although most are much shorter. In a few areas, they extend over hill and dale with apparent disregard for current topographic expressions. They are steep-sided in cross section, but the crest elevations are essentially level. Wisconsin and portions of Michigan have a number of eskers. In Europe, Finland has an especially large array of eskers, and some of them are quite impressive in size.

A similar feature, usually much smaller and typically less sinuous in appearance is a **crevasse fill.** These small ridge-like features apparently result when a crack in the ice has debris washed in from either side ultimately filling the crack, leaving a small ridge in postglacial terrain. Crevasse fillings are usually somewhat more numerous near the stagnant ice margin, and they are commonly only several score feet in length, whereas eskers usually have greater local relief, area, and length.

Kames are small conical hills of glacial debris that are apparently formed at ice margins when rivulets on the ice or in the ice lose their load as they issue from the snout of the glacier. Kames, which form well back of the ice front as debris accumulates at the base of a meltwater hole, are called **moulin kames.** They are roughly analogous to alluvial cones and frequently are associated with numerous kettle holes. Irregular, hummocky terrain associated with kames and kettle holes is known, logically enough, as **kame and kettle terrain (topography).**

Other features that result, in part, from glacial meltwater are **urstromtäler,** ancient stream valleys, where glacial meltwater has significantly augmented the waters in a preexistent stream causing the valley to be significantly deepened and widened beyond the capacity of the stream that formed them initially. When the ice has melted and the stream has returned to a normal size, it occupies a valley out of all proportion to what one would expect. It is a misfit stream. Although most glaciofluvial features are primarily the result of deposition, urstromtäler are principally the result of erosion. Some former overflow channels are not now occupied by streams. The valleys persist, but the streams may have disappeared as some streams returned to preexisting channels and others dried up as meltwater was exhausted.

Similarly, glaciers of continental scope have had major impacts on stream courses because of ice blockage. The existing courses of some major rivers were markedly altered. To illustrate, the Missouri River became an ice marginal stream because of the blockage of glacial ice. Before ice blockage, the upper Missouri and Yellowstone apparently drained northward towards James Bay, an arm of Hudson Bay. Preglacial rivers like the Grand, Moreau, and Cheyenne extended further east than present to join a northward flowing river in the James River basin. Links were made eastward along the ice margin and a joining with the preglacial Kansas River added the final link of the present course of the Missouri below Kansas City. Similarly, the lower portion of the Ohio is a composite of a proto-Ohio; its middle portion was a preglacial stream called the Teas, which flowed northwestward; and its upper reaches includes a portion of the Allegheny-Monongahela system, which flowed towards the St. Lawrence.

Periglacial Phenomena

Landforms or other effects of glaciation occurring beyond the margins of maximum ice advance are called **periglacial.** Urstromtäler, for example, might occur well beyond maximum ice advance. In northern Europe, the ice blocked the mouths of preexisting northward flowing rivers like the Elbe, Rhine, Oder, Weser, and Ems. These streams cut new courses along the margins of the ice, largely extending in a dominant east–west direction, until they could gain access to the sea. When the ice retreated, the original courses were reestablished, and the valleys cut during glaciation were left as depressions in the landscape. Since the present courses of many of these rivers are northerly and many of the urstromtäler have large east–west segments, they are readily amenable to canalization permitting easy east–west connections between northward-tending rivers. The interconnections of present and past stream valleys have produced an intricate water transport network, which is a major positive economic force in northwestern Europe. Man's current purposes have been augmented by the effects of past glaciation.

Marginal river systems like the Missouri and Ohio are periglacial features. Preexisting drainage patterns were radically altered, and some of the interesting crooks and bends of existing streams owe their origins to the connections and links with preglacial streams.

At the time of maximum ice extent, a prodigious amount of the earth's waters were locked up in the ice, and, as a result, sea level was a great deal lower than it is now. Consequently, preexisting streams continued their erosional effect across newly exposed land surfaces. When sea level rose again after the departure of the ice, these river valleys were drowned, and certain submarine canyons are clearly the extension of such past stream erosion. Some of these drowned extended river valleys now add to the irregularity of certain coastlines. Sea level continues to rise along many coasts, especially those of the United States. At least part of this rise appears to be due to the shrinking of the world's ice sheets. The effect of rising sea level is ominous in many places, since it subjects many of mankind's structures to the ravages of wave erosion. Large segments of the coastal plain margins of the United States would be inundated if the water now locked up in glacial ice were to melt, and the increased greenhouse effect certainly points to ice melting.

The enormous weight of ice several hundred feet thick of almost continental proportions caused the earth's crust to subside. After the ice melted, there has been a slow upward rebound. That rise in land surface of a few centimeters per year continues. In certain places, this rebound has been responsible for elevating lake beds and has resulted in accelerated shoreline erosion. This has been particularly true along the margins of several of the Great Lakes. In the late 1980s, Chicago was subjected to a great deal of shoreline erosion because of a combination of rebound, abundant runoff, and stormy conditions. Chances for continuing lakeshore damage in that city are high. Costs of protecting present shorelines by the construction of jetties and seawalls are enormous, but such efforts are necessary to protect valuable lake front property. Additional expense is likely to be incurred as natural rebound precipitates further wave damage. A temporary respite to a long-term problem occurred in the summer of 1988 as lake waters dropped in response to a prolonged and severe drought. It is likely that return to more normal precipitation patterns will exacerbate the problems of rising water.

Other periglacial phenomena are initiated by climatological changes. Storm tracks were obviously altered during the glacial epoch, and resulting changes in precipitation patterns affected marginal terrain.

The glacial period was somewhat wetter, climatologically, in middle and high latitudes for reasons not fully understood. In any case, some vast lakes were formed in areas where they no longer exist. These lake beds are interesting features of the current landscape, especially in areas like the Great Basin country of the United States, which is now arid or semiarid. The residual floors of such lakes are often composed of very fine-grained material, and many of them are exceptionally flat. These almost featureless surfaces have proven to be excellent sites for testing high-speed land vehicles and for landing experimental aircraft and the space shuttle.

Lakes like Great Salt Lake are only vestigial remnants of former lakes, which were much larger in area. Extensive flats in the Basin and Range country frequently represent lacustrine plains of lakes that disappeared thousands of years ago.

Many ice marginal lakes developed deposits of **varved clays** (i.e., clay deposits marked by an organic darker layer and a lighter layer with little organic material). Correlation of those layers from place to place has made it possible to date glacial features located some distance apart.

Figure 13.13
*The position of principal end and recessional moraines and
periglacial lakes in the American west.*
From Charles C. Plummer and David McGeary, *Physical Geology,
4th ed.* Copyright © 1988 Wm. C. Brown Publishers, Dubuque, Iowa.
All Rights Reserved. Reprinted by permission. After C. S. Denney,
U.S. Geological Survey and the Geological Map of North America,
Geological Society of America, and The Geological Survey of
Canada.

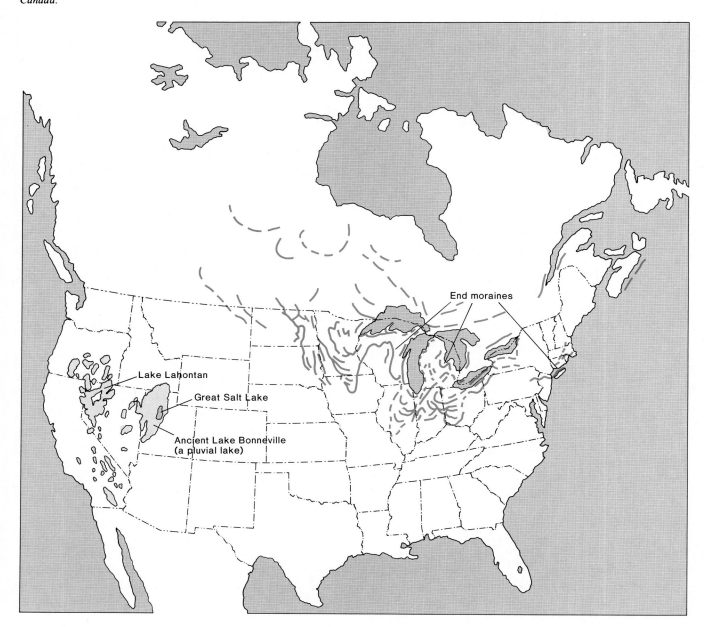

The glacial and periglacial features associated with
continental glaciation have certain counterparts in mountain and valley glaciers, but there are numerous features
unique to mountain and valley glaciation. A discussion of
mountain and valley glaciation and attendant features follows.

Mountain and Valley Glaciation

Virtually all of the high mountainous regions of the world
experienced significant glaciation during the Pleistocene,
and the higher mountains, especially in middle and high
latitudes, are still being modified by glaciers. The effects
of ice erosion in mountainous terrain is especially pronounced, and the tendency of mountain glaciers is to create
more angular and serrate forms than in preglacial times.

Figure 13.14
Bergschrund and development of a cirque.
From Charles C. Plummer and David McGeary, Physical Geology,
4th ed. Copyright © 1988 Wm. C. Brown Publishers, Dubuque, Iowa.
All Rights Reserved. Reprinted by permission.

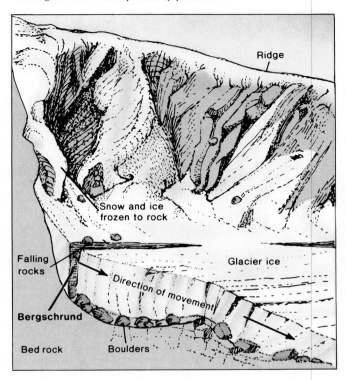

Figure 13.15
Horns, cols, arêtes, and cirque basin illustrated in the peaks
of Devil Paw Mountain, Tulsequah Glacier, British
Columbia.
Department of Regional and Industrial Expansion photo,
Government of Canada.

Figure 13.16
Horns, arêtes, cirque, and cirque glacier in Canadian
Rockies.

Slopes tend to be steepened, and local relief is increased. Some of the truly magnificent mountain scenery of the world has been created by the etching effect of mountain and valley glaciers. Several areas have experienced multiple episodes of glaciation, and the ice scour and depositional features in such regions may be quite complex.

Snow compacted to ice at high elevations yields the first glaciers in mountainous regions. These glaciers begin plucking at high slopes (in ways imperfectly understood) to produce broad amphitheater-like basins with steepened head- and sidewalls, and a floor, which may be flattened or even overdeepened. As the ice moves downslope, it tends to pull away from the high wall of the valley (figure 13.14). This gap between ice and headwall is called the **berg-schrund.** Glaciers occupying such basins are understandably called **cirque glaciers,** since the amphitheater-like basin is called a **cirque.** There are numerous regional names for cirque, but that term is now generally accepted as the term to use to describe the typical amphitheater-like feature. If they extend down into and modify preexisting valleys, they become **valley glaciers.** If they issue out of the mountain valley into surrounding, flatter terrain, they are called **piedmont glaciers.** The most persistent effect of these glaciers is scouring, which leaves the awesome angular forms so associated with alpine scenery.

Ice Scour Features of Mountain and Valley Glaciers

The plucking of headwalls and sidewalls by a series of cirques acting on opposite sides of a mountain ridge may produce a long serrate ridge called a cockscomb ridge, or an **arête** (figures 13.15 and 13.16). Cirques eating headward on opposite sides of an arête may cut a natural pass or notch through the ridge, called a **col.** A mountain knob eroded by glacial ice moving down from three or more sources may produce a single sharpened peak as the sidewalls of the cirques intersect. Such a glacially sharpened peak is called a **horn,** deriving its name from the famed Matterhorn of the Alps, which is the prototypical example of this geomorphological feature.

Figure 13.17
Ice scour features and ice depositional features in mountain and valley glaciers.
From Arthur Getis, Judith Getis, and Jerome Fellmann, Introduction to Geography, *2d ed. Copyright © 1988 Wm. C. Brown Publishers, Dubuque, Iowa. All Rights Reserved. Reprinted by permission.*

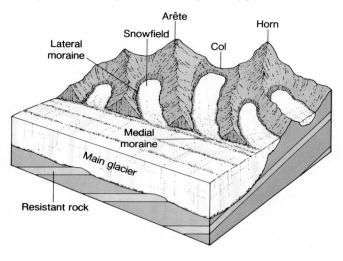

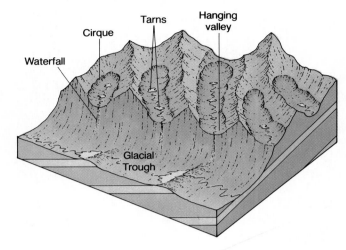

Figure 13.18
Hanging valleys, horns, cols, and arêtes in the Canadian Rockies.

As the cirque glacier extends itself into a valley, the erosion of the ice along the bottom and sides of the ice occurs both through abrasion and plucking. The typical effect is to produce a U-shaped trough from a preexisting V-shaped valley, which was created by water erosion. Frequently, there are intricate interconnections of glaciers in mountainous regions. Major streams of ice may receive tributary ice streams in considerable number. These tributary streams may be less effective at the business of erosion than the main stream of ice, resulting in the major valley being downcut more effectively than the tributary valleys.

As the ice moves over the valley floor, the rate of erosion varies, depending on the resistance of the exposed rock, the waxing and waning of movement, and varying thickness of the ice. The effect, typically, is to produce an irregular valley floor often proceeding step-like to lower elevations rather than descending in a smooth progression. These **cyclopean stairs** may, subsequent to the melt of the glacier, have a series of small lakes arranged step-like down the valley. These **paternoster lakes,** arranged like beads on a string, are essentially analogous to the **tarn lakes,** which may be left in an overdeepened portion of a cirque basin.

Many of the ice scour features in mountain-glaciated regions are brought to light only with the melting of the ice (figures 13.17 and 13.18). The **U-shaped valleys,**

Figure 13.19
The fjorded coast of South Island, New Zealand in the Milford Sound area.
New Zealand Tourist and Publicity Office.

Figure 13.20
How medial moraines develop as two valley glaciers merge.
From Carla W. Montgomery, Physical Geology. *Copyright © 1987 Wm. C. Brown Publishers, Dubuque, Iowa. All Rights Reserved. Reprinted by permission.*

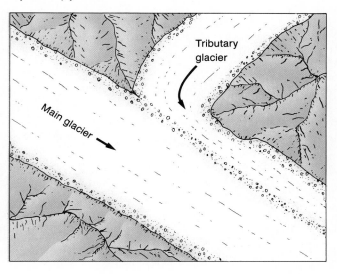

often intersected by the U-shaped valleys above the valley floor (such features are known as **hanging valleys**), are interspersed with current water courses often plunging from such hanging valleys in spectacular waterfalls. Where ice has overdeepened river valleys adjacent to seas, such valleys, after submergence subsequent to melt, often yield spectacular, deep, steep-walled inlets called **fjords** (figure 13.19). These fjords are especially prevalent in Norway, Greenland, and certain sections of southern Chile and New Zealand. Not only do they provide deep water anchorages almost adjacent to shores, but they contain some of the most magnificent scenery in the world.

Depositional Features Associated with Mountain and Valley Glaciers

Mountain glaciers also yield a full array of depositional features, usually of a smaller order of magnitude than counterpart features in continental glaciation. The panoply of moraines include the ground, terminal, and recessional moraines, in addition to **lateral moraines,** which form along the sides of the glacier and are left along the sides of the valley. **Medial moraines** form when two ice streams coalesce. Rocky debris is left at the midpoint of the ice stream, which has coalesced, producing a moraine oriented essentially parallel to the direction of ice movement, as compared to the almost right-angle orientation of terminal and recessional moraines (figures 13.20 and 13.21).

Figure 13.21
The development of medial moraines in the Northwest Territory of Canada.
Department of Regional Industrial Expansion photo, Government of Canada.

Figure 13.22
Moraine patterns in mountain and valley glaciers.
From Charles C. Plummer and David McGeary, Physical Geology,
4th ed. Copyright © 1988 Wm. C. Brown Publishers, Dubuque, Iowa.
All Rights Reserved. Reprinted by permission.

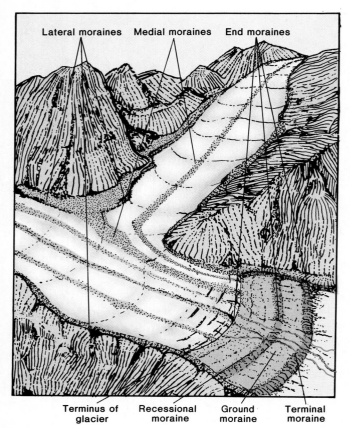

Figure 13.23
Valley trains (outwash plains) develop well out in front of maximum ice advance.

Figure 13.24
Glaciers are still active in mountainous areas in many portions of the globe as in the Columbia ice fields of Alberta, Canada.

Various small glaciofluvial features are also formed, but, with few exceptions, they are transient features soon obscured by subsequent effects of erosion (figure 13.22). The **kame terrace** may be formed in ice troughs adjacent to valley walls. Since the water course that produced them is relatively short, stratification of drift is quite imperfect. Kame terraces may also develop in troughs between ice margins and moraines in areas of continental glaciation. Outwash plains confined within valley walls are called **valley trains** (figure 13.23).

The overall pattern of glaciation in mountainous areas has been a steady ice retreat. Occasionally, however, for reasons not fully understood, mountain glaciers show sudden advances (figure 13.24). In late 1986 in southeast Alaska, the Hubbard Glacier advanced rapidly, blocking off an arm of the sea and creating flooding. The arm of the sea so quickly cut off by the ice dam became fresh because of ice melt, causing many trapped sea creatures to perish. The ice dam broke, and the lake again became an arm of the sea. That temporary retreat has been succeeded by another advance, however, and the same process shows signs of repeating itself.

Figure 13.25
Glaciers typically carry debris on, in, under, and in front of the ice margin as indicated by this photograph of Angel Glacier on Mt. Edith Cavell in the Canadian Rockies.
Department of Regional Industrial Expansion photo, Government of Canada.

Figure 13.26
High mountain scenery in glaciated Canadian Rockies.

Figure 13.27
The Tetons developed as the result of complex faulting, but they have been etched and sharpened by the effects of glaciation.

These glaciated regions become tourist meccas in virtually all seasons. Their outstanding alpine scenery is a delight to the eye, a challenge to the skier, and a joy forever.

Very large ice sheets, almost continental in scope, still cover almost all of Antarctica and more than 90% of Greenland. When mountains project above and are surrounded by these vast ice sheets they are called *nunataks*. Glacial ice at sea margins may break off to produce icebergs, which can be hazardous to ocean shipping in high latitudes.

The Antarctic icecap is a kind of giant refrigerator, which accentuates the effects of the polar high in the Southern Hemisphere. The unloading of ice, which could occur with glacial melt, would not only cause the rise of sea level around the world with probable harmful effects, but it would also result in considerable rebound of the land beneath the ice.

The power of existing glaciers hints at the awesome effects of ice sheets of continental dimensions and illustrates the efficacy of ice scour and deposition in the high mountain regions of the world.

Ice and running water are effective in sculpting the land and reducing it towards base level. Other elements of erosion will be considered subsequently. Each influences the intricate web of geomorphological change, as erosional and depositional forms create the furrows and wrinkles on the face of the earth. The never-ending struggle between diastrophic forces, and agencies of weathering and erosion continues.

Figure 13.28
High mountain lake in Colorado dammed by moraine.

Study Questions

1. List and briefly discuss theories of climatic change advanced as possible causes of glaciation.
2. What was the approximate maximum extent of Pleistocene glaciation in the U.S.?
3. Explain how glaciers advance. How they retreat.
4. How do glaciers erode?
5. List and characterize the various types of moraines in terms of their relationship to the ice front.
6. List and briefly characterize the various glaciofluvial features.
7. Explain how urstromtälers facilitate water transportation in northern and western Europe.
8. List the various ice scour features associated with mountain and valley glaciation.
9. Explain why portions of interior Greenland are below sea level.
10. Discuss periglacial phenomena.
11. Characterize kame and kettle topography.
12. Explain how the "driftless area" compares with surrounding glaciated regions.

Selected References

Butzer, K. W. 1976. *Geomorphology from the earth.* New York: Harper and Row, Publishers.

Dyson, J. L. 1962. *The world of ice.* New York: Alfred A. Knopf.

Embleton, C., and King, C. A. M. 1975. *Glacial and periglacial geomorphology.* 2d ed. New York: Halstead Press.

Matsch, C. L. 1976. *North America and the great ice age.* New York: McGraw-Hill Book Company.

Plummer, C. C., and McGeary, D. 1985. *Physical geology.* 3d ed. Dubuque, IA: Wm. C. Brown Publishers.

14

Groundwater and Solution Topography

Pohutu Geyser in Roturua, New Zealand erupts throwing groundwater heated by magma to a considerable height. New Zealand Tourist and Publicity Office.

Hence, loathed Melancholy, Of Cerberus and blackest Midnight born, In Stygian cave forlorn, 'Mongst horrid shapes, and shrieks, and sights unholy.

John Milton, *L'Allegro*

When rain falls, some of it runs off the surface, some is evaporated, and some soaks into the earth to become a part of the groundwater supply. The amount that soaks in at a given time and place hinges on several factors: (1) surface topography, (2) the rate at which the precipitation falls, (3) the permeability of the soil and rock materials, (4) the amount of water already contained in the soil, (5) the vegetative cover, and (6) the relative humidity.

On a very steep slope, there is usually very rapid runoff and less opportunity for water to percolate into the soil. Conversely, gentle slopes tend to inhibit runoff and provide greater opportunities for water to soak into the soil. Clearly, slope may inhibit or accelerate runoff, and the more rapid the surface runoff, the lesser opportunities for water to infiltrate into the soil.

Figure 14.1
Water budget of the conterminous United States in million acre-feet per year. Water flows off the surface and enters the subsurface realm.
From The Yearbook of Agriculture, *1983, U.S. Department of Agriculture, U.S. Government Printing Office, Washington, D.C., 1983.*

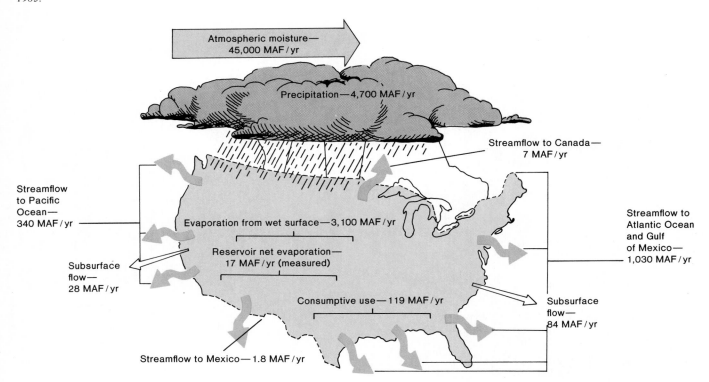

If rain falls in hard abrupt showers, there is a tendency for the earth to shed much of it to surface water courses, even if the surface is quite permeable. On the other hand, a gentle shower provides significant chances for much of the moisture to soak in, even when permeability is low. These generalizations tend to apply if we assume that other variables of soil texture and soil moisture are essentially the same. The interaction of rainfall amount, intensity, slope, permeability, vegetative cover, and other factors facilitate or inhibit percolation or runoff.

Soil with a great deal of **porosity** can hold a great deal of water. Sedimentary rocks, especially sandstone, certain limestones, and shales frequently have high porosities. Porosity is contrasted with **permeability,** which has to do with how water or other fluids move through rocks. Sandstone, for example, characteristically has high porosity and permeability (i.e., there is abundant pore space, which is large enough to facilitate the movement of water through it). On the other hand, shale has high porosity, but limited permeability. Some rocks, like granite, slate, and gneiss are almost impermeable. In the case of shale, the abundant pore spaces permit the absorption of a great deal of water, but the very small size of the pores inhibits the rate at which water moves through the rock.

Sandy soils may permit rainfall to percolate down very rapidly, whereas clay may resist water percolation. As a result, sandy soils may dry out rapidly after a rain, whereas clay may remain muddy for some time. As a matter of fact, this is the normal situation. The author's boyhood home in southern Illinois had a high percentage of clay in the soil, and rural unpaved roads became a quagmire for several days after a rain. His present home near Pensacola, Florida, has a high percentage of sand. The rural, unpaved roads quickly dry out after rains, and the wetted sand may actually prove to be a better surface for vehicular traffic than the unconsolidated sands are after a protracted dry period.

The author has had other intimate experiences with the limited permeability of clay. Living, as he does, in an area with septic tanks, he has watched with dismay as a small stratum of clay 7 feet (approximately 2 meters) beneath the surface has inhibited drain field percolation after periods of heavy rains have saturated the sandy layer at the surface. When percolation is thus inhibited, septic tank and drain field repairs may be necessary. The author was not naive about the conditions, but other attributes of the site overcame the handicap of slow percolation when a decision was made to build.

Obviously, soil that is nearly saturated with water can accommodate very little more. Conversely, soil that is very dry will literally drink available moisture, assuming that it is readily permeable. There are, of course, all gradations in between these two extremes.

Areas that are cloaked with vegetation inhibit runoff and allow more water to sink into the ground. Bare earth or rock has a greater tendency to shed water, hence increasing runoff and diminishing percolation. Indeed, one of the problems of urban development is the creation of large areas of pavement in addition to the surface areas of structures. These impermeable features shed water rapidly, which may add significant stress to existing drainage facilities and contribute to problems of flooding. By speeding runoff, these same features reduce percolation, which may contribute to lowering water tables. As lands are cleared for construction, rainfall is able to erode the exposed soils rapidly, which may increase the sediment burden of natural and man-made drainage systems and clog them with debris. Such overloading may change existing stream equilibrium and cause the establishment of a new equilibrium condition.

In very dry areas, more moisture evaporates after it falls, and there is less moisture available to sink in. If relative humidities are high, evaporation is inhibited and percolation is enhanced.

The ultimate source of any groundwater is, of course, precipitation that has fallen. That precipitation need not necessarily have fallen in immediate proximity to its present situation. Rain that falls or snow that melts in the Rockies, for example, may migrate through underground aquifers far out into the Great Plains. More humid sections of the highlands of Australia supply water that eventually feeds the aquifers, which are a part of the Great Artesian Basin in the eastern portion of the country.

Water Table

Groundwater seeps into all the cracks and pore space, and moves downward by the force of gravity. At great depths, water ceases to move downward, probably because there is limited pore space as a result of the enormous burden of material above. Water percolation is generally limited to the upper 3 miles (5 kilometers) of the earth's crust, although in very thick bedded sedimentaries, it may extend as deep as 6 miles (10 kilometers). When all the pore space is filled with water, the upper level of that saturated zone is known as the **water table** (figure 14.2). The water table tends to be located at shallow depths in humid areas, and it may be at considerable depths in arid regions.

Above the zone of saturation in an area where the pore spaces are partially filled with water and partially filled with air is a zone of aeration. Immediately above the water table is a capillary fringe where water moves by capillary action above the level of the water table. Groundwater moves in response to gravity and capillarity, although capillarity is usually quite slow.

Figure 14.2
The water table represents the top of the saturated zone where rock pore spaces are completely filled (see lower inset); whereas above that level rock pore spaces contain both air and water (see upper inset).
From Charles C. Plummer and David McGeary, Physical Geology, 4th ed. Copyright © 1988 Wm. C. Brown Publishers, Dubuque, Iowa. All Rights Reserved. Reprinted by permission.

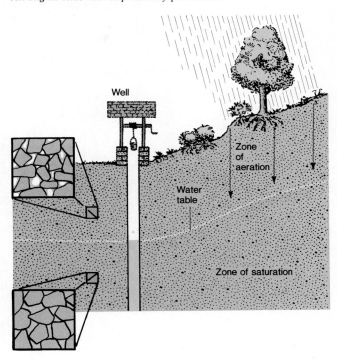

Figure 14.3
Examples of perched water tables where impermeable veins of shale preclude percolation of water through the shale to a lower level.
From Charles C. Plummer and David McGeary, Physical Geology, 4th ed. Copyright © 1988 Wm. C. Brown Publishers, Dubuque, Iowa. All Rights Reserved. Reprinted by permission.

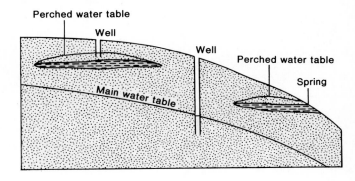

A **perched water table** may exist when an impervious layer of rock in a specific area resists percolation of water to lower regions where pore space is available to be filled (figure 14.3). A zone of saturation is perched above this impervious layer of rock; hence, a water table is found

Figure 14.4
Springs along canyon wall of the Grand Canyon of the Colorado River. © Peter Kresan

Figure 14.5
Pattern of an artesian structure. Water in a sandstone aquifer has a head of hydrostatic pressure that produces a flowing well below the level of the zone of saturation in the aquifer. From Charles C. Plummer and David McGeary, Physical Geology, *4th ed. Copyright © 1988 Wm. C. Brown Publishers, Dubuque, Iowa. All Rights Reserved. Reprinted by permission.*

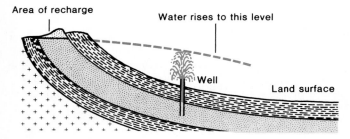

above where it would be if the impervious layer of rock was not present. Porous rocks, like sandstone, are able to hold large quantities of water, whereas nearly impervious rocks, like granite, usually have almost no water contained within them.

Wells are driven to the water table to serve all the usual human needs for domestic purposes, irrigation, or waste disposal. Natural springs occur where the water table is exposed as the result of diastrophic or erosional forces (figure 14.4). Some areas are especially known for their springs. In the thousand springs area of Idaho, for example, springs occur as water percolates through jointed basalt to be slowed by layers with fewer joints (figure 14.5). That zone has been exposed by erosion and springs issue from the valley sides high above the Snake River.

Groundwater Resources

Under quite special circumstances, water may flow from a well without the necessity of pumping. Such **artesian wells** exist where a tilting, porous, permeable, water-saturated bed is contained within two essentially impervious beds dipping at the same angle (figure 14.5). If the level of the saturated zone in the porous and permeable stratum is at a higher elevation than the spring or well, water will flow as a result of hydrostatic pressure. Such artesian wells exist in a number of places, notably in the Great Artesian Basin of east-central Australia. Overuse of such artesian wells may result in water discharge exceeding water recharge. If there is a steady drawdown in the saturated bed, or **aquifer,** the head of hydrostatic pressure may be lost, and what was once an artesian well may now have to be pumped. Continued overuse of water from the aquifer may result in complete depletion of the subsurface water resources. Subsurface supplies depend on recharge of water from the surface. When drawdown from pumping exceeds the rate of recharge, the water table drops. In many areas of the world, a rapidly expanding population using larger and larger quantities of water is causing the water table to drop alarmingly.

Of course, it is possible for aquifers to be depleted of available water resulting in the necessity of drilling deeper and deeper wells. Eventually, such an aquifer may be exhausted by excessive pumping and water use. When that happens, the wells run dry, and new sources of water must be found.

The Ogallala Formation, an aquifer that underlies much of the western Great Plains, is recharged from moisture that falls in and adjacent to the Rockies. Extravagant use of water in the Great Plains has resulted in an alarming drop in the level of the water table (figure 14.6). Since recharge does not begin to equal the amount of water used, there is the real danger that the Ogallala will cease to be a major water resource in several areas of the Great Plains early in the next decade. Because water must move great distances from the margins of the Rockies, the rate of recharge is slow in any case. Rapid drawdown and slow recharge accelerates the rate of depression of the water tables. Falling water tables are the rule rather than the exception in virtually all areas of the United States. Other problems also affect our groundwater.

An ominous increase in the amount of subsurface chemical or biological pollution has occurred as the result of excessive dumping of chemical wastes or the careless treatment of domestic or industrial sewage (figure 14.7). Since so many communities depend on wells for their water, this underground pollution is a mounting worry that matches the concern for surface water supplies. There seems little doubt that safe and dependable water supplies, especially for areas with a rapidly increasing population, will become a greater and greater problem in the years just ahead (figure 14.8). Even areas like Florida where precipitation is abundant must make plans for additional supplies of water. It is logical for central and south Florida to look towards north and west Florida for future water supplies. It behooves coastal regions to look seriously at the prospects of desalinization of sea water, since

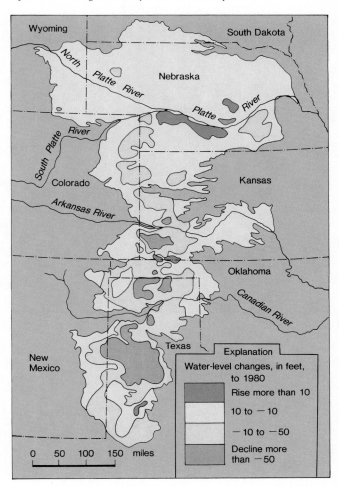

Figure 14.6
The Ogallala Formation is a major aquifer for the High Plains portion of the Great Plains. Water level drawdown, especially in the southern portion of the region, is reaching serious proportions.
After U.S. Geological Survey 1982 Annual Report.

population growth in all sections of the state militates against one region supplying another. The water supply problem can only become more acute over time.

The situation in southern California has been exacerbated by a burgeoning population and a subhumid climate. Californians have had to go further and further afield to the mountains of the north and east of the state to satisfy the mounting water need. Farmers in the Great Valley compete for some of the same supplies. Prudent planning for the area requires not only conservation of water, reuse of existing supplies, and the construction of more reservoirs to catch Sierra Nevada runoff, but the active exploration of desalinization of sea water. The pollution of water through the use of pesticides and herbicides is creating a problem for many species of animals, which habituate estuarine environments.

Groundwater and Solution Topography   239

Figure 14.7
Some ways pollutants may contaminate groundwater supplies.
From Charles C. Plummer and David McGeary, Physical Geology, 4th ed. Copyright © 1988 Wm. C. Brown Publishers, Dubuque, Iowa. All Rights Reserved. Reprinted by permission.

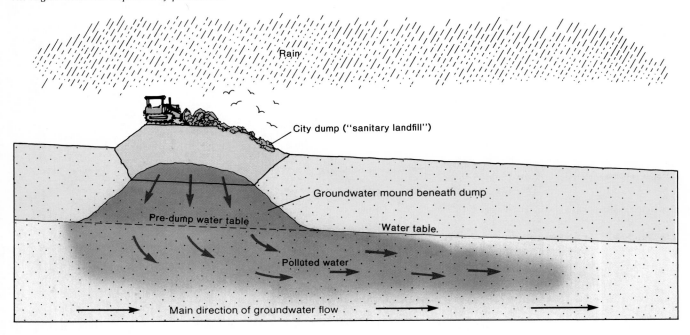

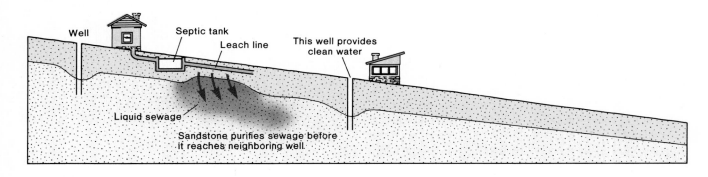

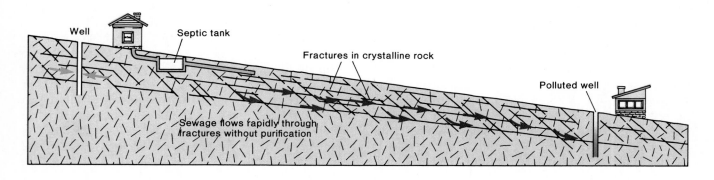

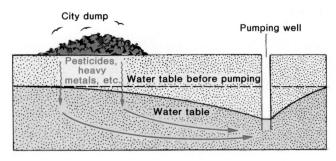

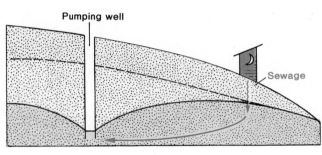

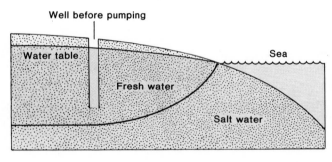

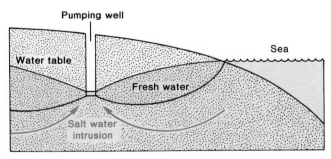

Mineral Waters, Geysers, and Hot Springs

Water, of course, dissolves soluble mineral materials as it percolates downward, and these mineral waters, or mineral springs, are often sought after as curatives for various physical ailments. Certain regions have become virtual spas (e.g., Hot Springs, Arkansas) because of the presumed therapeutic qualities of warm waters or mineral waters. Because of "miracle" cures at places like Lourdes in France, the economic livelihood of a region may depend on the people who come to take the cure. Other waters with few dissolved minerals or with "sparkling" qualities because of dissolved gas have been bottled for sale to the public. This sale of bottled water has generated a growth industry in the United States.

In areas of volcanic activity, groundwater may come in contact with magma or hot rocks and issue forth at another location as a hot spring. Under certain conditions, an underground chamber connected to the surface by a tube may accumulate enough water and steam for an explosive eruption of the hot water and steam. Such features are known as **geysers** (figure 14.9). Yellowstone Park is famous for its array of hot springs and geysers. Old Faithful, long known for eruptions at approximately one hour intervals, has become less reliable in recent years, apparently because underground chambers were disturbed or rearranged during earth tremors. Several volcanic areas of the world, like the Philippines, New Zealand, and Japan, are known for their hot springs (figures 14.10 and 14.11). Oftentimes, materials dissolved in such water are deposited at the place where the spring or geyser issues from the earth. Where the amount of water percolating downward to the hot rock may be quite limited so all the water flashes to steam, which issues at the surface in a kind of steam vent, a feature called a **fumarole** is produced.

The rising demands for energy and the precarious supply of hydrocarbons has led to the exploration of alternative energy sources. Steam produced by volcanic activity has led to the development of geothermal electric facilities using this steam (figure 14.13). The Pacific Gas and Electric Company has an operational facility in California. Other sites are being explored in the West, and several countries overseas such as New Zealand, Japan, and Guatemala have operational or planned facilities.

In Iceland, hot water from volcanic areas has been used to heat businesses and commercial establishments in Rekyavik and elsewhere for many years. It appears likely that greater uses for waters heated by nature will be developed in the future. Exploration for reliable geothermal areas continues as alternative sources for conventional sources of energy are sought.

Surface and subsurface water supplies are essential to life, and they must be managed prudently to insure that human needs are met. Running water at the surface is the most important erosional and depositional agent. In addition, however, there are important landforms, with both surface and subsurface ramifications, which are created principally by the effects of groundwater. A description and interpretation of such landforms follows.

Figure 14.9
How geysers erupt.
Source: Modified from W. R. Keefer, 1971, U.S. Geological Survey
Bulletin *1347.*

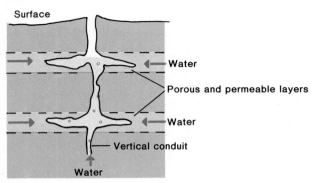

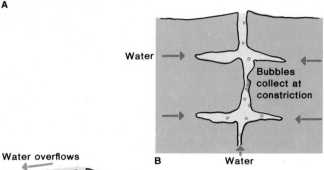

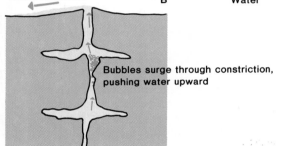

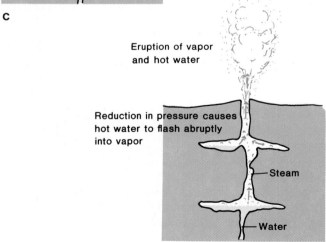

A

B

C

D

Figure 14.10
Steam issuing from volcanic vent on the slopes of Mt. Me-Akan, Hokkaido.
Japan National Tourist Organization.

Figure 14.11
Siliceous sinter deposited in Norris Geyser Basin, Yellowstone National Park.

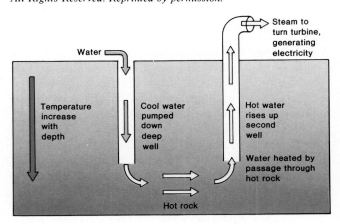

Landforms Created by Groundwater

Although pure water has a very limited effect in dissolving rock, groundwater has dissolved carbon dioxide and
oxygen in falling through the atmosphere, and the products of organic decay add more carbon dioxide to the water.
Although many rocks are quite resistant to what amounts
to a very dilute solution of carbonic acid, created from the
water and carbon dioxide, some rocks are not. Limestone,
dolomite, and gypsum are especially susceptible to solution by such water, and, in certain circumstances, marble
will succumb fairly readily as well. Halite and anhydrite
will also dissolve, but exposures are limited, and these materials are negligibly involved in the production of solution
topography. Solution adds another element of geomorphological change to the earth's landscapes as water dissolves away soluble rock to produce new surface and
subsurface forms.

Figure 14.14
Small sinkhole developed in gypsum in northwestern Oklahoma.

Figure 14.16
Sinkhole filled with water in the karst area of western Kentucky.

Groundwater seeping into cracks and joints in susceptible rocks begins to dissolve soluble materials away. Over time, the cracks may be enlarged into larger fissures and caves. The roofs of certain fissures and caves may collapse, creating sinkholes at the surface. The first such sinks to develop are usually small and frequently cone-shaped (figure 14.14). These features, called **dolines,** may be expanded as the underground caves develop into caverns of very large size. Enlarged sinks, often of significant dimensions, called **uvalas,** develop (figure 14.15).

Sinks may develop as percolating water along joints enlarges them to form sinkholes. The number of sinks at the surface is largely contingent on the number of joints, time of exposure, and quantity of precipitation. When sinks are plugged with debris, water may collect, and ponds or lakes may dot the surface (figure 14.16).

Eventually, much of the original surface may have been dissolved away, leaving behind only a few residual steep-sided hills, with summits at former surface levels variously called **haystack hills, hums,** or **mogotes** (figure 14.17). Some of these forms develop quite bizarre shapes. Some knobby, elongate forms, resembling large cucumbers, have been called **pepinos,** especially in Puerto Rico. Others with steep sides and pointed tops have produced the exotic forms so often characterized in Chinese paintings, which are representative of karst in southeast China and adjacent sections of Vietnam. In other areas, intricate fluting occurs as water seeks out and dissolves the least resistant portions of soluble rocks. All of these varieties of karst towers add pattern and character to old age karst regions.

The character and frequency of sinkholes and attendant subsurface forms varies depending on the thickness of the limestone member(s); its joint patterns; its dip; the amount, intensity, and distribution of precipitation; the character of surface vegetation; and the time of exposure. This means that although **karst** regions have certain features in common, great variations in forms occur because of the differences in lithological, biological, and meteorological conditions that exist from place to place. Obviously, a series of flat-lying limestone beds with a regular joint pattern will develop quite differently than a steeply dipping set of beds with an irregular joint pattern if both experience the same climatological and biological conditions. The regional variations in lithology, stratigraphy, and climate have produced a bewildering array of karst forms. The scientific literature is replete with detailed studies of karst features in different areas of the world.

Some have attributed a cycle of erosion to solution topography with youth being characterized by the beginnings of caves and a few small sinks at the surface, with most drainage still being affected by normal surface streams. Maturity is said to have been reached when there are many large surface sinkholes, much of the surface drainage has moved underground, and the whole area is honeycombed with an intricate array of caves, caverns, and galleries (figures 14.18 and 14.19). In old age, most of the

Figure 14.17
A haystack hill on the Luzon Plain being quarried for limestone.

Figure 14.18
Development of some of the forms characteristic of karst topography.
From Arthur Getis, Judith Getis, and Jerome Fellmann, Introduction to Geography, *2d ed. Copyright © 1988 Wm. C. Brown Publishers, Dubuque, Iowa. All Rights Reserved. Reprinted by permission.*

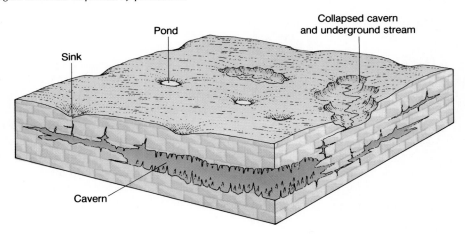

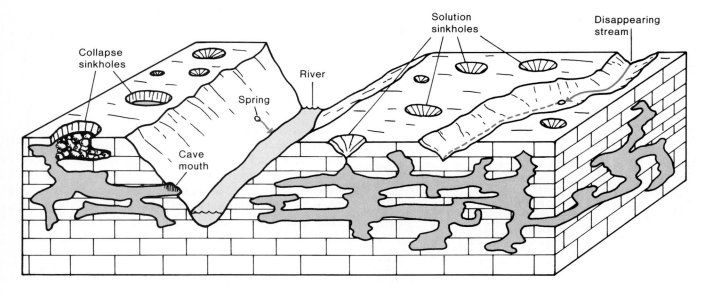

Figure 14.20
*Aerial photograph of the "Cockpit Country" karst area of
Jamaica.*

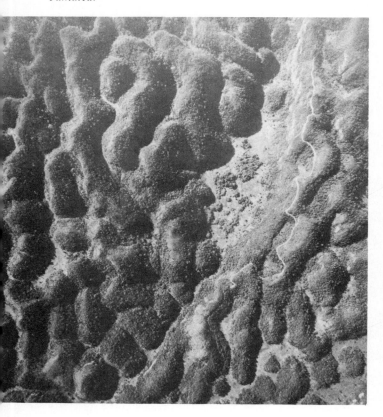

original surface has been reduced, and surface drainage
patterns have been established on a new, lower surface,
which is resistant to the effects of solution. This lower
surface is still surmounted by a series of steep-sided
honeycombed hills not removed by the process of solution.
This notion of a cycle of erosion as just described may not
be entirely credible, since it assumes a protracted period
of static conditions that probably does not exist, and it ig-
nores the significant differences in environmental conditions that exist in different karst areas.

Just as solution dissolves rock materials, so are those
materials deposited as the amount of dissolved carbon
dioxide in the water is reduced. Beneath the surface in
caves, an array of such depositional features called **speleothems** are deposited. Water dripping from the ceiling
of a cavern may produce small columnar structures on the
floor of the cave called **stalagmites,** and they are matched
by icicle-like protuberances from the ceiling called **stalactites** (figure 14.21). Eventually, these will fuse into **columns** reaching from floor to ceiling. Waters with a lot of
dissolved lime material may deposit these materials as they
issue from springs. These calcareous deposits are fairly
commonplace as waters issue from long underground trajectories in soluble rock areas.

The formation of caves has long been a subject of
considerable geomorphological debate. Some suggest that
subsurface streams cut caves above the water table. Still
others maintain that caves are cut essentially at the level
of the water table, and others believe that solution below
the saturated zone created them. It seems likely that caves
have been created in part from all these circumstances.

Figure 14.21
The development of stalactites, stalagmites, and columns.
From Charles C. Plummer and David McGeary, Physical Geology,
4th ed. Copyright © 1988 Wm. C. Brown Publishers, Dubuque, Iowa.
All Rights Reserved. Reprinted by permission.

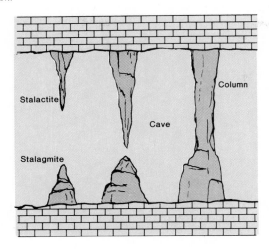

Figure 14.22
Turner Falls in the Arbuckles of Oklahoma deposits
travertine as it flows toward the plunge pool.

Figure 14.23
Section of topographic map in karst area of Kentucky.
From Mammouth Cave, 1:62,500 Quadrangle, USGS.

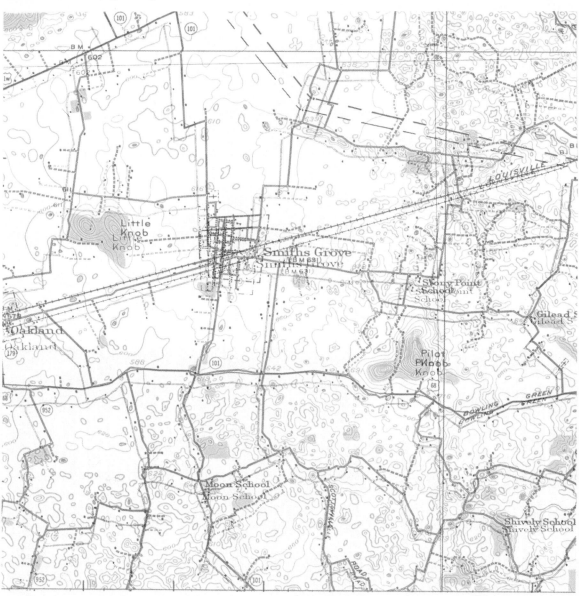

Some large caves, such as Carlsbad Cavern in New Mexico, exist where there are few surface manifestations of karst. Every cave is unique, and individual characteristics stem from a bewildering array of rock characteristics, stratigraphic circumstances, and climatological characteristics.

Solution topography is called karst after the Carso Plateau in Yugoslavia where sinkholes, caves, and caverns are omnipresent features. Karst is prominent in a number of areas in the United States, notably in west-central Kentucky, southern Indiana, central Florida, north-central Tennessee, and the Great Valley of Virginia and Tennessee. In addition to Yugoslavia, karst forms are especially significant along the Great Bight of Australia, southeastern China and adjacent Vietnam, southern France, Greece, the Andalusia region of Spain, northern Yucatan of Mexico, western Cuba, Jamaica, and northwestern Puerto Rico.

Although the total karst area in the world is not great, the landforms produced are spectacular and readily recognizable. In most areas, people's activities are restricted by karst forms, although tourism may mitigate some of the natural restrictions. Where collapse occurs, damage may be done to existing economic ventures or urban areas. On the other hand, caves are tourist attractions that enhance the economies of some karst areas.

Few landforms have created such an array of nomenclature for different forms. The differences in karst terrain from place to place is certainly as significant as similarities. Varying chemical characteristics of rock, different levels of hardness, heterogeneity of stratigraphic circumstance, different climatological regimes, and varying vegetation patterns create great varieties in karst forms.

In a sense, solution is relatively restricted in area to those places with exposed soluble rock. To that extent, solution is several orders of magnitude less extensive in its effects than running water or moving ice. Nevertheless, in certain locales, karst topography is *the* dominant aspect of a particular geomorphological milieu. In some, surface streams may be largely or completely absent, and the movement of water through joints and cracks may create an area of physiological aridity.

Groundwater adds another dimension to the forces that modify the face of the land. Other forces, like winds and waves, will be considered in subsequent chapters of this book.

Study Questions

1. What are the various factors that determine how much rain water soaks into the ground and becomes part of the groundwater supply?
2. Distinguish between porosity and permeability.
3. Compare and contrast water table with perched water table.
4. Explain how artesian springs or wells may develop.
5. Define aquifer. What is the significance of the Ogallala Formation in the Great Plains of the United States?
6. Compare and contrast geysers and fumaroles.
7. Compare and contrast dolines and uvalas.
8. Discuss a proposed cycle of erosion for karst forms.
9. List and briefly describe the principal forms of speleothems.
10. In which areas of the world is karst a particularly significant phenomenon?

Figure 14.24
Karst topography in central Florida is revealed on this infrared satellite image.
NASA.
From Carla W. Montgomery, Physical Geology. *Copyright © 1987 Wm. C. Brown Publishers, Dubuque, Iowa. All Rights Reserved. Reprinted by permission.*

Selected References

Baldwin, H. L., and McGuinness, C. I. 1963. *A primer on ground water.* Washington: U.S. Geological Survey.

Fahy, G. 1974. *Geomorphology.* Revised edition. Dublin: The Educational Company of Ireland Ltd.

Fernald, E. A., and Patton, D. J. 1984. *Water resources atlas of Florida.* Tallahassee: Institute of Science and Public Affairs of Florida State University.

Oberlander, T. M., and Muller, R. A. 1987. *Essential of physical geography today.* New York: Random House.

Plummer, C. C., and McGeary, D. 1985. *Physical geology.* 3d ed. Dubuque, IA: Wm. C. Brown Publishers.

Strahler, A. N., and Strahler, A. H. 1984. *Elements of physical geography.* 3d ed. New York: John Wiley and Sons.

15

Arid Landscapes

Bizarre forms are often characteristic of erosional features in desert landscapes. This natural bridge is found in Arches National Park, Utah.
Utah Travel Council.

And yonder all before us lie
Deserts of vast eternity.
Andrew Marvell, *To His Coy Mistress*

*A*rid landscapes occupy almost a third of the earth's terrestrial surface and they exhibit characteristics that appear to be alien in humid environments although a number of pieces of data suggest that similar forms are developed in humid regions. Some of these comparable features in humid areas may be masked by a thick soil or vegetative mantle, and particular forms may be more transient than in arid regions because of the speed with which weathering and erosion accomplish their work. A certain harshness—an angularity or a kind of abruptness—characterize desert landscapes, as opposed to the rounded forms, which are more prevalent in humid areas. Of course, diastrophic forces and mountain glaciation, especially, may introduce angular elements into humid landscapes. Except in dunes, the rounding, which is so characteristic of humid regions, is less frequently encountered in arid portions of the earth.

Figure 15.1
Some of the angular forms encountered in Monument Valley, Arizona.
Arizona Office of Tourism.

Figure 15.2
A portion of the Grand Canyon etched by the Colorado River.
Arizona Office of Tourism.

Permanent streams in deserts are rare. More often than not, streams that traverse desert regions have their origins outside of it. These are known as **exotic streams.** The Nile is a famous desert river, but it owes its life to the highlands of Ethiopia in its Blue Nile tributary and to the swampy Sudd and Lake Victoria, which gives birth to the White Nile in the wet and dry tropics. Monsoon rains swell the Blue Nile and until the High Aswan Dam backed up Lake Nasser, the annual Nile floods in timing and dimension depended principally on the monsoon pulse of water added by the Blue Nile. The timing and dimensions of the Nile flood greatly influenced the rhythms of planting and nurture of the agricultural crops in the Nile River valley and delta. The silt deposited by annual floods renewed soil fertility and increased the size of harvests.

The Colorado River in the United States has its source in the well-watered Rockies and wends its erosional way across the Great Basin deserts and Colorado Plateau before depositing its enormous sedimentary load in the Gulf of California. In its movement through arid country, it has etched the greatest terrestrial canyon (Grand Canyon) on the earth's surface (figure 15.2). Its tributaries and other rivers have cut some of the most interesting canyons and erosional features on the face of the planet. The waters removed from it for irrigation have produced fertile agricultural oases in areas that would otherwise be deserts.

Wherever desert rivers exist in middle- or low-latitude environments, they loom large in human use, since the water brings life to otherwise sterile regions. Both exotic streams crossing the desert and ephemeral streams within it have etched the surface in wondrous ways. Although surface water is generally scarce in dry regions, running water is by far the most important gradational agent in such regions.

Desert Landscapes Shaped by Running Water

Running water is the most significant erosional and depositional agent in most areas, and deserts are no exception. Although the amount of regolith (weathered rock material above the bedrock) in dry regions is typically reduced because chemical weathering is inhibited, the exposed surfaces, often unprotected by a mantle of close-growing vegetation, and the characteristically irregular, but hard showers, make transitory desert streams awesome agents of erosion and deposition. The permanent streams cutting across desert regions may be enormous agents of erosion and deposition.

The load of debris carried and the erosional impact of an ephemeral stream in flash floods are wondrous to observe. Travelers should heed warnings of flash flood dangers in desert regions. Unwary campers and travelers have been swept away when they have failed to consider the dangers of such streams and a cloudburst upstream has changed a dry river bed or a mere trickle of water into a raging torrent.

Figure 15.3
This river rising in the Zagros Mountains provides essential water to the adjacent desert.

Figure 15.4
This desert stream at low water has little erosive impact. In floods, the stone tools in the riverbed are very effective erosive agents.

Figure 15.5
A portion of Canyon de Chelly in Arizona. Arizona Office of Tourism.

The Grand Canyon of the Colorado River is an example of a powerfully erosive stream that was able to cut a tremendous canyon as the Colorado Plateau was slowly uplifted across the river course. The entire canyon area of southeastern Utah and northern Arizona represents a myriad of landforms created by a slow uplift of land across streams actively cutting downward in areas of strongly jointed sedimentaries, especially sandstones. The array of forms, including steep-sided canyons, mesas, buttes, arches, and needles produces a landscape of stark beauty. Hundreds of western movies have capitalized on beautiful scenery and ideal weather for photography in these desert environs. Tourists flock to such scenic wonderlands, and in many such regions, tourism may be the most important industry.

The erosional and depositional effects of desert streams can only be called erratic, both temporally and in terms of characteristics of erosion or deposition after a given storm. A downpour in areas with steep slopes may enable the ephemeral (short-lived) stream to move massive amounts of material including rocks of considerable size. The cloudburst precipitating such flow might be so short-lived, however, that materials moved a few hundred feet might be redeposited in the river bed as the water sinks into its own bed and yields to the desiccation of low humidities, permeable soil, and frequent winds. Erosion and fill that is in and adjacent to desert streams creates patterns significantly different from those commonly encountered in humid regions. Most of those differences, but not all, may be accounted for by differences inherent in permanent and intermittent streams.

Figure 15.6
Typical pattern of an arroyo in the American Southwest.

On gentler slopes, there is a tendency for the short-lived stream to be quickly choked with the debris it carries. This creates a braided pattern in many of these streams. As we have seen in earlier sections of this book, braiding occurs more frequently in streams with a high percentage of sand and larger debris in its load, whereas meandering is more typical of a stream carrying mostly clay and silt. Because streams in dry areas frequently drop their load in their own bed, they are characterized by steep-sided valley walls and relatively flat valley bottoms. Such stream valleys are called **arroyos** in the southwestern United States, **dry washes** in the American Northwest, **barrancas** in much of Latin America, and **wadis** in the Middle East and North Africa. The smaller amount of slope wash along valley sides tends to maintain the steep valley sides.

Basins of Interior Drainage

A few arroyos make their way to permanent streams, but most do not. Some reach basins of interior drainage at certain times, but not at others. Such basins of interior drainage may be filled with ephemeral or permanent lakes. Typically, these are salt lakes. The dissolved minerals carried by all streams are concentrated by the effects of solar evaporation in a lake with no outlet. Certain such lakes have brines so concentrated as to preclude most forms of life. Great Salt Lake in the western United States and the Dead Sea of Israel have such high concentrations of dissolved salts that bathers are readily floated by the high specific gravity of the water. The Great Salt Lake has expanded markedly in response to more pluvial (wetter) conditions in the decade of the 1980s. This has made it necessary to move some transportation facilities and has created some economic difficulty. In 1987, water was being diverted to an adjacent dry basin by a canal in the hope of lowering the water level in Great Salt Lake.

Figure 15.7
A basin of interior drainage in Iran with a salt-encrusted playa surface.

Figure 15.8
A plethora of angular forms in Arizona's Monument Valley.
Arizona Office of Tourism.

Some intermittent desert lakes appear to be sterile. Yet, after they're filled by periods of above-normal rainfall, the lake may soon be filled by brine shrimp and other creatures whose eggs lay dormant in lake sediments awaiting more favorable times. A person unfamiliar with such dry lakebeds may be surprised to see a lakebed teeming with life a few days after it has been temporarily filled by a desert cloudburst.

Lake basin flats where water has been evaporated are called **playas.** Such playas may be occasionally filled with water. In the Great Basin country of the United States, many dry lakebeds from earlier, wetter times exist. It's clear, too, that certain existing lakes are simply small remnants of much wetter times. Larger lakes and more lakes existed during the period of Pleistocene glaciation. These periglacial remnants from more pluvial periods are significant landscape features.

Some of these playa flats assume a peculiar importance. The very level surface of the Bonneville salt flats in Utah, for example, has been the site of frequent trials for high-speed land vehicles. In fact, the world speed records for land vehicles have almost always been established on the Bonneville salt flats. The dry lakebed around Edwards Air Force Base in California provides enormous expanses for landing high-speed experimental aircraft or space vehicles like the space shuttle. It's no accident that much of the flight testing of exotic designs and simulated combat between high-speed aircraft are carried on in such fortuitously situated regions. The large expanse of landing area is forgiving of some design imperfections; the broad expanse provides extra runway for unpowered vehicles like the space shuttle; and a high incidence of clear weather makes it possible to fly most days of the year.

The smooth surfaces of playas contrast sharply with the angular forms or sharp edges that exist in so many desert landscapes. Where there is little soil mantle, the exposed bedrock responds dramatically to differences in hardness. This is especially true in areas with a series of sedimentary beds of differential hardness. The weak and resistant beds are obvious because the softer rock is eroded away first. The walls of the Grand Canyon in Arizona exhibit this characteristic where resistant rocks stand out as distinct escarpments, and less resistant strata have been notched back.

Angular edges have led quantitative geomorphologists to measure such things as the number of crenulations (wavy edges) in an exposed edge, ratios between crenulations and length of exposed edge, and so on. There is something immensely satisfying about recording known quantities and attempting to develop relationships in an effort to produce operant theories. The testing of such theories is made enormously complex by the assumptions that must be made and the enormous amounts of time, by human standards, that are required to produce existing forms. Computer simulations, which may speed observed processes exponentially, afford some additional opportunities to test assumptions. A major problem in developing comprehensive theories will always be a changing set of climatological, biological, and geological variables, which may upset a destined result prior to completion of an erosion cycle. The circumstances of climate, vegetation, stratigraphy, slope, and lithology, which may exist in one place may never be duplicated elsewhere. It is difficult to develop the generalizations that can represent plausible theory when so many variables of time and circumstance are involved.

Figure 15.9
Erosional retreat can leave isolated forms like mesas and buttes.
From Charles C. Plummer and David McGeary, Physical Geology,
4th ed. Copyright © 1988 Wm. C. Brown Publishers, Dubuque, Iowa.
All Rights Reserved. Reprinted by permission.

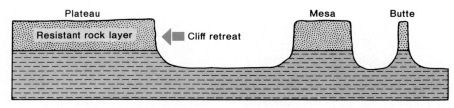

Figure 15.10
Mesas are common erosional remnants in the American Southwest.

Figure 15.11
A mesa and a needle.

Erosional Landforms

In several locales, a resistant cap rock has protected less resistant rock beneath it. These tabular surfaces protect the steep sides of a hill called a **mesa** (figure 15.9). Mesas are eventually reduced in surface area to smaller features called **buttes,** and buttes, in turn, may be reduced to very small features known as **needles.** Various efforts have been made to establish quantitative limits between mesas and buttes. Any such distinction is arbitrary and probably irrelevant. The ultimate fate of the needle, of course, is destruction, leaving behind a plain-like surface. Mesas and buttes are typical of erosional features encountered in desert (or certain semiarid) regions, but numerous other forms give character to the landscape (figures 15.10 and 15.11).

Along the margins of many, perhaps most, arid mountains are broad, gently sloping skirts of erosional surfaces called **pediments.** These pediments have developed as a consequence of the degrading of rock surfaces over time by lateral erosion and the normal consequences of fluvial erosion. These erosional surfaces are usually masked with the debris of deposition and rockfall, but they do form sloping surfaces, which grade gradually from plains to the steeper mountain slopes around the periphery of most desert mountains.

Running water is generally less effective in reducing the landforms to grade in desert regions than in humid areas because most streams are ephemeral, and the stream systems are generally less well-developed. Water in channels and sheet wash both appear to be less effective because of the lack of moisture, except in exotic streams. Nevertheless, running water is the most important erosional agent in most desert regions.

Figure 15.12
Alluvial fans develop in desert areas as streams leave steep slopes and flow into adjacent plains.
From Charles C. Plummer and David McGeary, Physical Geology, 4th ed. Copyright © 1988 Wm. C. Brown Publishers, Dubuque, Iowa. All Rights Reserved. Reprinted by permission.

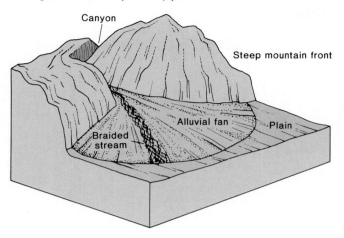

Figure 15.13
Alluvial aprons skirt these hills in the desert of Iran.

Figure 15.14
A miniature alluvial fan along a road cut in the western part of Florida. Such features are transient.

Water-Deposited Forms

On the other hand, the scarcity of close vegetative cover exposes surfaces to rapid erosion when erratic rains come. Certain depositional forms of desert water courses, however, have no observable large-scale counterparts in humid regions.

Alluvial fans are deposited at the base of mountains in arid landscapes (figure 15.12). These forms, a bit like a steep sided delta deposit on dry land, frequently coalesce when the fans of two adjacent ephemeral watercourses overlap and produce **alluvial aprons** (figure 15.13). These aprons frequently mask a pediment surface beneath. If these forms develop in humid areas to any extent, they tend to be removed by subsequent slope wash, or they are cloaked with such a mantle of vegetation as to mask their profile and character.

Although running water is generally less effective as a grading agent in arid regions than in humid areas, it seems abundantly clear that water is by far the most significant of all the gradational agents. Other agents of erosion contribute to the modification of the physical landscape, and, in a few areas, their effects are dominant, but usually they add elements of detail in the changing face of landscape.

Such an agent is wind. In the past, too much significance was attributed to its work. It is still considered a potent force, but is several orders of magnitude less significant than running water.

Figure 15.15
Possible origins of some of the basin and range topography of the western part of the United States.
From Charles C. Plummer and David McGeary, Physical Geology, 4th ed. Copyright © 1988 Wm. C. Brown Publishers, Dubuque, Iowa. All Rights Reserved. Reprinted by permission.

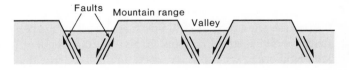

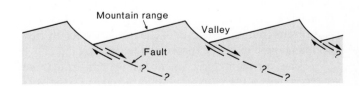

Figure 15.16
A portion of the basin and range area of Utah. Note stony surface in foreground and sparse vegetation.
Utah Travel Council.

Figure 15.17
A rocky desert surface left by deflation.

Work of the Wind

The work of the wind is more significant in arid landscapes than in others because the effects of abrasion have direct impact on the unprotected regolith. The general absence of vegetation, or the sparse cover, over broad areas means that particles carried by the wind can work more effectively than in humid areas cloaked by vegetation.

Wind erosion and deposition is characterized as **aeolian** (or eolian). Wind erosion occurs as materials are picked up and moved from one place to another. This removal of fine particles is termed **deflation** (figure 15.17). Blowouts, or depressions, may occur as a result of deflation. It also seems clear that such removal of fine materials may be responsible for the surface character of rocky deserts, called **regs** in North Africa, although water erosion probably has also had a hand in the creation of regs.

In addition to the removal of materials through deflation, the wind may also abrade surfaces using the particles carried as grinding tools. Grains of sand frequently are angular and tend to grind, etch, and polish areas they come in contact with. In certain regions, desert bedrock surfaces may be polished to a sheen by this constant abrading action. Where this occurs, **desert varnish** may be observed. It also seems clear that the patina on rock surfaces results from the presence of chemicals, especially iron and manganese oxides, exposed at the surface.

Wind carries materials in modes very similar to those we recounted in the case of running water (i.e., the very fine particles are carried in suspension, the somewhat larger grains are bounced along the ground by saltation, and somewhat larger particles still are rolled along the surface by traction). It seems clear that, except in rare circumstances, sand can only be lifted a few feet off the surface. Normally this lift is less than 3 feet (1 meter) above the surface. Simultaneously, the entire surface may

Figure 15.18
The joint pattern in this sandstone plus the work of water and wind has produced this peculiar form.

move downwind as a result of the constant impact of grains moved by saltation. This movement is called **creep,** but it differs from soil creep in being generated by the impact of wind-borne grains as opposed to soil creep, where water saturated soil moves downslope in response to gravity (figure 15.19).

Figure 15.19
Dune migration occurs as the result of countless individual grain movements.
From Carla W. Montgomery, Physical Geology. Copyright © 1987 Wm. C. Brown Publishers, Dubuque, Iowa. All Rights Reserved. Reprinted by permission.

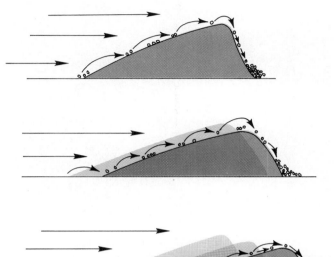

Figure 15.20
Patterns of common dune forms with prevailing winds indicated by an arrow: (A) *barchan,* (B) *transverse,* (C) *parabolic,* (D) *longitudinal.*
From Charles C. Plummer and David McGeary, Physical Geology, 4th ed. Copyright © 1988 Wm. C. Brown Publishers, Dubuque, Iowa. All Rights Reserved. Reprinted by permission.

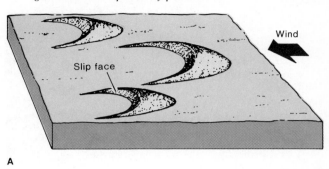

A

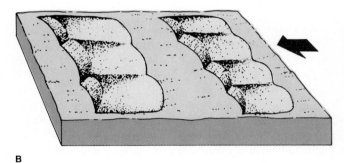

B

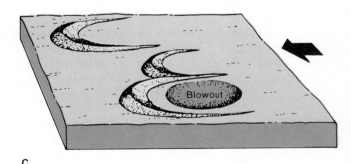

C

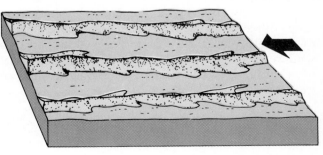

D

In certain areas of the desert, notably the Sahara and the Arabian, large expanses of the landscape are covered by sand that has been transported and deposited by the wind. These **ergs,** or sand seas, may literally cover hundreds or thousands of square miles. Smaller examples of sand deposits include an array of different kinds of dunes, usually formed because of prevailing wind directions, velocity of the wind, rock or vegetation obstruction, or availability of sand source.

Dunes

The common dune types include **barchans,** which are crescent-shaped in plain view with the horns or cusps of the dune pointing downwind (figure 15.20). These dunes tend to march over rocky surfaces, singly or in groups, in areas where the amount of sand available is limited, and the winds come from a strong prevailing direction perpendicular to the long axis of the dune.

Transverse dunes are somewhat similar to barchans; individual dunes extend over more area and occur in regions where the sand supply is greater than for barchans. They have convex faces to the windward, like the barchan, and they may degenerate into barchans if the sand supply is diminished.

The **longitudinal dune** occurs in long parallel ridges in great abundance. The exact orogeny of these dunes is in question, but it appears that they have developed in an intermediate position between two dominant wind directions.

Arid Landscapes   259

Parabolic dunes are sometimes referred to as blowout dunes. The horns point upwind and are usually anchored by vegetation. Such dunes typically occur in areas with abundant sand.

Most dunes are represented in most deserts, although barchans appear to be missing from the Great Australian Desert, and longitudinal dunes are rare in the deserts of North America. Active dunes continue to advance as grains of sand move up over a windward slope and move down the leeward slope to assume an appropriate angle of repose on the slip face. Many other dune forms are recognized, reflecting the complexity of deposition attending quantity of available sand, wind directions, and underlying structure or available vegetation. Various attempts have been made to establish elaborate classifications of dunes according to ground plan, height, and mobility or lack of it. The usefulness of elaborately classifying dune types is of questionable value, and, in any event, is beyond the scope of this book.

Of course, dunes also develop along many of the world's coastlines. The pounding waves grind rock fragments into smaller and smaller pieces. In addition, materials are transported from the land by proximate streams. These small particles, typically sand, are tossed on the shore or moved along the shore. Once on shore, the wind works and reworks the sand, forming dunes. In some locales with abundant sources of sand and prevailing onshore winds, dunes may cover a significant area. In southwestern France, for example, the dunes are pushed inland as far as 6 miles because the prevailing winds are from the west.

At periods of especially high tides and during storm surges, waves and tides may push sand substantially further inland than is normally the case. Sandy beaches are part of the charm of seaside resorts, but unstabilized dunes often create costly encroachments. Dunes are often stabilized by vegetation, either natural or introduced. Since sand has little readily available nutrients or water for plants, it is difficult for plant colonizers to take hold. In coastal regions, it is even more difficult because salt spray or salt water encroachment may limit further the number of plant species available for colonization. In many areas such as Florida, the sea oats, which are dune stabilizers, are protected by law. Not only do the plants add an element of stability to the dunes, but they are also picturesque additions to the beach landscape. Periodically, people make efforts to enhance dune stabilization by planting sea oats. These attempts have had mixed results.

Efforts are also made to stabilize dunes by the use of fences like snow fences. Such fences are used in arid regions as well as in beach environments. Such effort to stay the encroachment of dunes has often created more problems than have been solved. Dredge-and-fill operations where dunes have been eroded away are temporary solutions that usually do not have any long-lasting positive effects. Indeed, it is often true that the modification of natural processes is counterproductive because the dynamics of the natural forces are not understood.

Particles smaller than sand also are wind-transported. In certain circumstances, they assume considerable local significance. Sometimes the deposits may be consolidated over time, and they may become a useful agricultural resource. In other instances, the blowing silt of a dust storm may create conditions of lowered visibility, and the material may become a nuisance as it filters into a house or clogs operating machinery.

Figure 15.23
Loess bluffs near Vicksburg, Mississippi.

Figure 15.24
Loess bluffs at Missouri Valley, Iowa.
Wallace E. Akin.

Figure 15.25
Distribution of loess in the United States. Darker color represents thicker deposits.
Source: U.S. Bureau of Reclamation, 1960, and various other sources.

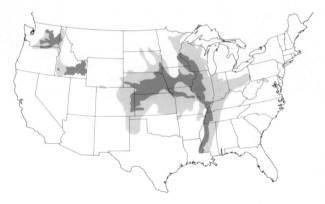

Loess

Loess is an exceptionally fine-grained aeolian deposit, which is not specifically related to desert regions, although one source of such materials may be in arid areas. It does appear that much loess was deposited during the Pleistocene. Vast areas of glaciofluvial material were left exposed to winds blowing across expanses of territory unprotected by vegetative cover. The early absence of vegetative cover in such exposures provided a ready source material for wind. Playa areas in desert regions may have served as a source for the fine-soil materials. It seems clear that floodplains in semiarid or subhumid landscapes, especially, have also served as a source for the silty material to be transported by the wind.

In any case, loess, which is composed dominantly of particles of silt and clay, has a number of interesting characteristics. First, it is composed of very fine particles, which feel like powder when rubbed between the fingers. Second, there is no discernible stratification (i.e., the material is essentially homogeneous throughout). Third, there is often an abundance of calcareous material present. Finally, loess tends to stand in almost vertical banks when cut through by erosion or some human agency. Even after prolonged exposure, this tendency to maintain an almost vertical

profile persists. In humid areas, steep-walled valleys are etched through loess quickly, and an intricate network of streams frequently develops where loess is exposed.

The enormous load of fine silt in the Hwang Ho in northern China is significantly related to the contributions of rivers like the Fen and Wei, which have eroded the great loess deposits of that area. The fact that loess maintains itself in almost vertical banks has led to the construction of caves and houses, especially in the loess regions of northern China. There may be catastrophic consequences when earthquakes occur in such regions because loess tends to collapse even with light tremors. The loess deposits of northern China seem to have originated, at least in part, in the drier areas of Mongolia. The strong winds of the outblowing monsoon carried along the finer materials from the surface of the Gobi Desert.

Figure 15.26
The desert is marked by beauty where wind and water create a sometimes bizarre landscape.
Daniel Ehrlich.

Some of the great loess areas of the world include portions of the Mississippi River drainage system; eastern Washington and western Idaho; the Great Plains; the Soviet Ukraine; along the Rhone, Danube, and Rhine River Valleys; in Soviet Turkestan; and northern China. Loess generally exists as a mantle over preexisting landscapes. It is thicker close to source regions and thins out rapidly away from such areas.

The persistence of the vertical banks so characteristic of loess deposits were poignantly revealed to the author during a recent visit to the Civil War battlefield at Vicksburg. The outlines of trenches, artillery emplacements, and earthworks ramparts are still plainly obvious more than 125 years after men of the Blue and Gray bled and died there during that hot summer of 1863.

Running water and wind are omnipresent erosional and depositional features affecting virtually every geomorphological landscape on the planet; solution and glaciation affect fewer and less extensive areas. The effects of waves and currents are essentially limited to the sea bottoms and margins. These influences will be examined in the following chapter. Of course, seas occupy the greatest portions of the planet, but our access to ocean bottoms is limited, and we must concern ourselves principally with the terrestrial landscape at the margin of the seas.

Figure 15.27
The desert has its own kind of stark beauty.

Study Questions

1. About what percentage of the terrestrial globe is arid?
2. What is an exotic stream?
3. What are the essential characteristics of arroyos, barrancas, and wadis?
4. What are the essential characteristics of playas?
5. Explain the relationships between and among mesas, buttes, and needles.
6. Compare and contrast pediments and alluvial fans.
7. Compare and contrast the erosional and depositional effects of the wind.
8. What kinds of efforts are made to stabilize sand dunes?
9. What are the essential characteristics of loess?
10. Where are some of the principal deposits of loess in the world?

Selected References

Bloom, A. L. 1978. *Geomorphology: A systematic analysis of late cenozoic landforms.* Englewood Cliffs: Prentice-Hall.

Gabler, R. E.; Sager, R. J.; Brazier, S. M.; and Wise, D. L. 1987. *Essentials of physical geography.* 3d ed. Philadelphia: Saunders College Publishing.

Plummer, C. C., and McGeary, D. 1985. *Physical geology.* 3d ed. Dubuque, IA: Wm. C. Brown Publishers.

Ritter, D. F. 1978. *Process geomorphology.* Dubuque, IA: Wm. C. Brown Publishers.

Strahler, A. N., and Strahler, A. H. 1984. *Elements of physical geography.* 3d ed. New York: John Wiley and Sons.

Thornbury, W. D. 1969. *Principles of geomorphology.* 2d ed. New York: John Wiley and Sons.

16

Oceans, Seas, Lakes, and Associated Landforms

The restless sea pounds the Oregon coast.
Daniel Ehrlich.

> *Myriad laughter of the ocean waves.*
> Aeschylus, *Prometheus Bound*

Only about 29 percent of the earth's surface is above sea level. If all the land above sea level were deposited in the seas, only about one-eighteenth of the total volume of water would be displaced. If the earth's crust were ironed out and reduced to a perfect sphere, ocean waters would cover the whole surface to depths of almost 8,000 feet (about 2,500 meters). In a very real sense, ours is a water planet. The availability of free water is *the* distinguishing characteristic of earth among planets in the solar system. Water is, essentially, the giver of life as we know it. Indeed, if, in some future time, extraterrestrials do visit or observe our planet, they are likely to give it their name for water rather than for earth.

The vast oceans contain about 300,000,000 cubic miles (1,230,000,000 cubic kilometers) of water. From this vast amount, about 80,000 cubic miles (328,000 cubic kilometers) are sucked into the atmosphere by evaporation and returned by precipitation and drainage to the sea. More than 24,000 cubic miles (98,000 cubic kilometers) of water fall as precipitation over the continents. This vast amount is essential to replenish surface and groundwater supplies, which nourish our ponds, lakes, rivers, streams, swamps, and underground aquifers. The remaining 56,000 cubic miles (230,000 cubic kilometers) of water fall directly back to the sea. This rhythmic cycle of evaporation, condensation, and precipitation is a kind of never-ending rhythmic dance, which has within it the pulse of life. This **hydrologic cycle** is the system that dominates climate and is responsible for the posttectonic face of the landscape. Failure of this cycle, initiated and perpetuated by the sun, would quickly result in the extinction of terrestrial life. Life in the sea would persist for a longer period, but eventually it, too, would flicker out as the oceans evaporated away. With the disappearance of life, earth would join the biological silence, which characterizes our sister planets in the solar system. Water is the giver and sustainer of life as we know it; without it, life is impossible.

Not only is water essential to life as we know it, but the liquid is endowed with several characteristics that make it unique among liquids. One such anomaly is the fact that water expands by about 9 percent when it freezes, whereas most liquids contract as they cool. As a result, ice appears on the surface of the water. Otherwise, it would sink, resulting in freezing at a depth that would effectively eliminate many forms of life in most bodies of water.

Another interesting characteristic of water is that it has the highest heat capacity of all liquids and solids except for ammonia. This characteristic permits oceans and seas to be giant heat sinks (i.e., they are able to absorb and store vast quantities of heat, which tends to minimize climatological extremes). A waterless planet would experience far greater extremes of temperature than is the norm for our good blue-green earth. We have already seen how water moderates the temperature extremes, which characterize land masses. The interplay between land masses and water bodies also affect pressure, wind, humidity, and other climatic elements.

Further, water dissolves more substances than any other liquid. It is almost a universal solvent. It is this characteristic that makes oceans and seas great storehouses for minerals, which have been carried into them by the great rivers of the world or have issued from vents in the ocean floor.

The Seas as a Resource and Waste-Disposal Site

In many areas of the world, these minerals have been extracted for human use. Salt is evaporated from seawater along many coastlines, especially in the developing societies. It is almost a universal truth that coastal societies that develop in an area that has some period of dry weather will extract salt through the solar evaporation of seawater. Potash is extracted from the Dead Sea by the Israelis, and magnesium is extracted from seawater along the American Gulf Coast.

It seems reasonable to expect that in the years ahead, more and more mineral resources will be extracted from the sea. The manganese nodules that litter the ocean floor in a number of regions are tempting targets for future mining. In fact, the reduction in the quality of ore deposits from the terrestrial sources and the increased demands of a burgeoning world population seem certain to increase mineral extraction from the sea. While the seas hold out the prospect for additional minerals, it should not be assumed that they will supply all our future needs. Similarly, although seas have an enormous capacity to absorb terrestrial wastes without apparent harm, there are limits to this absorptive capacity. Controversy rages as to what those limits are.

Seas have long been used as dumps for the world's wastes. Small numbers of people and simple products posed no significant threat to life in the sea, since the sea was able to cleanse itself readily of modest burdens of filth; however, a burgeoning population and a widening array of toxic wastes, including radioactive materials, portend great danger to and damage of the ocean environment. We are in imminent danger of so overloading the system as to create irreversible damage. We are in great peril if we fail to recognize this danger and to respond to it properly.

We are obviously concerned with the sea and the creatures in it, and we are also interested in crustal stirrings beneath its surface and especially with its relentless modification of the land where the two meet. Those characteristics of the oceanic realm add spice and variety to the mosaic of earth patterns. Expanding knowledge makes it possible to give a more complete description of the aquatic realm than was possible just a few decades ago.

Characteristics of the Lands beneath the Sea

Much of the mystery that formerly characterized the areas beneath the sea has been dispelled by more and more sophisticated exploration techniques and more active exploration programs, but numerous secrets remain. Indeed, we are treated to some new discovery on an almost weekly basis. Our knowledge of ocean depths and seafloor topography was meager until about 1920. At that time, sonic soundings were developed, and it became possible to map vast segments of the ocean floor heretofore unexplored. Submarine and countersubmarine warfare in World War II added a significant store of knowledge. As Axis and Allied submarines alternately played cat and mouse in a life-and-death struggle, new information about subsurface configurations, ocean currents, and physical characteristics of ocean waters was learned by military planners and later became a part of the public domain. Drilling along the continental shelf and scientific borings elsewhere have added information about the ocean bottom and the continental shelf. Deepwater submersibles, both manned and unmanned, with remote cameras, have helped to increase knowledge about the ocean.

In the late 1970s and early 1980s, for example, vents issuing hot water and dissolved minerals at great depth were discovered. Even more exciting at these depths was the discovery of new life forms depending on a food chain based on bacteria that are able to use the minerals from those vents directly. In turn, they are consumed by an array of tubeworms, small crustacea, and mollusks. Apparently, these vents are more commonplace than was once thought, and the base of the biotic pyramid depends on the direct use of dissolved minerals rather than photosynthesis, which heretofore was assumed to be the basis of all life on earth.

The constant games of maneuver and countermaneuver between American and Soviet submersibles and antisubmarine warfare units are yielding new stores of information about the sea and underseas topography. Unfortunately, a great deal of such information is not yet in the public domain because of its potential military significance. Both sides move about the silent business of riding unseen and, heretofore, unknown density currents, and hiding behind the sonar veils of thermoclines. They hug the ocean depths in submarine canyons, which do not yet show on unclassified bathyscapic maps. They observe mudslides at depths never before seen by the eyes of man. Their sophisticated sonar decodes the babble of mammalian and piscatorial voices never before properly identified.

Deepwater soundings, photographs, and drilling provided the essential evidence for seafloor spreading, and led to the underpinnings necessary to validate plate tectonics as a useful theory. Each passing day seems to reveal new data to support the theory. Further mysteries will doubtless be revealed as our exploratory methods become even more sophisticated. Geophysicists have considerable interest in drilling through the earth's crust. When such an effort occurs, it is likely to reveal a great deal of useful information about processes involved in plastic subsurface flow and the movement of tectonic plates. Drilling beneath the sea would reach through the crust at a substantially shallower depth than drilling on land. It is devoutly hoped that funding to permit such drilling develops before the end of the twentieth century.

Ocean Basins

Although the ocean basins differ from one another topographically as much as continents, with high mountains and great abyssal plains, they share certain common features. Each ocean basin can be divided into three parts: continental shelves, continental slopes, and ocean floors. The **continental shelves** are, essentially, the submerged edges of the continents. During past geologic times, they have been alternately exposed and submerged with the fall and rise of sea level. Most continental shelves were exposed within the past 1,000,000 years or thereabouts during Pleistocene glaciation. The vast quantities of water locked up in glacial ice caused sea level to fall markedly. Subsequent ice melting and concomitant rise in sea level are responsible for the present form and extent of most continental shelves. In a sense, because of glacial ice melt, essentially all of the world's shorelines currently show the effect of submergence.

From the edges of most continents, the sea depths descend gradually from shore to about 600 feet (about 200 meters). Continental shelves vary in width from 10 miles (16 kilometers) to almost 800 miles (almost 1,300 kilometers), although in a few instances, very deep water is encountered almost immediately offshore.

Continental slopes begin at the edge of the continental shelf and plunge to the ocean floor. The slope is greater than that encountered on continental shelves by a ratio of three or four to one, but diagrams illustrating the slope exaggerate the slope markedly. In a few areas adjacent to subduction zones, the plunge to ocean depths may be as much as 30,000 feet—more than 9,000 meters. Such a steep plunge to ocean depths does not exist except in areas that have oceanic trenches adjacent to the continental shelf.

Ocean floors, or basins, occupy almost one-half the total area of the earth. In a sense, the word *floor* is misleading, since only a few areas are really flat. Most ocean floors are ribbed and corrugated with innumerable rises and declivities, adding to the complexity of submarine relief. Their margins are often seamed with oceanic deeps, cut by subsurface canyons, wrinkled with subsurface mountain ranges, and interrupted with isolated mountain peaks. Indeed, in most ocean basins, the variations in relief are at least as significant as those that exist on the terrestrial globe.

Deep trenches exist in areas of active crustal subduction (figure 16.1). The deepest known trench is the Mariana Trench in the Pacific, where oceanic depths exceed 38,000 feet (11,600 meters). The Mindanao Trench, off the Philippines, is almost as deep.

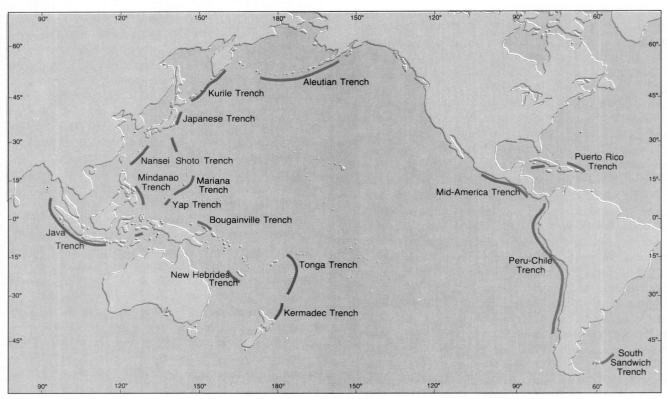

The Mid-Atlantic Ridge is characterized by its great north–south extent (figure 16.2), and it is punctuated here and there with peaks, such as Ascension and the Azores, which project above the surface of the sea. Submarine canyons, which etch the ocean floor, and curious flat-topped submerged mountains called **guyots,** which rise above it, add diversity to the ocean basins (figure 16.3). If the seas were drained away, the observer would be treated to a landscape at least as diverse as now observed on the terrestrial surface, in terms of the geomorphological features encountered. The absence of a cloak of vegetation would yield exposed landscapes very similar to bare ridges and surfaces encountered in arid landscapes. A dried-up ocean basin would be a geomorphologist's paradise, since exposed surfaces could be readily observed in terms of lithology, structure, process, and stage.

Ocean floors everywhere experience a "rain" of suspended particles of materials and the skeletal remains of untold billions of sea creatures. Near shores, vast loads of inorganic debris are contributed by streams. People continue to add an increasing load of effluent to the oceans' suspended and deposited load. Even the deep ocean remoteness has not escaped pollution, as we have dumped our particularly noxious materials, including toxic and radioactive materials, on the naive assumption that the oceans have a limitless capacity to tolerate wastes. It doesn't. Our out-of-sight, out-of-mind attitude will rise up to haunt us unless we change our ways.

Within the ocean, a combination of factors leads to a never-ending set of motions. These motions of waves, tides, upwellings, and currents affect landforms, climate, and other elements of the physical environment as well as producing opportunities for and limitations to human endeavors.

A description and analysis of the principal oceanic movements and their environmental implications will be forthcoming in the following sections. Some of those motions exhibit subtle influences, whereas others are responsible for dramatic environmental impact.

Figure 16.2
The position of the Mid-Atlantic Ridge for the North Atlantic Ocean.
From Richard A. Davis, Jr., Oceanography: An Introduction to the Marine Environment. *Copyright © 1987 Wm. C. Brown Publishers, Dubuque, Iowa. All Rights Reserved. Reprinted by permission.*

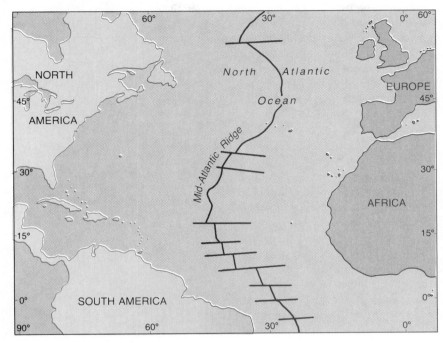

Figure 16.3
Relationships among major features in ocean basins. The seamount and guyot are greatly exaggerated in size.
From Richard A. Davis, Jr., Oceanography: An Introduction to the Marine Environment. *Copyright © 1987 Wm. C. Brown Publishers, Dubuque, Iowa. All Rights Reserved. Reprinted by permission.*

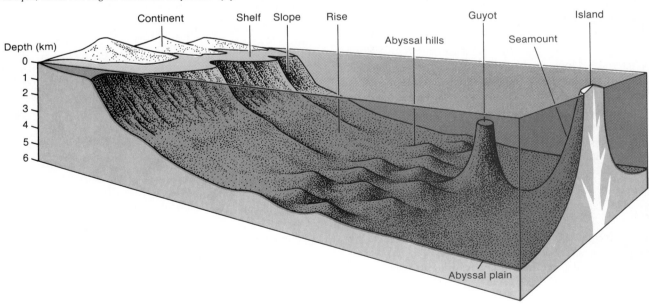

Water Movement

Five factors are principally responsible for the pattern of movement of ocean waters. They are (1) the pull and drag of the prevailing winds; (2) the differences in density of seawater; (3) the rotation of the earth; (4) the gravitational attraction of celestial bodies, notably the sun and the moon; and (5) the shape of the continents and the configuration of the ocean basins. Since water is a liquid, the oceans constantly exhibit three principal motions: waves, currents, and tides. Each of these movements is important in a variety of ways. Some have climatological effects; all of them, in one way or another, exhibit geomorphological impacts.

Waves

Waves are generated principally by the wind. Waves tend to move water particles within them in a kind of circular oscillating pattern. On the other hand, when a wave reaches shallow water, the ocean bottom interferes with the oscillatory pattern, the wave height and slope increases until it breaks or crashes on the shore. **Wave height** is the vertical distance between the crest (high point) and trough (low point) of the wave (figure 16.4). The **wave length** is the horizontal distance between two succeeding wave crests or intervening troughs. Waves vary from the kind of benign lapping of water on shore during a quiet day to pounding breakers during storms. At the height of a major storm such as a hurricane, waves may be as high as 50 feet (about 15 meters), although they are usually less than that, even during a major storm.

This water moving on shore has considerable hydraulic and pneumatic force, and the wave may force apart cracks in rocks and use existing sand and rock material to abrade the shore. These waves rarely strike the coast at right angles; rather they approach the shore at an oblique angle (figure 16.5). This oblique approach also tends to set up longshore currents which have the effect of moving rock particles along the coast. Particles, which may be carried by the waves, move in a kind of parabolic path, which tends to move materials along a shore. As the water moves back to the sea, debris is carried along and redeposited. Waves and surf are persistently and pervasively modifying and reworking deposits of materials along shorelines. No shoreline has exactly the same configuration on two successive days.

Tides

Unlike waves, tides are moved principally by gravitational attraction, especially of the sun and moon. This attraction causes a rise and fall of water levels at about twelve hour

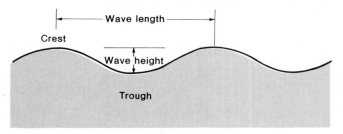

Figure 16.4
Wave height and wave length.
From Charles C. Plummer and David McGeary, Physical Geology, *4th ed. Copyright © 1988 Wm. C. Brown Publishers, Dubuque, Iowa. All Rights Reserved. Reprinted by permission.*

Figure 16.5
Wave refraction and longshore currents.
From Richard A. Davis, Jr., Oceanography: An Introduction to the Marine Environment. *Copyright © 1987 Wm. C. Brown Publishers, Dubuque, Iowa. All Rights Reserved. Reprinted by permission.*

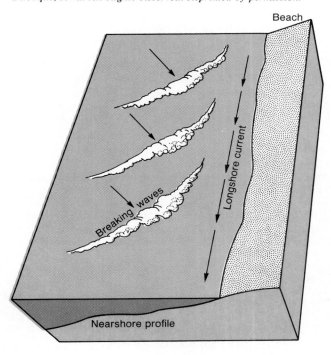

and twenty-five minute intervals on most of the world's coasts. This means that most coastal areas experience two high tides and two low tides a day. A few areas like the American Gulf Coast experience a rhythm of only one high and one low tide per day, reflecting the effects of the semi-enclosed nature of the Gulf and a resulting modification of the tidal pattern.

Figure 16.6
How tides develop. Tidal ranges are greater during spring tides and less than normal during neap tides.
From Richard A. Davis, Jr., Oceanography: An Introduction to the Marine Environment. *Copyright © 1987 Wm. C. Brown Publishers, Dubuque, Iowa. All Rights Reserved. Reprinted by permission.*

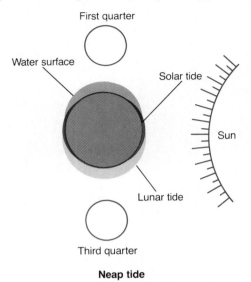

Neap tide

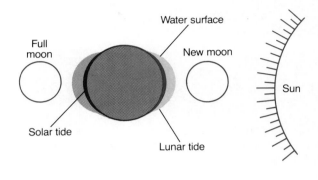

Spring tide

Tides have the effect of rearranging some materials along the beach, and in a few areas where tides are tightly constricted in coves or between islands they are significant erosive agents. When a major storm, such as a hurricane, makes landfall during high tide, the effects of water damage are increased. Generally, however, tides have quite modest erosive impact.

The possibility for harnessing tides to produce power is slowly being recognized. Such a tidal power project for the United States was proposed by Franklin Delano Roosevelt more than half-a-century ago. He proposed a site in Maine in an area with a narrow inlet and particularly high tidal ranges. There was no American follow-through, but the French have an operating tidal power facility in the Rance Estuary in Brittany.

At certain times of the month when sun, earth, and moon are essentially in line, a condition called **syzygy** exists. Tides are particularly high at such times (**spring tides**), and the effects of erosion are likely to be more pronounced (figure 16.6). When the earth, moon, and sun are at right angles, lower-than-normal tidal ranges are called **neap tides.**

A condition that occurs every several years compounds the effects of syzygy (i.e., when the sun is in perihelion and the moon is in perigee (closest to the earth), and when those two bodies are in line with the earth). All of these factors raise abnormally high tides, which subject normally protected beach areas to the ravages of wave erosion. Such an occurrence in 1987 led to serious beach erosion in several posh Malibu locations in California.

Major ocean currents appear to have little influence on marginal ocean landforms, but localized currents, especially longshore currents, may transport sand for some distances along a coastal edge. In some areas, density currents are apparently responsible for transporting significant amounts of materials along continental slopes or in submarine canyons.

In the end analysis, however, waves are most important in creating and modifying landforms at the margins of seas. A discussion of coastal landforms in terms of origin and evolution follows.

Coastal Landforms

Various aspects of water movement result in erosional and depositional forms, especially along coastal margins. In areas where rocky headlands are exposed to the fury of pounding waves, the erosive power may produce some striking features. Water forcing itself into cracks exerts a force, and air that is trapped in rock interstices is compressed. When waves depart, the air expands. Both influences expand cracks and work to pry rock materials apart.

Figure 16.7
How wave-cut terraces evolve.
From Carla W. Montgomery, Physical Geology. *Copyright © 1987*
Wm. C. Brown Publishers, Dubuque, Iowa. All Rights Reserved.
Reprinted by permission.

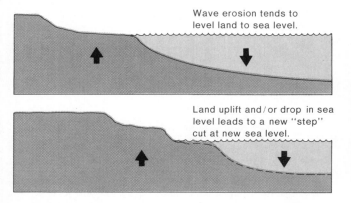

Wave erosion tends to
level land to sea level.

Land uplift and/or drop in sea
level leads to a new "step"
cut at new sea level.

Figure 16.8
*Exposed ocean terrace along the northwestern coast of Puerto
Rico.*

Figure 16.9
*As waves attack headlands, the cliffs, stacks, and arches are
formed.*
From Charles C. Plummer and David McGeary, Physical Geology,
4th ed. Copyright © 1988 Wm. C. Brown Publishers, Dubuque, Iowa.
All Rights Reserved. Reprinted by permission.

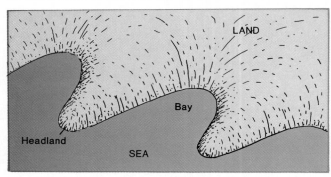

LAND

Bay

Headland

SEA

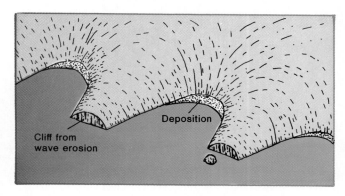

Cliff from
wave erosion

Deposition

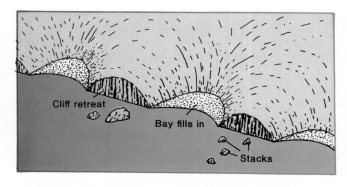

Cliff retreat

Bay fills in

Stacks

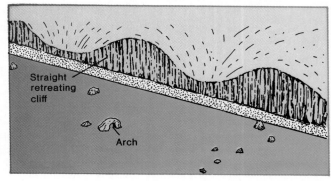

Straight
retreating
cliff

Arch

Waves pounding directly on such shores pry inexorably into the cracks in the rock, break it away, and leave behind a **wave-cut cliff,** often with a notch at the base closest to the waterline. This notch undercuts the cliff face (which collapses periodically keeping the face at a high angle of repose) and provides a relatively flat surface near and slightly below the water line known as a **wave-cut bench,** or **wave-cut terrace** (figure 16.7).

The debris from the cliff and bench is incessantly pounded by the waves, and broken and ground into smaller and smaller pieces. Eventually, these pieces are small enough to be transported by the backwash where they are usually deposited immediately seaward of the wave-cut terrace as a **wave-built terrace** (figure 16.8). The wave-cut and wave-built terraces are responsible for a relatively gentle progression from the beach to deep water along most such coasts. Changes of sea level or diastrophic uplift may expose these terraces along certain coastal fringes.

Wave crashing onto a headland from slightly different angles may expose a zone of rock weakness and isolate a segment of the cliff offshore as a **stack** (figure 16.9). The particular joint patterns in coastal cliffs may be responsible for the varied patterns of cliffs, caves, and stacks that may mark such a shore. Or, over a protracted period, a cliff may retreat some distance from its initial position leaving very resistant islands of rocks as stacks. Some stacks may be cut through by wave action producing picturesque **sea arches.**

Figure 16.13
Offshore bar development along a gently sloping coast.
From Charles C. Plummer and David McGeary, Physical Geology,
4th ed. Copyright © 1988 Wm. C. Brown Publishers, Dubuque, Iowa.
All Rights Reserved. Reprinted by permission.

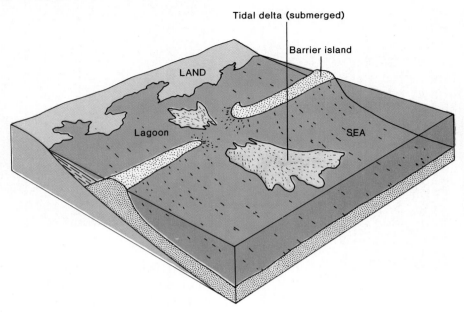

Tidal delta (submerged)

Barrier island

LAND

Lagoon

SEA

In areas with relatively shallow water and abundant sources of sand, offshore depositional features tend to be the most conspicuous aspects of the wave action. A family of bars, especially **offshore bars,** are omnipresent along much of the Atlantic and Gulf Coastal Plains of the United States (figure 16.13). Waves breaking in shallow water offshore have built up deposits paralleling the coast as offshore bars. Some of these bars may be quite long, but they are, typically, very narrow, and usually they rise only a few feet above the sea. Many of these bars have become popular seaside resorts, and more and more of them have become heavily populated as condominiums and apartments have risen to take advantage of the surf and sand. A disaster could occur if a hurricane strikes these areas.

Between the bar and the shore, there is usually a lagoon of quieter water. Over time, such lagoons may be filled by material washing from both the landward and seaward sides. Such filling may be facilitated by the development of a salt marsh with all the attendant aquatic vegetation. If such filling occurs during periods of relatively stable water level, the bars may be tied to a new shoreline and become part of it.

Many sections of the intracoastal waterway of the Atlantic and Gulf Coast lie in these lagoons between offshore bars and the mainland. These waterways, which are relatively protected from all but the most severe storms, carry enormous tonnages as strings of barges carrying bulky cargo, especially, ply these waters in a steady procession.

In several sections of peninsular Florida such bars are located well inland far from the present coastline. Obviously, the land surface has risen over a protracted period of time in a shallow sea. Bars formed offshore along with former lagoons were left high and dry as new bars were formed offshore in an oft-repeated process as the land was steadily uplifted or sea level declined.

The sandy debris characteristically found along the seacoast is moved along by longshore currents and redeposited as projections in adjacent waters. These **spits** may project across the mouth of an inlet to an adjacent headland where they are called **baymouth bars** (figure 16.15). Because of the vagaries of wind and current, such bars or spits are also formed at **midbay** and **bayhead** positions. The materials in all of these is essentially similar, and the distinguishing characteristic is largely the position of the bar. Uncommonly, waves and currents moving from two directions may tie an offshore island to shore with a sandspit. Such a feature is called a **tombolo** (figure 16.16). The capricious nature of wind and current may cause the projecting ends of certain spits to be curved or even recurved. Such features add interesting variety to coastline form. These curved spits are often called **hooks.** Not infrequently, such features enter into the lexicon of geographic place names (e.g., Sandy Hook).

Figure 16.14
Offshore bar (island) near Florida–Alabama line.
Jerome Coling.

Figure 16.15
Sand drift can cause the development of a variety of bars and spits.
From Charles C. Plummer and David McGeary, Physical Geology, 4th ed. Copyright © 1988 Wm. C. Brown Publishers, Dubuque, Iowa. All Rights Reserved. Reprinted by permission.

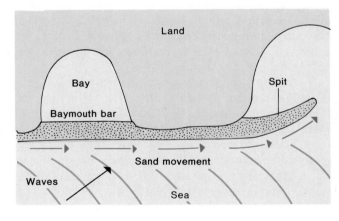

Beaches

Beaches, or zones of sediment, which extend from the low-water line inland to an upland surface, have several parts that are modified to varying degrees by surf, tides, and storms (figure 16.18). The beach **foreshore** is regularly covered and exposed by rising and falling tides. The **beach face** is steepened by wave action, especially at high tide. The **backshore** is located landward of the high-water line.

Figure 16.16
A tombolo forms when a former rock or island offshore has been connected to the coast by a bar.
From Charles C. Plummer and David McGeary, Physical Geology, 4th ed. Copyright © 1988 Wm. C. Brown Publishers, Dubuque, Iowa. All Rights Reserved. Reprinted by permission.

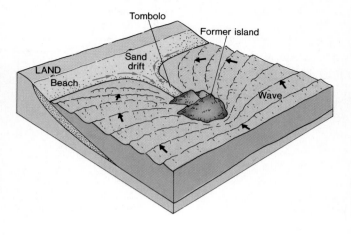

Figure 16.17
A tombolo along the Washington coast.

Figure 16.18
Parts of a beach.
From Charles C. Plummer and David McGeary, Physical Geology, 4th ed. Copyright © 1988 Wm. C. Brown Publishers, Dubuque, Iowa. All Rights Reserved. Reprinted by permission.

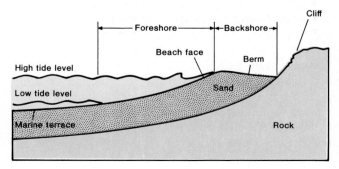

Figure 16.19
The sugar-white sand of a Florida Gulf Coast beach with ripple marks.

Figure 16.21
A lava flow on this Oregon beach has been excavated by the erosive action of waves and surf.
Daniel Ehrlich.

Figure 16.20
The black sands of a volcanic beach in the Philippines.

Beaches are frequently composed of quartz particles and fragments of coral in tropical environments or of volcanic debris in areas where vulcanism is prevalent. Larger gravel or pebbles exist in a few areas where wave energy is very high and effectively removes smaller particles.

Beaches vary from the sugar-white colors of the Florida Gulf Coast to black volcanic shores in areas such as Hawaii. All of these sands are constantly shifted and reworked by waves and surf.

Types of Shorelines

Just as the effects of waves and currents can modify the nature of the coastline and associated beaches, in plan and profile, so can fluctuations in sea level modify, bury, or expose certain features caused by waves. Not only do diastrophic forces raise and lower segments of coasts, but epochs of glaciation and postglacial periods can also change coastal relationships.

Attempts to classify shores in terms of their relationship to rising or lowering sea level are of limited usefulness, although in a few instances they may have some value. As indicated earlier, the great majority of the world's coasts exhibit characteristics of submergence currently as the result of post-Pleistocene glacial melt. During the height of glaciation, of course, essentially all the world's shorelines were emergent.

Shorelines of submergence occur when the sea level has risen relative to adjacent land surfaces, or, conversely, the land may have become depressed as the result of diastrophic forces. The effect generally is to increase irregularity of such coasts, since river mouths and adjacent valleys are drowned by the rising waters, and lands at higher elevation stand as promontories. Coasts with an abundance of such drowned river mouths creating a very irregular shoreline are called **ria coasts** or **ria shorelines** (figure 16.22). The area of northwestern Spain is especially noteworthy for this ria pattern. Irregularity of shoreline by itself does not imply subsidence, however.

Figure 16.22
A miniature example of a ria shoreline.
Jerome Coling.

Shorelines of emergence exist where the land has risen relative to sea level, or the sea level has fallen relative to the shore. Frequently, broad expanses of coastal plain of very low relief testify to the fact that the land recently emerged from the sea. Wave-cut and wave-built terraces now situated well above sea level testify to past emergence. Again, however, regularity of coastline is no guarantor of recent emergence. **Compound shorelines** exhibit characteristics of both emerging and submerging coasts. Because of glacial melt, many of the world's coasts exhibit this characteristic.

Neutral shorelines show little evidence of either emergence or submergence. Delta coasts and coral coasts exemplify many neutral coasts. The relationships of coastline to rising or falling sea level is of limited concern in most current geomorphological study. Rather, interpretations of the dynamics of construction or destruction of the coastal margin are made in terms of energy leading towards equilibrium conditions. Again, it is probably more useful to describe the salient characteristics of coastal form rather than to attribute existing forms to variations in sea level. Some examples of such forms are considered in the sections that follow.

Coral Reefs and Atolls

Coral organisms in warm tropical waters frequently build landforms by extracting and depositing calcareous materials from seawater, which add new solid material that may ultimately be exposed at or above sea level. Polyps may be the principal organisms adding materials to reefs, but other contributions are made by certain calcareous-depositing algae, the skeletal remains of reef dwellers, and other sources. Fringing reefs are built out from near land to some distance offshore. The skeletal remains of countless generations of coral polyps serve as the platform on which new generations add their stone load. Reefs hidden beneath the surface are hazards to navigation. Those projecting above the surface which develop a soil and vegetative cover are usually settled promptly.

Coral growth is fostered by water temperatures between 77° and 86° F (25° and 30° C), water depths of 150–200 feet (45–60 meters), and water salinity of between 27 and 38 parts per thousand. Turbidity is disruptive to coral growth. Essentially, coral is restricted to regions with a tropical climate and to areas with relatively clear waters. Many reefs have been destroyed when dredging or construction has contributed a heavy silt burden to the water. Silt essentially smothers the coral organisms and diminishes their ability to gather food.

Barrier reefs develop offshore and parallel to it. The coral heads are often exposed at high tide, and the modifying effect of the waves may ultimately build a reef above the level of high tide. Between this reef and the shore, a quiet water lagoon stands in marked contrast to the open sea.

The Great Barrier Reef off eastern Australia extends for 1,500 miles as the greatest expression of reef building on the planet. It is at once a navigational hazard and home of the greatest variety of coral species and reef creatures in the world. As a result, it has become a mecca for scientists and tourists as well.

Atolls are ring-shaped coral islands that stand in the open ocean well away from existing land. It is believed that they begin development as a fringing reef around isolated mountain peaks, such as volcanoes (figure 16.23). The mountain may begin to subside, or sea level to rise, or both, and the polyps are able to maintain their growth even when the mountain has long since vanished beneath the waves. Obviously, the rate of submergence would have to be very slow, since coral cannot grow in very deep water because of diminishing light and cooler temperatures. For atolls to exist, the water must be sufficiently warm and shallow to support the coral polyps.

Figure 16.23
The Darwinian theory of atoll formation.
From James L. Sumich, An Introduction to the Biology of Marine
Life, *4th ed. Copyright © 1988 Wm. C. Brown Publishers, Dubuque,
Iowa. All Rights Reserved. Used by permission.*

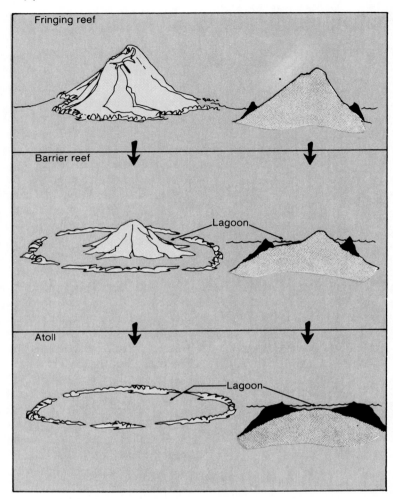

Many scientists have addressed the question of atoll formation, including Charles Darwin and William Morris Davis. The details may vary from island to island, but many of the basic premises of Darwin appear to be correct. Darwin's insights in this area of scientific investigation are almost as remarkable as his conceptual development of an evolutionary plan. The voyage of H. M. S. Beagle provided the impetus for a remarkable scientific *tour de force,* which has changed the thinking of the scientific world. The declaration of a theory of evolution has forever changed the thinking of the scientific world about the earth's biology. Although his theory about atoll formation is less momentous, it does reveal another facet of his brilliant powers of deduction. It is noteworthy that no better explanation for atoll growth and development has evolved.

Lakes

The size and salinity characteristics of large bodies of water are not very useful characteristics for distinguishing between lakes and seas, since there are size and salinity overlaps between and among them. Generally, however, we think of most lakes as being fresh and typically somewhat smaller than seas.

Several very large lakes assume extraordinary significance in both the physical and cultural landscape. Most continents have one or several such lakes. Australia lacks any very large lakes, and Europe has some picturesque lakes, but nothing of any great size.

A few of the largest of those with special characteristics merit some discussion because of the significance of their physical relationships and impact on other parts of the environment. In North America, the Great Lakes (Superior, Huron, Michigan, Erie, and Ontario) are of great economic significance because they serve as important transportation arterials in the heart of the continent. They also have a major role in the climate and hydrology of the area. Their margins tend to modify temperatures somewhat, and Erie and Ontario, especially, serve as sources of moisture, which increase snowfall along their eastern and southeastern shores. They are modulators of runoff to and within a significant part of the east coastal portion of North America.

In South America, Lake Titicaca has the distinction of being the world's highest large lake. Its margins and surface support unique types of reed vegetation which, in turn, evoke interesting human settlement responses. The ecology of the lake is also unique.

The great rift system of East Africa is occupied in part by large lakes like Victoria, Nyasa, and Rudolf. These lakes provide important protein food sources for the people who live along their banks, and they are major hydrological features. Lake Victoria is the source of the White Nile which, when joined by the Blue Nile, supports the exotic Nile, which crosses the Sahara and gives life to Egypt.

In Asia, Lake Baikal has the distinction of being the world's deepest lake. Its waters support some unique forms of life, and it, too, is of considerable economic significance to the people who live along its banks.

Tonlé Sap in Kampuchea is interesting because it serves as a kind of safety valve for the Mekong during floods. Because of low gradients when the Mekong rises, water flows into Tonlé Sap greatly expanding it in size, but modulating the effects of flooding along the lower Mekong. During the low-water season, water reverses its flow and moves from Tonlé Sap into the Mekong. The lake diminishes greatly in size during the dry season, but the water flowing to the Mekong equalizes water levels in the lower course and reduces the danger of flooding in the delta.

Life History of Lakes

No two lakes have precisely the same origin and no two lakes experience the same life cycle or destiny. Surface area, depth, latitudinal position, altitude, inlets and outlets, and surrounding rock types all have a bearing on the life history of a particular body of water. A few generalizations do, however, tend to apply to certain classes of lakes.

Most lakes without outlets are salt. Since dissolved materials are carried in, but not out, their concentrations are increased as the result of additional accumulations and evaporation. Some, such as the Dead Sea (really a lake), and Great Salt Lake are so saline as to preclude most forms of life.

Figure 16.24
This lake is so overgrown with vegetation that the surface of the water can't be seen.

All lakes, including those with outlets, tend to fill over time as debris is washed in and deposited. Aquatic vegetation takes hold and adds organic debris to the lake's burden. Some will fill quite rapidly, moving to swamp form and finally to dry land.

Eutrophication occurs in waters that are enriched by dissolved minerals. This tends to enhance vegetation growth and results in oxygen depletion in the water (figure 16.24). This causes a change in the animal population, and results in fish kills and degeneration of water quality. People tend to speed eutrophication when our phosphate-containing detergents are carried into lakes by sewage, and fertilizers from our fields or lawns are carried into lakes by runoff. This enrichment causes galloping eutrophication to affect lakes.

Industrial nations have also contributed to the deaths of hundreds of lakes through the exhaust from industries, which carry sulphur dioxides and trioxides. These materials, combined with atmospheric moisture, fall to the earth's surface as acidic solutions. In areas where basic rocks are unable to buffer or neutralize the solutions, such rain makes lake water become so acid that it kills off life in the lakes.

Europe and the eastern half of the United States have been affected in increasingly dramatic ways, as more and more lakes are dead or dying. According to *The State of the World, 1988,* published by Worldwatch Institute, lakes in the United States are threatened because of the effects of acid rain and pollution. Unless prompt action is taken soon, there will be an increase in the number of dead and dying lakes.

Although we create new artificial impoundments to supply us with domestic water supplies for irrigation or for power, our increasing numbers; our use of fertilizers, pesticides, and herbicides; and our clearing of land will accelerate the rates of lake filling and destruction of both natural bodies of water and artificial reservoirs. The world population growth and increasing assaults on the terrestrial environment will also result in the deterioration and destruction of lakes, both natural and man-made.

The lakes and seas, like other elements of weathering and erosion, are in a constant state of flux between the agencies that are grinding down the coastal fringes and those that are reworking and redepositing materials elsewhere. The inexorability of wave striking shore, the movement of currents, and the ebb and flow of the tides with concomitant modification of existing landscapes demonstrate the accuracy of the adage of Heraclitus, "The only constant in the universe is change."

The dramatic effects of storm-driven waves acting in sharp counterpoint to a gentle surf and the significant effects of strong longshore currents compared to the subtle effects of subsurface currents exemplify the complicated influences on land at the margins and beneath the seas. People's fascination with the sea and its coastal margin may somehow reflect an unseen tie leading back to the time when all life resided in the sea. Although our ancestors left the sea millions of years ago, in a sense, we carry around miniature replicas of seawater in our blood and body chemistry. The restless waters of the seas add another dimension to the sculpting of the land.

Ultimately, earth materials, diastrophic forces, climate, vegetation, and the various elements of weathering and erosion (including those of the sea) add their influences in a complex matrix, which helps to create the mantle of soil that cloaks the earth's surface. The anatomy, character, and distribution of those soils will be considered in the following chapter. It is important to recognize that the seemingly ordinary soil makes our existence possible. Treating our soil resources cavalierly is perilous.

Study Questions

1. Approximately how much moisture is evaporated from the seas and returned to the earth each year as precipitation?
2. Draw a schematic diagram of the hydrologic cycle.
3. Briefly explain the hazards of the continued dumping of waste materials into the world's seas.
4. Explain what factors have motivated people to gain a greater understanding of and appreciation for the seas, and associated landforms and features.
5. What is a guyot? How were such features probably formed?
6. What are the principal factors responsible for the movement of ocean waters?
7. Explain syzygy.
8. Draw a diagram illustrating the relationships between wave-cut terraces and wave-built terraces.
9. Enumerate and briefly characterize some of the landforms that develop in areas where waves break directly onshore in regions with moderately deep water immediately offshore.
10. List and characterize the array of depositional features at and adjacent to the shoreline.
11. What effect does rising or falling sea level have?
12. How are coral reefs formed? Atolls?
13. Where is the Great Barrier Reef?
14. Why is channel dredging a danger to proximate coral reefs?

Selected References

Bascom, W. 1980. *Waves and beaches.* Rev. ed. New York: Doubleday Anchor Books.

Briggs, D. 1977. *Sources and methods in geography: Sediments.* London: Butterworth.

Davis, R. A., Jr. 1987. *Oceanography: An introduction to the marine environment.* Dubuque, IA: Wm. C. Brown Publishers.

Komar, P. D. 1976. *Beach processes and sedimentation.* Englewood Cliffs: Prentice-Hall.

Shepard, F. P., and Wanless, H. R. 1971. *Our changing coastlines.* New York: McGraw-Hill Book Company.

Strahler, A. N., and Strahler, A. H. 1984. *Elements of physical geography.* 3d ed. New York: John Wiley and Sons.

17

Soils

An exposed soil profile exhibiting columnar structure.

The earth is cover'd thick with other clay, which her own clay shall cover, heap'd and pen, . . .
Lord Byron, *Childe Harold's Pilgrimage*

The topsoil, which blankets the terrestrial globe, is, next to water, people's most precious resource. Like potable water, it is under unremitting assault from deforestation and destruction of vegetation; poor agricultural practices, which speed runoff and accelerate erosion; and substantial overgrazing, which lays bare the topsoil to the erosive ravages of water and wind. This topsoil, which averages about 8 inches (20 centimeters) worldwide, is the key to the survival of humanity. Eons of time were required to produce the legacy of soil, which civilizations began to use significantly with the invention of agriculture. Soil lost through vegetative denudation, and attendant erosion or imprudent use is, for all intents and purposes, lost forever, although in very limited areas it may be possible to rebuild soil to make it productive through massive additions of soil conditioners, fertilizers, and organic materials. Even in those cases, additions of material and the use of energy add to the cost of soil use. As population grows, more pressure is placed on our soil resource, and as it is asked to provide more and more for a hungry world, it is exposed to increasing erosion hazards.

Soil science (pedology) is essential to an understanding of and an appreciation for the physical environment. Soils are inextricably linked to other facets of the physical environment, and they are particularly and peculiarly subject to modification through human intervention. The vulnerability of the earth's soil resources is not generally understood, and we continue to remain ignorant of this important fact at our peril. The soils that support us also support the green vegetative mantle, which is a complement to the blue of the earth's seas. The vegetation renews our oxygen supply so that animal life is possible.

Soils are worked and reworked countless times as they are eroded away by one or several of the agents of erosion, and are subsequently deposited at a new location. Soils may be buried and reexcavated. The loss at a given locale may create catastrophic problems of agricultural impoverishment for the inhabitants of a region where erosion is severe. The eroded topsoil of one area may, however, become the rich alluvial deposits in a floodplain downstream. Since nature is value neutral, the shift of soil materials is simply a transient event in the never-ending cycle of landscape modification. The effects on human beings and their works are no part of the inexorable continuation of ever-present geomorphological processes.

Soils that have developed in the past, **paleosols,** may be buried by subsequent deposition. In some instances, such soils may be reexcavated and exposed by a rejuvenated stream. Preexisting soil regimes may also be buried by lava flows, by ash falls, or beneath the sands of migrating dunes. Often these paleosols remain hidden from the eyes of man until construction or scientific excavation reveals them. Their burial probably eliminates them from productive use for any reasonable period of time.

Soil is the stuff of life. At once, it supports life and teems with life. What a complex and precious material it is. It is important for us to value it, preserve it, and conserve it for ourselves and our posterity.

Nature of Soil

It is no simple matter to define soil, a highly complex mineral and organic material, although an attempt will be made to describe its essential characteristics. Basically, soil may be considered a material that is composed of minerals, organic compounds, living organisms, air, and water in interactive combinations. Those combinations vary markedly in kind, association, and percentages from place to place over the globe. The soil zone is that portion of the terrestrial globe that supports, or is capable of supporting, high-order plant growth. Certain low-order plants such as varieties of lichens and mosses may exist on bare rock surfaces, but trees, shrubs, and grasses depend on soil for support and sustenance.

Each of the constituent elements, their combinations, and their interactions are essential in determining soil characteristics, erodibility, fragility, inherent productivity, and utility. Soil is a dynamic interactive system undergoing constant change through persistent physical and chemical actions and reactions. Changes in soil may be accelerated or inhibited by human intervention. Some of these influences may be subtle, as in the case of transient human occupance, whereas others may be dramatic as the result of devegetation of a region, the fertilization or augmentation of soils, or through imprudent agricultural practices. Few corners of the earth have failed to feel the impact of man on soil, and certain very closely settled regions may have seen the soil so modified by protracted and intensive use that it has little physical or chemical relationship to the soil formed by natural processes in the region. The human modifications in a relative instant of geological time may mask or obscure the effects of protracted periods of geomorphological processes.

Soil may be considered as teeming with life, while, at the same time, it is the originator and sustainer of life. An examination of its constituent elements tends to explain differences and similarities between and among soils from place to place over the globe.

Minerals

The kinds and character of minerals in soil vary greatly from one place to another. Many factors control the minerals present in a given soil type. One of the most significant of these is the **parent material** (i.e., the substance from which the soil was derived). To illustrate, a soil developed on limestone ($CaCO_3$) bedrock will quite obviously be composed of different materials than one that has developed on sandstone (SiO_2). The mineral content may be relatively uniform if the soil has developed on a single bedrock supporting a homogeneous biota. On the other hand, there may be a complex array of minerals if soils have developed on complex of heterogeneous rocks or if other materials have been transported from remote locales. Most soils possess a substantial array of minerals deriving from both organic and inorganic sources. Some of the minerals may be tightly bonded to one another within soil particles, whereas others are more loosely incorporated. Under certain peculiar and special circumstances, soil minerals may be so concentrated by natural processes as to yield prospective commercial ores (e.g., iron and aluminum). In others, harmful substances may preclude the development of agriculture (e.g., salt or alkali). In some, mineral content, form, and combinations may foster intensive agricultural production.

A thorough knowledge of the mineral content of a soil is essential if its inherent productivity is to be understood or if prescriptive fertilization is to be undertaken. Plants require a large number of minerals for proper growth, but only a few are consumed in large amounts by most plants. Those most commonly depleted from the soil are nitrogen, phosphorous, and potassium. These minerals are frequently replaced in the soil by the addition of mineral or organic fertilizers, and manures, although nitrogen may be readily replaced by the use of **legumes** (nitrogen-fixing plants) in the crop rotation. Many leguminous plants exist as part of natural vegetative associations. Experiments in plant genetics are attempting to develop nitrogen-fixing characteristics in plants that do not have that natural capacity. Whether such plants will be developed to take advantage of the limitless quantities of nitrogen in the air remains to be seen. Obviously, if a corn plant could be manipulated genetically to cause it to extract nitrogen from the atmosphere, it would be very advantageous for the plant and the farmer. Fertilizer and fuel costs for cultivation would be dramatically reduced. The promises of and prospects for genetic engineering are exciting indeed. A few worry about the release of genetically altered plants into the environment for fear that unknown characteristics may pose a danger to other plants or animals. It seems likely that proper controls will prevent such danger, and the prospects for good results clearly seem to outweigh the potential for bad effects.

The clay minerals and quartz are the two common residual materials remaining after most rocks weather. The clay minerals tend to hold water in soil, whereas the quartz materials provide opportunities for aeration and water percolation. Most clay minerals develop a negative charge on flat surfaces, which holds water and plant nutrients in the soil (figure 17.1). Obviously, this is essential directly in the growth of plants and indirectly for the sustenance of animals. Clays may absorb large quantities of water. They are porous, but not very permeable. They have great volumetric changes between wet periods and dry periods.

In addition to the materials that are used in greatest abundance by plants, certain trace elements may be essential for plant growth. Some of these materials may serve as catalysts to assist plants in the absorption and use of other materials. Trace minerals like copper, zinc, boron, and antimony may be essential to the healthy growth of plants and/or the creatures that subsist on the vegetation, even when minerals that are most used by plants are present in abundance. It should be recognized, too, that the presence of a mineral does not necessarily mean that it is readily available for plant use. Some are so tightly bonded within the soil that they are unavailable for plants, and others may exist in an unusable form for plants.

Figure 17.1
Negative charges on clay minerals attract positive charges on ends of water molecules.
From Charles C. Plummer and David McGeary, Physical Geology, *4th ed. Copyright © 1988 Wm. C. Brown Publishers, Dubuque, Iowa. All Rights Reserved. Reprinted by permission.*

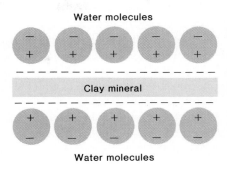

Figure 17.2
Ionic exchange between clay mineral and plant root.
From Charles C. Plummer and David McGeary, Physical Geology, *4th ed. Copyright © 1988 Wm. C. Brown Publishers, Dubuque, Iowa. All Rights Reserved. Reprinted by permission.*

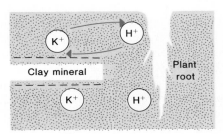

A careful soil analysis is essential in determining a fertilization regimen that may be necessary to insure agricultural productivity. Fortunately, in the United States, citizens may have their soil tested by county agents found in every corner of the country. In significant part, the skill and dedication of county agents and the scientific message they bear, along with the acceptance of American farmers, are responsible for the enormous productivity of American agriculture. These county agents and the land-grant universities that educated them deserve enormous credit for a job well done.

Organic Compounds

Humus, which breaks down into organic compounds, is composed of semidecayed remains of plant and animal life. Humus contains colloidal substances, which facilitate the transfer of nutrients to plants. In addition, organic materials are extremely useful in soils, since they improve **tilth** (workability), promote water retention, provide a home for living organisms, and replenish minerals. As a rule, soils with a high humus content are chemically active, whereas those lacking in humus tend to have slower chemical actions and reactions. Soils with a high organic content tend to be more fertile, reflecting the lush vegetative mantle that contributed to the organic compounds initially. Not all areas with a thick vegetative mantle have soils rich in organic matter, however. For example, soils developing in tropical rainforest areas are low in organic material because of accelerated rates of decay and the quick use of organic residue by living plants.

There is a natural symbiosis (i.e., vegetation contributes to the organic content of the soil, and, in turn, vegetative growth is facilitated by the high organic content of the soil). These same colloids and organic materials foster the presence of living organisms in soil. Biological and chemical activity are facilitated by the presence of organic compounds.

Living Organisms

When soil is examined closely, it is found to be the habitat of literally millions of organisms. Larger animals, **macro-organisms,** such as earthworms, ants, beetles, and grubs are quite useful in soil, since some, like the earthworm, pass organic materials through their bodies, digesting useful substances and producing fecal materials, called **casts,** which change the soil texture and perhaps increase the opportunities for mineral absorption. Certain authorities state that although earthworms are present in soils that are rich in organic material, they contribute little to it. In other words, according to these experts, they are there because the organic material is there, and they are minimally involved in modifying the soil. On the other hand, anyone who has observed earthworm casts on the surface of soil will testify as to the mixing of soil materials. It seems likely that passage of earth materials through the digestive tract of the animals modifies those materials in subtle physical and chemical ways. In addition, the insect remains contribute to the organic mix of the soil. It does seem clear that the burrows of larger organisms provide avenues for the movement of water and air through the soil. As a rule, soils with abundant micro- and macro-organisms appear to be the most agriculturally productive. Whether this reflects the attraction of organic materials for organisms, rather than the contribution of the organisms to the soil is not clear.

Micro-organisms

Animals, plants, and fungi not visible to the naked eye, or micro-organisms, cause the decay and disintegration of organic material, and contribute to the break-up of mineral particles. This process helps to release nutrients to plants and is an essential factor in soil formation and development. Until very recently, it was assumed that living organisms existed only in the top few feet of earth materials, but it's now clear that such organisms may exist at considerable depth if water is available. It is a general rule that organisms diminish in numbers at greater depths.

Herbert Franz* determined that one cubic foot (.026 meter³) of organic woodland soil in Europe contained 45,000,000 one-celled protozoa; 4,000,000 subinsects (including eelworms and rotifers); 60,000 other insects including millipedes, woodlice, larvae, ants, beetles, and so on; and 150 earthworms. These creatures had a total collective weight of about 4 ounces (120 grams). He also estimated that there were up to 10 ounces (300 grams) of fungi, molds, and algae, and up to 30 ounces (900 grams) of roots of higher plants per cubic foot and 1,300,000 bacteria per ounce. Other soils might contain lesser or greater amounts of micro- and macro-organisms, but these data suffice to illustrate that most soils are teeming with life. Soils are both harborers of life forms and sustainers of life. As indicated earlier, soil is a living substance that is constantly undergoing change and alteration. The abundance of these organisms, or their absence, usually reflects the health or lack of health of a particular soil.

In some environments, where temperatures and humidity conditions permit, such as in the rainforest, the micro- and macro-organisms reduce leaf litter and other organic materials very quickly. In fact, the rates of decomposition in such regions is little short of astonishing. On the other hand, in cold environments, the rate of decay is very slow.

Other major constituent elements in soil include air and water. Both are found in varying amounts in all soils depending upon the climate of the region and the porosity and permeability of the soil.

Air and Water

Air and water are as essential to plants as to animals. Since plants are rooted in soil, they are dependent on it for most of their water requirements, although in some species, absorption of rain or dew through the leaves makes some contribution to the plants' needs. A soil's capacity for holding water depends principally on its porosity (amount of pore space in the soil). If the soil is excessively porous

*Franz, Herbert, *Feldbrodenkunde,* Vienna: Fromme, 1960, pp. 134–36.

Figure 17.3
How water is held in soil.

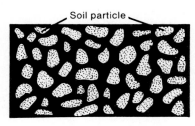

Gravitational

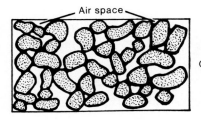

Capillary

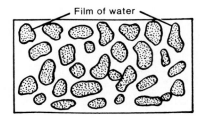

Hygroscopic

Capillary water is the water that is held by surface tension on and within soil particles and in portions of intervening pore spaces. Under varying climatic conditions, capillary water may move downward, upward, or laterally in much the same way that oil in a kerosene lamp moves up the wick or water in a paper towel moves in the direction of greatest capillary tension. Capillary water is used by plants, and the movement of such moisture through fine textured soils in periods of drought is of great significance. Since most agricultural plants are shallow rooted, the movement of water upwards by capillary action is especially significant to their continued survival during the intervals between rains. Capillary water disappears from soil only after a protracted dry period. Capillarity is most effective in soils with a loamy texture.

Hygroscopic water is held as a microscopically thin layer around each individual soil particle. Hygroscopic water is the last moisture to be removed from a soil. It has little or no use to plants, and sometimes may be removed only by baking soil in a high-temperature oven. In a sense, it is a kind of exotic presence that is of greater significance chemically than biologically.

The various ingredients of soil combine to produce an array of physical and chemical characteristics that are useful in categorizing soils as well as determining their biological activity and utility. These characteristics are as variable as the myriad geomorphological, climatological, and biological environments that make up the earth's mosaic.

The physical and chemical nature of soil are essential elements in classifying soil in terms of its spatial distribution. A discussion of some of the significant physical and chemical relationships in soil follows.

Physical and Chemical Characteristics of Soil

Soils possess certain obvious physical characteristics including texture, structure, depth, and color. These characteristics, singly and in combination, contribute to or detract from the vegetation a soil may support or the prospective utility it may have as a forest, grassland, shrub, or an agricultural resource. These physical characteristics may inhibit or facilitate the cultivation of soil. They tend to determine soil fragility and dictate conditions of wise use. They are an integral part of the distinguishing characteristics of various soil types and are among the principal characteristics used in soil classification.

and permeable, water may percolate through it rapidly, resulting in rapid desiccation. If, on the other hand, the soil is very compact, water may have difficulty in percolating through the soil, although a great deal of water may be held in the soil interstices. In extreme cases, the soil may be so compact as to be essentially impervious to percolating water. Clearly, soils with an appropriate balance between porosity and permeability are most amenable to water-holding capacity and transport. Soils with this happy mix of characteristics are usually most suitable for cultivation.

Water is found in the soil as gravitational water, capillary water, and hygroscopic water. **Gravitational water** percolates downward in the soil after precipitation has occurred in response to the pull of gravity (figure 17.3). Such water moves between individual soil particles, or aggregates of soil particles, completely filling all of the pore spaces in the soil. Gravitational water is present in soil in greatest abundance after a rain and is the first water to be depleted from the soil in dry periods. It moves rapidly through permeable soils and more slowly through soils with limited permeability.

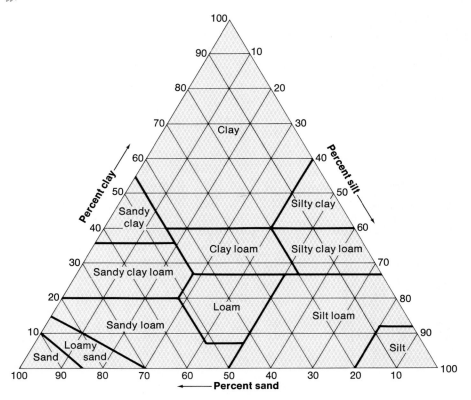

Soil Texture and Structure

Soil texture, which refers to the size of individual soil particles, may vary from very fine clays and silts to coarse sands and gravels. Clay particles are less than .002 of a millimeter in diameter; silts are .002 to .05 of a millimeter in size; sands range from .05 to 2.0 millimeters. Larger materials are gravel, cobbles, and larger rock, which are essentially inert materials in the soil, although as they weather, they contribute mineral substances to the soil. As they are broken down, they may assume particle sizes that fall into the sand, silt, or clay category. Soil texture is of considerable significance, since it influences the tilth of the soil; the supply and availability of plant nutrients; and the absorption, retention, and transport of air and water. For most types of plants, a loamy texture (20 percent or less of clay, 30–50 percent of silt, and 30–50 percent of sand) is the most satisfactory textural medium for plant growth. Such loamy soils are easily worked; they facilitate water retention and transport; and they enhance root penetration and nutrient absorption.

A convenient method to classify soils according to texture is the soil texture pyramid (figure 17.4). Applicable descriptive terminology derives from the percentages of various soil particles present in a given sample.

Soil structure refers to the way individual soil particles are aggregated or clumped together. Because of variations in climatic conditions, the nature and character of particle size, and presence of certain organic and mineral materials, soils may be aggregated in numerous ways, which affect erosion susceptibility, root penetration, permeability, water absorption rates, and tilth. Aggregates may assume numerous forms including **granular, platy, lens-like, columnar,** or **prismatic** form (figure 17.5), among several others. Together, soil texture and structure significantly affect the tilth of the soil as well as the prospective plant habitat. Aggregates of soil particles may hold together quite tenaciously, even in the face of significant physical disturbance. Clay particles may adhere very tightly together and what is a sticky, viscous substance when wet may become almost rock hard clods when dry. Indeed, most soils do not break down into individual soil particles, unless there is a physical effort to reduce them to this size.

The depth of soil in different layers varies from place to place depending on an assortment of physical conditions. Soil depth is important in determining extant natural vegetation or prospective agricultural use.

Figure 17.5

Major types of soil structure: (A) *prismatic,* (B) *columnar,*
(C) *angular blocky,* (D) *subangular blocky,* (E) *platy,*
including lens-like, (F) *granular.*
Source: Soil Survey Staff, 1951.

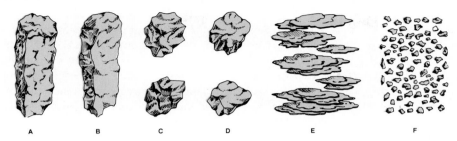

Figure 17.6
Soil profile showing relationship to underlying bedrock.

Figure 17.7
Deep soil profile.
Jerome Coling.

Soil Depth

Deep soils are normally better from an agricultural stand-point than thin soils, since plowing may expose the subsoil in areas with a thin topsoil. Topsoil is usually more fertile and fragile than subsoil; hence, it provides the greatest

opportunities for agriculture. In addition, deep soils provide greater opportunities for root penetration and a greater reservoir for water and nutrients for plant use. Under most natural conditions on flat or gently sloping land, soils tend to increase in depth as more and more of the bedrock becomes a part of the soil. In fact, in certain humid tropical areas where weathering is very rapid, soils may develop great depth. Conversely, in very dry areas where weathering tends to be slowed, the soils may have very shallow depths. On very steep slopes, soil tends to remain thin, since sheet erosion constantly removes surface materials. Some sheet erosion occurs on virtually all surfaces, but more material is removed from steeper slopes. As man cultivates crops, the soils tend to lose some of their depth, since removal of surface cover makes them more susceptible to erosion, and, as erosion proceeds, a portion of the soil horizon is stripped away. In addition, cultivation inevitably leads to some compaction and resulting diminished depth of the topsoil. In some areas of the world, the ravages of imprudent grazing or agricultural practices have resulted in the removal of alarming amounts of topsoil by water and wind.

Soil Color

The color of soil varies from place to place. Parent material, minerals, and organic material are principally responsible for variations in soil color. Subtle and dramatic chemical reactions are principally responsible for the different spectra revealed.

As a general rule, however, the darker colored soils contain the greatest quantity of organic material. Certainly there are exceptions to that general dictum as in young soils derived from recently exposed dark-colored volcanics or in certain of the intense reds characteristic of many humid tropical soils. In middle latitudes, soils typically range from browns to black in humid environments. They tend towards lighter browns and grays in subhumid to arid regions. As a rule of thumb, the soils with the darkest colors are inherently the most fertile. As with all generalizations, however, there are exceptions.

Red and yellow colors predominate in humid tropical and subtropical environments resulting largely from the presence of iron and aluminum oxides, and hydroxides. High latitudes in humid environments tend to have light-colored topsoils reflecting the paucity of organic material. Grayish or bluish colors in middle latitude soils usually reflect reduced iron content of the soil in areas of poor drainage. Of course, there are other factors that are responsible for other colors. Certain red soils, called Rendzinas, are characteristic of a particular limestone parent material.

Soil temperatures may be affected by soil color. Darker colors may absorb somewhat more heat, whereas light-colored soils may reflect more heat. The temperature of the soil, in turn, may influence the productivity of certain crops. For example, white potatoes do better in cool, moist soils, whereas corn responds positively to high soil temperatures. Soil temperatures are also important in the growth patterns of other plants. In every instance, the soil temperature must be sufficiently high to permit plant metabolism if the soil supports natural vegetation or is amenable to agriculture.

Soil colors are often indicative of principal minerals and those minerals may be involved in a variety of chemical reactions within the soil. The types and rates of chemical reactions are salient aspects of different soil types.

Chemical Reactions in Soils

The chemical reaction of the soil is significant, particularly as it relates to its prospective agricultural utility. Serious efforts may be made to change the chemical balance or reactivity of soil through the addition of fertilizers or soil conditioners.

Soil pH

Colloids, the small gelatin-like particles of minerals and organic matter, can attract and hold ions. Certain basic ions are used by plants in a process known as **base exchange.** Hydrogen ions in the soil produce conditions of acidity. The accumulation of hydrogen ions in the soil provides a measure of chemical acidity known as **pH.** A pH of 7.0 is considered neutral, and low pH numbers indicate soil acidity, whereas high pH numbers are indicators of alkalinity. Most commercially important agricultural commodities grow best in neutral to slightly alkaline soil, although the absolute range of soil pH for high order plant growth is between pH 4.0 and pH 10.0. Plants with a tart, acidic taste (e.g., gooseberries and cranberries) usually indicate their affinity for and use of acidic soil. Most of the world's principal food crops grow best in soil pH conditions close to 7.0. Some common substances will give an impressionistic idea of pH relationships. For example, distilled water has a pH of 7.0, whereas ammonia, a strong base, is at 12.0, and sodium bicarbonate (baking soda) is about 8.5 pH. Milk averages about 6.5, bananas are about 4.5, tomatoes are slightly above 4.0, and vinegar has a pH of 2.2. Vinegar is, of course, a weak solution of acetic acid.

It may be necessary to treat soils with neutralizing materials to improve their potential productivity. Limestone or dolomite is frequently added to certain acidic soils to "sweeten" them. The addition of the limestone or dolomite increases the pH and improves the productivity of low pH soils. Less frequently, acidic materials may be added to highly alkaline soils in dry regions to lower the pH readings and enhance the productivity of such soils. Such chemical treatments must normally be repeated, since exposure to the elements has a way of diminishing the impact of a single application of materials, whether as mineral fertilizers or as soil neutralizers. The normal leaching of materials by rainwater will have a constant subtle effect on the pH balance in soils.

There is increasing concern about acidification associated with the burning of high sulphur coals for industry and power generation. There appears to be an absorption of sulphur dioxide and sulphur trioxide, which results from the combustion of the hydrocarbons, especially coal, by rainfall producing dilute solutions of acid. Over time, these "acid rains" have very harmful effects on vegetation, including agricultural crops, and they have lowered the pH of some lakes sufficiently to kill certain aquatic life. Similarly, this acidification process may lower the pH of the soil. Studies concluded in the summer of 1988 revealed that the streams in the eastern and northeastern parts of the United States had a lower pH than previously surmised. A continued lowering of pH values in soils and waters seems certain to have deleterious effects on plants and animals.

Soils in irrigated areas tend to accumulate salts over time. Irrigation waters contain a certain amount of dissolved salts. As these waters are used by plants and evaporated from the soil, the salts are left behind. Eventually these salts may concentrate to toxic levels. Several areas in the Salt and Gila River Valleys of Arizona have become so toxic as to eliminate commercial agriculture in the affected regions. Increasing accumulations of these dissolved salts threaten the continuing productivity of the Imperial Valley of California. Costly chemical treatment of the soils or flushing with sweet water (water with few dissolved minerals present) with adequate drainage facilities in place may be necessary if these soils are ever again to be brought into sustained agricultural production. Of course, having enough sweet water to flush such soils is a problem, and it is likely that toxic salts will continue to accumulate. This difficulty must be countered by the judicious use of water, adequate drainage facilities, and sound agricultural practices.

The inherent danger in irrigated soils in dry regions is the ultimate accumulation of saline or alkaline materials, which are antithetical to plant growth. The remains of past civilizations in the Tigris-Euphrates Valleys and elsewhere stand in mute testimony to the unfortunate effects of unwise irrigation practices and associated salt encroachment. Numerous examples of salt accumulation in irrigated regions around the world can be cited, and this problem is steadily worsening. At a time when a burgeoning world population requires more and more food,

we can ill afford to allow an unintended conspiracy between the existing forces of nature and the folly of man to shrink the world's irrigated acreage. At the very least, reduction in irrigated lands in the United States will result in higher prices for agricultural commodities.

The combination of physical and chemical characteristics are involved in the development of horizons in soils. These horizons serve as major markers in the differentiation of soil types. Their nature and characteristics are described in following sections of this chapter.

Soil Profiles

Most of the earth's soils are developed from weathered rock material. Some have developed new horizons and characteristics from alluvium, which has been transported from other regions. As weathering results in rock disintegration and decomposition, plants and animals are able to establish a biological beachhead. The evolution of soil begins with this biological invasion of weathered rock materials. The continuing weathering, and the biological and chemical reactions over time cause soils to develop their inherent characteristics, which tend to vary considerably from place to place.

The principal result of soil-forming processes is the development of recognizable layers in the newly forming soil materials (figure 17.8). These layers (**horizons**) are different physically and chemically, but they all evolve from the same rock materials, which are known as parent

Figure 17.8
Generalized relationship of soil horizons in soil development.
From Charles C. Plummer and David McGeary, Physical Geology,
4th ed. Copyright © 1988 Wm. C. Brown Publishers, Dubuque, Iowa.
All Rights Reserved. Reprinted by permission.

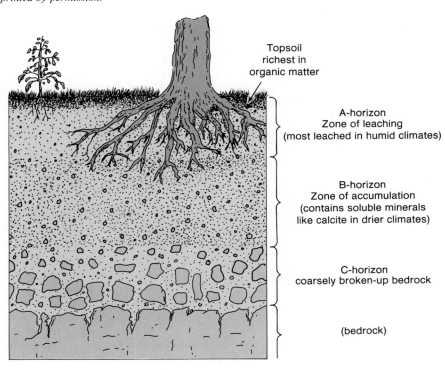

Topsoil
richest in
organic matter

A-horizon
Zone of leaching
(most leached in humid climates)

B-horizon
Zone of accumulation
(contains soluble minerals
like calcite in drier climates)

C-horizon
coarsely broken-up bedrock

(bedrock)

materials. A cross section through those layers is known as a **soil profile.** Only mature soils exhibit a completely developed soil profile. Immature soils or transported soils may have incomplete horizons, and some of them may be quite homogeneous throughout the profile (i.e., they may exhibit only a single horizon).

The top layer of the soil is known as the topsoil, or **A-horizon.** The A-horizon is the zone of greatest accumulation of organic material, most biological activity, and is typically the most fertile portion of the soil. It is also the zone subject to the greatest loss of material through the downward percolation of water and as a result of erosion. Solution (**leaching**) dissolves out the soluble materials, whereas small particles are removed in suspension by a process referred to as **eluviation.** Both processes occur in all soils, but the process is retarded in dry regions and accelerated in wet environments.

Immediately below the A-horizon is a zone of accumulation of materials, which were removed from the topsoil. This zone, the **B-horizon,** often called the subsoil, is typically more compact than the A-horizon and normally contains little organic material. Fine particles of soil, especially clay, which are removed from the A-horizon and deposited in the B-horizon create the more compact conditions. The deposition of these materials carried in suspension is called **illuviation.** The A- and B-horizons together constitute the soil proper, or solum.

Fragmented pieces of weathered bedrock immediately below the B-horizon make up the **C-horizon.** As soil-forming processes continue and weathering occurs, upper reaches of the C-horizon become lower reaches of the B-horizon. In areas of minimum slope the A-, B-, and C-horizons become thicker over time increasing the depth of the soil profile. Conversely, on steep slopes, the A-horizon may be thinner because erosion continues to remove topsoil material at a fairly rapid rate. The **D-horizon,** or R-horizon as it is now called, is the unweathered rock material below the C-horizon.

Boundaries between soil horizons may be quite sharp, or, on the other hand, may blend almost indistinctly into the adjacent horizon. In a particular soil profile, a given horizon may be missing entirely. Erosion and misuse may have caused the A-horizon to be destroyed or removed, or the C-horizon may be missing in areas where the weathering processes are particularly vigorous. This situation occurs especially in flat areas in the rainy tropics. Further, individual horizons may be subdivided into parts based on subtle differences between different levels within the horizon (figure 17.9). For example, portions of the A-horizon might be labeled A^0, A^1, A^2, or A^3, or now simply as O_1, O_2, and so on. Each of those subhorizons might exhibit very small chemical, biological, and textural differences. For example, A^0 or O_1 might be a zone of undecayed leaf litter or organic debris.

Figure 17.9
Idealized soil profile illustrating subdivisions of horizons.

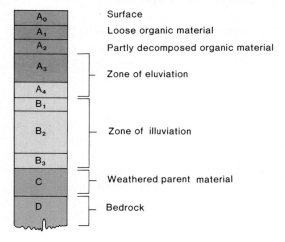

The gross divisions in A-, B-, C-, and D-horizons are of greatest significance to the physical geographer. They give further character to soils and assist in their classification. Basically, soils exhibit different horizons because of different soil-forming processes. Some of those principal processes are described in the section that follows.

Soil-Forming Processes

Soil is formed from **regolith** (weathered bedrock), and a number of factors acting in complex interrelationships produce a particular soil type. The most important of these factors are **parent material, climate, natural vegetation, living organisms, slope,** and **time.** Obviously, a particular factor(s) may be dominant in a given milieu, but all of them, acting and reacting in concert, are important in creating the essential and recognizable characteristics of a particular soil.

It is the most bitter of ironies that the soils with the greatest inherent fertility tend to be situated in subhumid to semiarid areas, whereas those found in humid areas are characterized by lower inherent fertility. The very humid conditions that supply the quantities of water vital to most commercial crops dissolve or erode away the soluble materials in soils so essential to plant growth, whereas the lowered rates of leaching and eluviation in subhumid, semiarid, or arid areas, which leave an abundance of plant nutrients, at the same time, lack the essential moisture for plant growth.

Because of the variations of soil-forming processes from place to place, the resultant soils show similar variability around the globe. A soil formed on a ferruginous (iron-rich) sandstone, under tropical rainforest climatic conditions, beneath a canopy of selva vegetation, and with

Figure 17.10
Idealized soil profiles illustrating laterization, podsolization, and calcification.

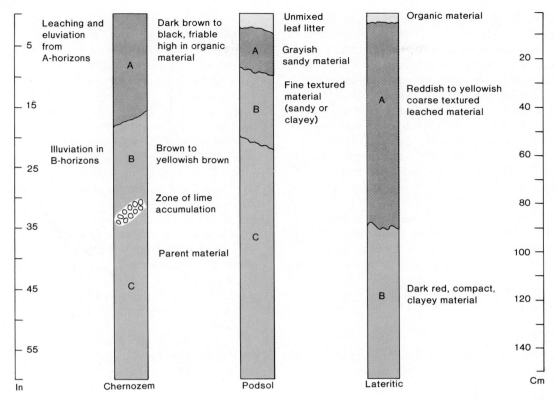

many living organisms working on a flat surface over a protracted period of time will be significantly different from one that has developed on limestone, in a desert climatic regime with sparse vegetation, a scarcity of living organisms, and on a steep slope in a short period of time. Clearly, the variety of geological, topographic, biological, and climatological conditions at both the macro and micro levels creates a variety of soil-forming conditions that literally beggar the imagination. In detail, the mosaic of soils that has developed is exceedingly complex. On the other hand, the broader patterns can be ascertained fairly readily. The generalizations, which are enunciated, fall victim to the particulars of a localized circumstance.

The myriad combination of factors producing different soils may be reduced, essentially, to three principal soil-forming processes: **laterization, podsolization,** and **calcification** (figure 17.10). These processes serve to explain the distribution of mature soils on the surface of the earth. A succinct description of their salient characteristics is included in the following sections.

Laterization

Laterization occurs most commonly in humid tropical and humid subtropical regions. Leaching and eluviation are developed to a high degree because of the abundant precipitation that characterizes such regions (i.e., soluble materials and fine particles are removed rapidly from the A-horizon, resulting in a coarse, granular topsoil and a very compact and fine-textured subsoil). This results in the removal of most soluble materials from the topsoil, leaving a residue of iron and aluminum compounds. Organic material is in very short supply in the A-horizon as well as all others. This may be considered anomalous when one observes the luxuriant vegetation mantle characteristic of such soil regions. On the other hand, it is useful to recognize that an enormous amount of organic material is tied up in the living plant, and the organic litter falls quick victim to living organisms such as insects, microorganisms, and fungi. High temperatures and abundant moisture speed physical and chemical disintegration and decomposition. Hydrolysis and oxidation operate at very high rates in these humid areas. This quick decomposition and disintegration markedly reduce the amount of organic material that might become an integral part of the soil horizon.

Lateritic (latosolic) soils are often very deep in the complete horizon, although the A-horizon is not extraordinarily thick. Shades of reds and reddish yellows characterize soils that have been produced by laterization, reflecting the presence of iron and aluminum compounds. Such soils have low inherent fertility and are subject to rapid destruction when natural vegetation is removed. As a matter of fact, they are among the most fragile soils if vegetative cover is removed. There is considerable cause for alarm about soil destruction in lateritic soil regions because the forest cover, which cloaks most of these soils, is being removed at the rate of about 2 percent per year. The ravages of erosion accelerated by cultivation in these regions foretell an unhappy future for these fragile soils. Rapid runoff and attendant flooding can also be attributed to the clearing of forest vegetation. Unfortunately, too, many of the developing societies in the tropics have lateritic soil as their principal soil legacy, and a poor legacy it is.

Podsolization

Podsolization occurs in cool and moist environments, which may have either mild or severe winters under forest vegetation. Animal life in the soil is meager and, as a result, there is little mixing of organic material with the mineral matter in the soil. A retarded rate of decay, because of the cool temperatures and long periods of below-freezing weather, results in the accumulation of a peaty mat of acidic organic material on the surface of the soil. Leaching and eluviation of the topsoil produces a sandy or granular texture in the A-horizon. Most of the soluble bases are removed from the A-horizon, and there are accumulations of aluminum and iron compounds in the B-horizon. The B-horizon is often compact and clayey as the result of illuviation of the subsoil.

Podsolization is characterized by shallow, acidic soils with minimum organic material and organic life. Extreme podsolization results in soils of generally low productivity, although along the warmer margins of this process, where the effects aren't so extreme, soils of moderate fertility develop. Along the humid subtropical and continental climatic interface, the effects of podsolization and laterization interdigitate, and both processes influence soil development.

Calcification

Unlike laterization and podsolization, calcification occurs in subhumid, semiarid, and arid environments. The dominant vegetation cover is grass or xerophytic shrub. Since precipitation is light, leaching and eluviation are minimal.

Calcium and other bases accumulate in the upper horizons because materials taken into solution by the scanty precipitation are redeposited as the moisture evaporates and, of course, there is a constant addition of calcareous materials, especially from the dead grasses or shrubs, which form the vegetative cover. This zone of lime accumulation is found at some depth beneath the surface in the more humid areas of calcified soils, whereas it reaches to or almost to the surface in the extremely dry areas. There is a direct relationship between the amount of precipitation and the depth of the calcareous nodules (i.e., in the more humid sections, the nodules may be 2 to 3 feet beneath the surface, whereas in the very dry areas, they may be found at the surface). The gradations in between generally reflect varying rainfall amounts. Organic material and living organisms are abundant in the tall-grass areas, although organic material and living organisms diminish in the arid areas where vegetation is sparse. Calcification results in the production of neutral or basic soils in contrast to the acid pH encountered in areas of laterization or podsolization. In the very dry areas, the pH may be so high as to inhibit the growth of all but alkali-loving plants. The extreme effects of calcification produce conditions that are the antithesis of the extreme effects of podsolization.

The various soil-forming processes produce identifiable soil characteristics. The constant desire of scientists to detect and describe order in the environment has led to a number of logical classification systems for soils. Some of the classification systems will be described in the section that follows.

Soil Classification

Mature soils exhibit specific physical and chemical characteristics when developed under discrete combinations of the several soil-forming factors. These physical and chemical characteristics give character to soils and make it possible to develop rational classification schemes. Major categories of soils are the end-products of the soil-forming processes over a long period of time in a given area. In a general classification scheme, soils are divided into three orders, nine suborders, and forty great-soil groups. The three principal soil orders are zonal, intrazonal, and azonal.

Zonal soils are those that have well-developed characteristics in particular response to climate and natural vegetation, and, as such, are widely distributed in areas with similar climates and vegetative cover. These soils usually have a fully developed soil profile. Although there is broad homogeneity within different zonal groups, this similarity masks an enormous complexity as the soil is studied in detail.

Intrazonal soils have developed primarily because of parent material or age, and, therefore, may transcend major climatic and vegetation regions. Indeed, that is a significant characteristic of the intrazonal soils, which appear as enclaves within the pattern of zonal soils.

Azonal soils are immature soils that have not yet developed a full soil horizon. Common examples of azonal soils are loess and alluvium. Such soils are widely spread throughout different climatological and biological regions.

Climax Soils

Some soil experts hold that under identical climatic and other physical conditions, given time and irrespective of the nature of the parent material, there will be no significant difference in the kind of soil formed. This contention is not likely to be true and would be extremely difficult to test, since it would be impossible to tell when (if) soils do, in fact, reach equilibrium conditions. And, in the overwhelming majority of situations, the time required to develop a soil would preclude study of the processes and results. Soils quite probably continue to evolve and change so long as they are exposed to the soil-forming processes.

The classification of soils has been an enormously complex task and has undergone significant evolution within the last century. The dramatic and subtle differences of environment from place to place are reflected in the heterogeneity of soil types encountered on earth. To bring some order out of apparent chaos, a large number of classification schemes have been developed. As knowledge has increased, more and more sophisticated classification systems have evolved. Doubtless, future developments will result in new iterations of existing classification systems or the development of still other systems.

The Russian–American Classification

A scheme developed by Russian soil scientists stressed the relationship of soil types to vegetation regions, which, in turn, reflect climatic regimes. Indeed, the Russians were pioneers in soil science, and they have continued to make significant contributions to an evolving field. This classification was modified by the United States Department of Agriculture (USDA) in 1938, and it continues to be a useful integrative scheme, for it reinforces the significance of the interrelationship of patterns on the earth's surface. *The Yearbook of Agriculture* for 1938 (Soils and Man) was a seminal work that guided pedological investigations until about 1960. It continues to be a pedological classic, and it has use more than fifty years after publication and in the face of new classification approaches and systems.

The major soil orders of the USDA system (i.e., zonal, intrazonal, and azonal) are, of course, further subdivided. Within the zonal order, two gross categories, **pedalfers** (aluminum- and iron-accumulating soils) and **pedocals** (calcium-accumulating soils) are recognized (figure 17.11). Pedalfers occur in humid environments, and pedocals are found in subhumid, semiarid, and arid environments. Pedalfers are products of laterization and

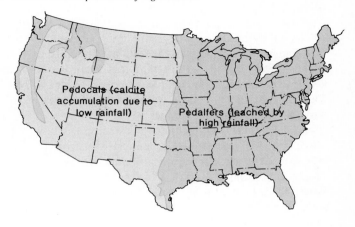

Figure 17.11
Distribution of pedalfers and pedocals.
Source: U.S. Department of Agriculture.

podsolization, whereas pedocals are products of calcification. Obviously, there are transition zones where several soil-forming processes operate alternatively, depending upon the vagaries of climate. Clearly, too, there are blendings of soil-forming processes near their points of contact and isolated local areas with one soil may be surrounded by a larger area of different soil that may exist because of some microclimatological or biological circumstance.

It is beyond the scope of this book to categorize all of the great soil groups in the USDA system. On the other hand, it is useful to describe the important characteristics of a few of them to illustrate climatological and vegetative interconnections with associated soils.

Great soil groups within the pedalfers include lateritic, red and yellow podsolic, gray-brown podsols, podsols, and tundra soils. Essential characteristics of these great soil groups are described in the sections which follow.

Lateritic Soils Lateritic soils are the most weathered, eluviated, and leached, since they occur in areas of abundant precipitation where temperatures are uniformly high. Decay and weathering proceed constantly and rapidly because there is no cold season to suspend such activity. The rapidity of weathering is almost staggering. Plant and animal remains are reduced by the bacteria of decay, the action of fungi, and by macro-organisms of many types. Such soils evolve in Af, Am, and Aw environments under rainforest, savanna, or certain kinds of monsoon forests.

Lateritic soils are made up of the end products of weathering such as iron compounds and hydrous oxides of aluminum. The topsoil is usually coarse textured, porous, and subject to rapid oxidation. The reddish and yellowish colors reflect the abundance of iron and aluminum compounds. There is very low organic content because of the intense activity of micro-organisms and fungi. The subsoil is compact because of illuviation. Much of the clay is removed from the A-horizon and carried to the B.

Inherent fertility of these soils is very low, and the soils are quite fragile. If the vegetative cover is removed, they quickly lose the remainder of their low residual fertility, and they are subject to extensive erosion damage. The streams flowing through areas with lateritic soils are often heavily burdened with sediment. The Amazon, for example, which drains a vast area of lateritic soils, carries such a heavy burden of silt that the ocean waters are stained by the sediment load scores of miles out to sea. Turbidity in adjacent ocean waters is clearly revealed by various remotely sensed images.

Under certain conditions of extreme leaching and eluviation, the topsoil may be almost like a brick. This true **laterite** is sometimes literally quarried as road surfacing or building material. The very word laterite stems from the Latin meaning brick. The author has observed the cutting of laterite blocks to be used for construction in areas of southeast Asia (figure 17.12). True laterites, however, are quite restricted in area and scope, whereas lateritic soils are found in large areas of the humid tropics.

Red and Yellow Podsolic Soils
The red and yellow podsolic soils may be found in certain of the drier margins of the Am and Aw environments and in the Cfa and Cwa climatic areas. At their equatorward margins, they are closer to the lateritic soils; whereas in their middle latitude expressions, they more closely approximate the gray-brown podsols. These soils typically occur under certain monsoon forests, savannas, and pine and certain mixed pine-deciduous forests. These soils have a higher mineral content than the lateritic soils, and they are richer in organic materials. Nevertheless, they still exhibit relatively small amounts of organic materials compared to pedocals. The high percentage of aluminum and iron compounds account for reddish and yellowish soil color. They are still low in bases and inherently marginal in productivity, although they are more amenable to fertilization and continuous cultivation than the lateritic soils.

Although they are subject to deterioration and severe erosion when exposed, they are not so fragile as the lateritic soils. With reasonable management, such soils have supported considerable population pressure. Especially with fertilization, these soils in the American southeast are fairly productive. In many areas of southern China, the soils have been so modified by centuries of cultivation, deliberate flooding, and the additions of large quantities of organic waste (including human waste) that they bear little resemblance to the soil developed in response to nature's dictates.

Gray-Brown Podsols
Gray-brown podsols are found in Dfa-b and Dwa-b climatic environments under mixed hardwood and coniferous forests. They are shallower, less leached, have more organic matter, and are inherently of

Figure 17.12
Laterite dug for use as a road surfacing material in Thailand.
Joseph Castelli.

greater fertility than the lateritic, or red and yellow podsolic soils. They are moderately fertile, and with appropriate fertilization and proper care are amenable to long-term agricultural production. The A-horizon is gray-brown in color (tending toward brown when wet and gray when dry), moderately leached, and of medium to coarse texture. The B-horizon is more compact and typically yellow to red-brown in color. Substantial sections of middle latitude agriculturally productive areas depend upon the gray-brown podsols. They are less fragile than the lateritic, and red and yellow podsols. Gray-brown podsols usually support close settlement and quite dense populations. This is especially true in sections of the Ohio Valley and Great Lakes states of the United States as well as portions of western and central Europe.

Podsols
Podsols are found largely in the Dfc and Dwc-d climatic regions under coniferous forest vegetation. There tends to be an acid-reacting mat of vegetable material at the surface, but there is little mixture of organic material in the A-horizon, which is gray in color. The B-horizon is yellowish or reddish in color because of the relocation of the aluminum and iron compounds. Living organisms in the soil are limited, since the acidic organic mat and cool conditions inhibit their activity, and inherent fertility is low. A severe climate and low fertility limit agricultural production in regions with podsols. Typically, areas with podsol soils have smaller populations and lower population densities than areas with a more healthful climatic and edaphic environment. Shorter growing seasons and lower soil fertility inhibit crop choice and restrict agricultural production. The low fertility of the soil and the severe climate argue against dense settlement in such regions.

Tundra Soils Tundra soils are of negligible agricultural value for a number of significant reasons. First, they are inherently low in fertility because of the reduction of bases. Second, they are waterlogged during the summer season and strongly acid in pH. Third, they exist in an area where the growing season is very short. Tundra soils are located in ET or EM climatic regimes and develop under a low-order vegetation of grasses, mosses, lichens, and shrubs. The surface color is usually grayish and there is a fair amount of organic matter present because of the slow rate of decay. Tundra soils are almost always underlain by a zone of permanently frozen ground known as permafrost. Permafrost produces a kind of perched water table accounting for the fact that many tundra soils may be waterlogged during the frost-free season. The zone of permafrost restricts root penetration and limits the kinds of vegetation that can grow. Tundras are essentially treeless environments. Normal agricultural practices are precluded, and they are used as grazing sites, usually for nomadic animals such as the caribou and musk ox, or as homes for migratory birds.

Prairie Soils The prairie soils are transitional between humid and subhumid environments, and they exhibit the effects of podsolization and calcification. Typically, they are found in Dfa-b and Dwa-b climates, although certain patches of such soils exist in Cfa climates. These soils evolved under tall prairie grass. They are rich in minerals and organic material, possessed of an excellent texture and structure, and they are highly productive when put to the plow. The dark brown to almost black color when wet, which characterizes these soils, may be stained to another hue by underlying bedrock as in the case of the Permian Redbeds in Oklahoma. Prairie soils test neutral to slightly acidic and are essentially at the boundary between the pedocals and pedalfers. Prairie soils usually support significant rural population densities. At their drier margins, agricultural risks increase.

Prairie soils serve as a transition between representatives of the pedalfers and the pedocals. Some representative examples of pedocals include chernozems, chestnut and brown soils, sierozems, and desert soils. These calcium-accumulating soils contrast markedly with the pedalfers in generally having a much higher pH, more abundant organic materials, and soluble minerals. Their inherent fertility is generally quite high, although the amount and reliability of precipitation may reduce their agricultural utility.

Chernozem Soils The chernozem, or black earth, soils are located at the drier margins of the Cfa-Cwa, Dfa-Dwa, and in the more humid margins of the BS environments. They have evolved under shorter grass than the prairie soils, but the grasses have provided a thick mat and continuous cover of vegetation. The topsoil is very dark, almost black, reflecting the high organic content of the soil. The topsoil is shallower than that of the prairie soils, and there are calcium carbonate nodules at the base of the A-horizon. Chernozems are pedocals that test slightly basic in pH. The soluble minerals are present in abundance, and the soils are highly productive. The high fertility of such soils is balanced by the elements of risk in the subhumid to semiarid climate. Wetter margins of the Great Plains and the steppes of the U.S.S.R. have large slices of the chernozems. They are usually characterized by extensive grain production, especially wheat.

Chestnut and Brown Soils The chestnut and brown soils occur in the BS environment under short steppe grass cover. The lighter color of the topsoil, which grades from darker to lighter as progression is made from more to less humid margins of the steppe, reflects the diminishing quantity of organic matter. The shorter, less continuous growth of grass is responsible for the smaller quantities of organic material. These soils are replete with soluble minerals, and zones of lime accumulation are found closer and closer to the surface as one moves to drier climates. Inherently, the chestnut and brown earths are quite fertile, although agricultural risks are magnified as precipitation amount and reliability diminish. Chestnut and brown soils are pedocals that have a basic pH. These soils have supported wheat agriculture, especially in North America and Eurasia. During periods of protracted drought, such as occurred in the 1930s in the United States, plowed soil lay bare to the ravages of wind erosion, which produced dust clouds that blackened the sky and led to the Dust Bowl. Grazing is a more compatible pursuit in most such regions, but care must be exercised to avoid overgrazing and subsequent destruction of the grass cover.

Sierozems and Desert Soils Sierozems and desert soils occur in BW climatic regimes with sparse vegetation. Soluble mineral matter is high, and those pedocals often have calcium or nitrogenous accumulations at or very near the surface. Occasionally, an almost impermeable layer of rock material called **caliche** is formed. The grayish color is reflective of the reduced organic matter, which, in turn, reflects the lack of vegetative cover. Without irrigation, agricultural possibilities are minimal, but irrigation may produce abundant crops. Since water is not available in quantity in most desert regions, the agricultural production of sierozems is extremely limited. Sierozems epitomize the soil-forming process of calcification, and they have the most basic pH of the pedocals. Areas with sierozem soils tend to be sparsely settled unless water is available for irrigation.

Other Soils Numerous other examples of great soil groups could be discussed, and a whole range of intrazonal soils such as those that have developed in boggy conditions like the **Weisenboden soils,** or soils that have developed because of bedrock control like the **Rendzina** on certain kinds of limestone could be described. In addition, there are types of azonal soils such as **lithosols** (rocky soils), alluvium, or loess, which could be further described; however, in a book of this kind, certain limitations must be imposed.

Another soil classification system has become more significant to most pedologists, and it is assuming more importance among geographers. This classification system is described in the following sections.

The Comprehensive Soil Classification System

All soil classifications involve generalization to a greater or lesser degree since there is an almost infinite variety of soil types if they are examined in great detail. **Soil orders** (i.e., zonal soils, intrazonal soils, and azonal soils) represent the greatest level of generalization in the USDA scheme. The orders are then subdivided into nine to twelve suborders. Further division of the suborders is made into great soil groups. Several of those great soil groups have been described in the preceding discussion. That level of generalization is appropriate to our purposes, but pedologists use a much more detailed breakdown. Great soil groups are divided into **soil families,** the families into **soil series,** and the series into **soil types.** The thousands of soil types are further divided into **soil phases** on the basis of slope, stoniness, or some other physical feature, which might have an important bearing on agriculture.

Clearly, great detail is important to allow people like county agents to provide practical advice to farmers. That level of detailed mapping and accurate scientific advice as to preferred crop type, tillage practices, and fertilizer application has made American agriculture the most productive in the world. In no small measure, that productivity relates to accurate soil mapping, astute professional advice by county agents, and educated and skilled farmers who have made use of the advice offered.

A new classification system introduced in the United States in the 1960s was designed to describe and characterize soils on the basis of their inherent physical and chemical characteristics without reference to their location in a particular vegetative, geomorphological, or climatological region. That system is briefly discussed in the sections that follow. It has become the dominant schema for continuing soil classification and mapping in the United States.

The Comprehensive Soil Classification System (CSCS), which was introduced in 1960 and subsequently modified, identifies six levels of classification (i.e., 10 soil orders, 47 **suborders,** 185 **great soil groups,** an expanding number of **subgroups,** families, and series. There are more than 10,000 recognized divisions within the United States, and the number expands if the rest of the world is considered.

CSCS classification describes the physical characteristics and chemical composition without reference to soil-forming process or environment. In a sense, the system is analogous to the Linnaean system of classification of plants based on physical characteristics rather than association or evolutionary development. The CSCS classification adds heterogeneity to the landscape, and the kaleidoscopic character of earth's patterns is further illustrated. It should be recognized that it is like any classification scheme of any sort in trying to develop order, sense, or pattern in observed phenomena. Any number of classification schemes may be used, but all have in common an attempt at a rational codification of extant phenomena.

Of course, there is a body of opinion that rejects order (i.e., there is an assumption by some that chaos rules, and the order described by scientists is perceptual rather than real). To most, however, there appears to be order in the universe, and the human mind can cope with order far more effectively than with chaos.

The CSCS system examines horizons carefully to classify soils. The surface layers are called **epipedons,** and the subsurface layers are termed horizons.

It's difficult to correlate the USDA and CSCS systems, but there is some association at the greatest levels of generalization. For the purposes of correlation, a number, comparable to the great soil groups described in the USDA system, will be described. The environmental interconnections will be considered, although it must be borne in mind that the soils were classified on the basis of observed physical and chemical characteristics and not environmental associations.

Figure 17.14
Oxisol (latosol, lateritic).
Reproduced from the Marbut Memorial slide set, set of 85 slides, 1968, SSSA, by permission of the Soil Science Society of America, Inc.

Oxisols The oxisols are essentially analogous to the lateritic soils. They are heavily leached soils high in aluminum and iron compounds. They are essentially of the same extent and have developed in the same climatic regimes and under the same vegetation as the lateritic soils. Obviously, they suffer the same fertility shortcomings and manifest the same fragility characteristic of lateritic soils. They are the meager legacy of the humid tropics and illustrate again the age-old irony of an inverse relationship between soil fertility and amounts of precipitation. That is, soils developing in perhumid or humid areas tend to be less productive, whereas those evolving in semiarid or arid conditions tend to be high in fertility. Those possessing the greatest inherent fertility, generally, are limited in terms of productivity by the paucity and unreliability of precipitation.

Figure 17.15
Ultisol (red and yellow podsol).
Reproduced from the Marbut Memorial slide set, set of 85 slides,
1968, SSSA, by permission of the Soil Science Society of America,
Inc.

Figure 17.16
Alfisol (gray-brown podsol).
Reproduced from the Marbut Memorial slide set, set of 85 slides,
1968, SSSA, by permission of the Soil Science Society of America,
Inc.

Ultisols The ultisols have approximately the same distribution as the red and yellow podsolic soils, which means that they extend from the wet-dry tropical regimes into the humid subtropics. The colors in the red and yellow ends of the spectrum derive from the aluminum and iron compounds. Although these soils are acidic, with relatively low fertility, their organic content is greater than the oxisols, and they are amenable to cultivation if fertilized and carefully managed. With expanding populations, these soils, which have been conditioned and fertilized, are supporting increasing population densities. Undoubtedly, a growing world population will place even greater productivity demands on these soils in the future.

Alfisols Alfisols show an enormous range of climatic and vegetative distributions, although they typically occur under forest vegetation. The epipedon is typically gray-brown in color and the B-horizon is usually clayey. They are similar to the gray-brown podsol; they contain a fair amount of organic material; and they are only mildly acid in reaction. They are productive agricultural soils and, with reasonable management, rotation, and fertilization, are very productive. They support quite dense human populations and, with care, can continue to do so for the foreseeable future.

Figure 17.17
Spodosol (podsol).
Reproduced from the Marbut Memorial slide set, set of 85 slides, 1968, SSSA, by permission of the Soil Science Society of America, Inc.

Figure 17.18
Mollisol (chernozem).
Reproduced from the Marbut Memorial slide set, set of 85 slides, 1968, SSSA, by permission of the Soil Science Society of America, Inc.

Spodosols The spodosols are very much like the podsols and are usually found in subarctic environments under coniferous forest vegetation. The light-colored epipedon is low in organic material and important plant nutrients, and it has a low pH. The clayey subsoil is yellowish to brownish, reflecting the relocation of aluminum and iron compounds, and the illuviation of clays eluviated from the A-horizon. These soils are of limited agricultural value, because they are low in inherent fertility, and they are found in climatic regimes with very short and unreliable growing seasons.

Mollisols The mollisols embrace, essentially, the prairie, chernozem, and chestnut-brown soil regions. A dark brown to almost black topsoil reflects the abundance of organic material. Their presence in subhumid to semiarid climates accounts for the abundance of soluble mineral matter present. These soils have high inherent fertility and, if the rains come or irrigation is provided, they are highly productive. As a matter of fact, they are among the world's great surplus grain producers. In an increasingly crowded world, it is important to retain their structural integrity and fertility.

Figure 17.19
Calcium carbonate nodules near the surface of a mollisol.

Figure 17.20
Mollisol (chemozem) in central Texas.
Jerome Coling.

Figure 17.21
Aridisol (red desert soil).
Reproduced from the Marbut Memorial slide set, set of 85 slides, 1968, SSSA, by permission of the Soil Science Society of America, Inc.

Aridisols Aridisols are desert soils, and they are characterized by high mineral content, not infrequently at toxic levels for many plants, and low organic content. The epipedon is typically light in color, reflecting the paucity of organic material in the soil. With appropriate chemical treatment, in some instances, and irrigation water, in all instances, these soils may be reasonably productive. These soils are essentially coincident with the world's desert climates. Perhaps as much as a third of the world's surface area is covered by aridisols.

Vertisols Vertisols really have no USDA classification equivalent. They are widely spread in tropical and subtropical regions subject to drought. They really have no horizons in the usual sense. They are composed of a high percentage of clay minerals, which absorb copious quantities of water during wet periods, but shrink and crack during drier times. They are high in organic materials and base minerals since they typically develop under grass. They are quite productive, although they may be difficult to work during wet periods because of the viscous clay content. Such soils in eastern Texas were termed the black,

waxy prairies by early settlers reflecting the color and cultivation characteristics. The polygonal pattern of surface cracks during dry periods on exposed vertisols may be an important identifying characteristic.

Entisols Entisols are the equivalent of azonal soils in the USDA classification. They have no specific horizons, since their relative youth precludes such development. Alluvial or aeolian deposits are common examples of entisols, and they may occur in any climatic or vegetation region. Their utility varies greatly depending on the nature of the alluvium or loess, and the climatic region where they have developed. Drainage patterns shown on maps and areas of thick loess accumulation reflect the localization of entisols. Frequently, zones of loess and alluvium have proven

to be agriculturally productive, since they contain the nutrients of the topsoils from which they were derived. They may also be quite fragile and subject to the ravages of severe erosion when the natural vegetation is removed.

Inceptisols Inceptisols are also immature, and typically they have very limited horizon development. They are common in tundra regions where they have developed on glacial deposits, and they are often found in volcanic regions developed on volcanic ash. Clearly, their productivity is determined by the nature of the parent material and the climatic conditions, which exist in the locale of their development. They typically exhibit shallow horizons and poor horizon development. Basic volcanic parent

material in humid regions usually develops into productive agricultural soil, whereas acidic volcanic rocks are less inherently fertile, and typically they are so coarse textured as to be subjected to physiological aridity. Obviously those inceptisols, which have developed in tundra regions, are severely restricted in utility because of the harsh climate.

Histosols Histosols are examples of intrazonal soils in the USDA classification system, and they develop in swampy areas. The waterlogged soil inhibits aerobic bacteria, and anaerobic species do not significantly limit the accumulation of abundant organic matter. The organic remains contribute to an acidic pH. The effects of anaerobic bacteria may produce foul-smelling gases in the clayey horizons. These soils may be made productive directly by using certain acid-loving plants such as cranberries as cultivated crops, or they may be drained and, after appropriate soil treatment, are often areas of intensive horticulture, such as certain drained peat and muck areas along the margins of the Everglades in Florida. Here, again, is a conflict between the short-term value of draining such mucklands for agricultural purposes and their long-term value in helping to modulate the hydrological system while serving as a wetlands nursery for countless species of living creatures. Whether long-term effects are desirable or undesirable, it seems certain that more areas will be drained as hunger lurks on the horizon. The Everglades have shrunk in size and become less fecund in biological terms as a result of the draining of lands for crops or human settlement. Again, these soils are fragile when disturbed.

Other soil classification systems are employed in different parts of the world. Neither the USDA nor the CSCS system are considered the *sine qua non* (absolutely essential) for soil classification in the world-wide situation. The complexity of soil types across the world continues to test the ingenuity of human beings to develop fully rational and universally applicable classification systems.

Mankind continues to be intimately tied to soils, and the species continues to modify soil characteristics. Conservation and the long-term survival of the species dictate an increased understanding of and positive response to our soil resources.

Soils and Man

Irrespective of the system of classification, the soil resources of the earth continue to be *the* resource, along with water, which supports *Homo sapiens* on this planet. The thin veneer of soil, which stands between people and starvation, is a fragile shield that must be constantly reinforced and protected. Our past record of reckless extravagance in the use of all of our resources, including soil, and the expanding numbers of people are hardly good omens for the long-term survival of the human species. In

Table 17.1
pH Scale of Soil Acidity and Alkalinity

CSCS	pH value	Acidity–alkalinity	Soil groups	
	4.0			
	4.5	very strongly acid		
histosols			podsols	
	5.0		gray-brown	
spodosols		strongly acid	podsolic soils	
			tundra soils	
	5.5			
		moderately acid		
	6.0			
			brown forest soils	
	6.5	slightly acid		
	6.7		prairie soils	tropical black earths
			latosols	
			chestnut and brown soils	
alfisols				
ultisols	7.0	*neutral*		
mollisols		weakly alkaline		
oxisols	8.0			
		alkaline		
	9.0			
aridisols		strongly alkaline		
	10.0		black alkali soils	
		excessively alkaline		
	11.0			

spite of the short-term successes of some countries, using the scientific fruits of the "Green Revolution" to become self-sufficient in food (e.g., India and China), a continuing growth in world population of almost 2 percent per year threatens the long-term survival of people. Soil erosion, leaching, and lack of fertilizer in many areas of the world have led to a kind of chronic famine. Indeed, these chronic conditions represent a kind of never-ending cycle of poverty yielding poor soils, meager crops, and hungry people.

Africa is especially prone to famine conditions because of the rapid increase in population, the limited soil legacy, austere climatic conditions, lack of fertilizer, low availability and high costs of nonhuman energy sources, an imperfect understanding of modern cultivation practices, and an unstable political condition in many regions. Indeed, Africa has actually had a per capita decline in food production since the 1960s. There appears to be no reasonable prospect for significant improvement of the situation within the foreseeable future. Interventions from developed societies can only be viewed as temporary solutions. Longer term solutions are in the hands of Africans who must learn to practice conservation and, above all, to reduce their population growth rates. Ethiopia, Mozambique, Mali, and other states in the Sahel district of Africa experience chronic food shortages, which frequently assume the dimensions of famine. Other African states are in danger of slipping into the same pattern of chronic food shortages. Failure to take remedial actions can only lead to economic and political chaos with deleterious ramifications for the entire world.

Soil erosion is expanding in virtually all regions, and effects of flooding; siltation of reservoirs, waterways, and harbors; damage to transportation facilities; and reduction in permeability of soils are all undesirable consequences. Rapidity of runoff and lowered soil permeability, along with accelerated drawdown of subsurface water by pumping, are causing the water table to drop alarmingly in many areas of the world.

Figure 17.26
Generalized CSCS soil map of the world.

World Soil Orders

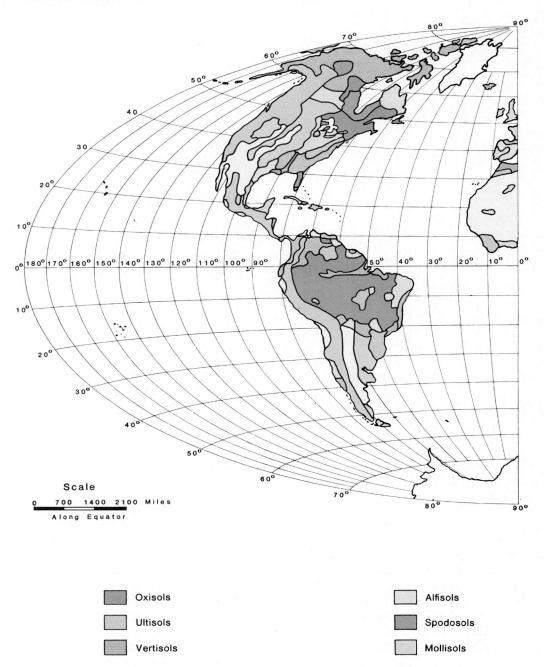

Scale

0   700   1400   2100   Miles

Along Equator

Oxisols	Alfisols
Ultisols	Spodosols
Vertisols	Mollisols

Aitoff's Equal Area Projection

	Aridosols		Histisols
	Entisols		Highlands
	Inceptisols		

Figure 17.27
Generalized Russian-American soil map of the world.

World Zonal Soils

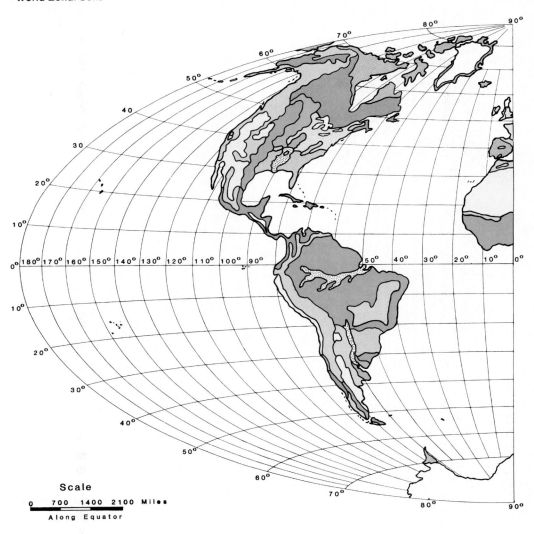

Scale

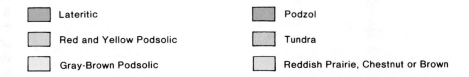

	Lateritic		Podzol
	Red and Yellow Podsolic		Tundra
	Gray-Brown Podsolic		Reddish Prairie, Chestnut or Brown

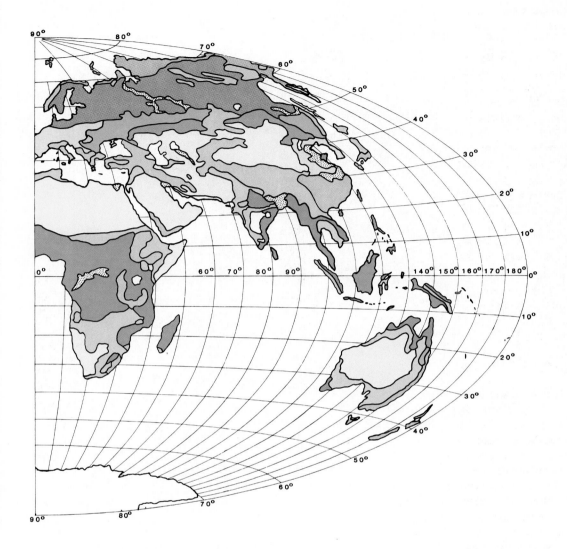

Aitoff's Equal Area Projection

	Prairie—Chernozem		Mountain Soils
	Chestnut—Brown		Alluvium
	Desert		

Figure 17.28
Generalized map of major CSCS soil orders in the United States. Boundaries are approximate and tentative.
Source: Adapted from a USDA map.

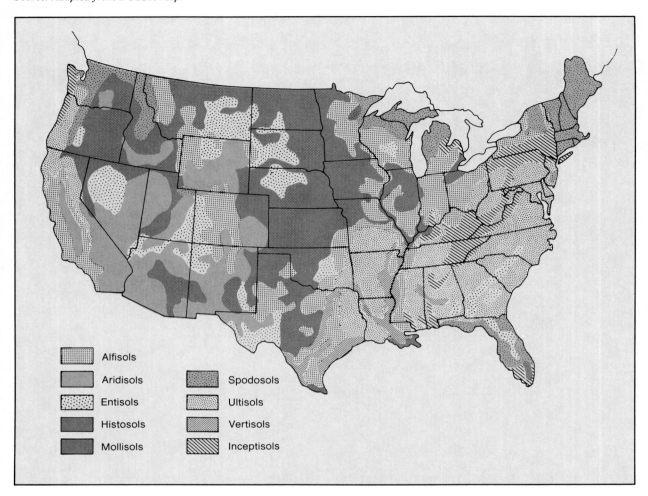

Alfisols

Aridisols

Entisols

Histosols

Mollisols

Spodosols

Ultisols

Vertisols

Inceptisols

Unwise practices of deforestation, overgrazing, and agriculture have resulted in more loss of soil in the last few decades than nature can replace in thousands of years. The devastating effects of soil erosion have caused farm abandonment, rural migration, debts, poverty, and bankruptcy for many farmers in developed countries and outright famine in many developing countries. Soil resources must be husbanded; soil conservation must become an international commitment; and rational population policies must be established if this good blue-green earth is to maintain its attributes as a haven for man. Anything less seems certain to result in an expansion of a brown and less healthy environment, suitable only to lower forms of life, or to no life at all. Without proper care of the earth's soils, a benign world may become hostile. People's home may become their grave.

Soil resources serve as a habitat for plants and animals. Many of those plants and animals have direct economic value for people, and virtually all of them contribute to the aesthetic aspects of the world. The destruction of soil has an indirect influence on plant and animal communities. In many cases, these indirect assaults on the world's biota may be as significant as the direct denudation of vegetation cover or the purposeful destruction of animal species.

Extant vegetative and animal patterns and associations are considered in the following chapters. Relationships to soil associations and climate patterns are obvious and pervasive.

Figure 17.29
Soil erosion in the United States. The situation is even more desperate in many developing societies, but the data to produce accurate maps of such areas are often lacking or unreliable.
Data from USDA Soil Conservation Service.

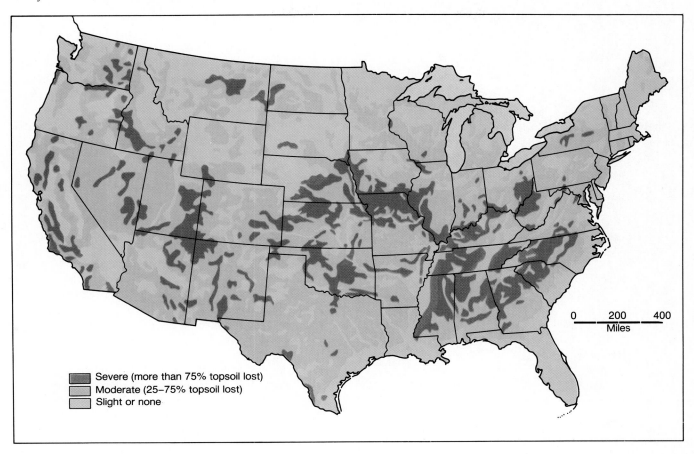

■ Severe (more than 75% topsoil lost)
■ Moderate (25–75% topsoil lost)
□ Slight or none

0   200   400
Miles

Figure 17.30
Sheet erosion is occurring even in areas where gullying is not obvious.
Jerome Coling.

Figure 17.31
Gullies strip away countless tons of topsoil, and cultivation accelerates erosion.
Jerome Coling.

Figure 17.32
Erosion has exposed these tree roots. It is likely that the tree will soon topple.

Study Questions

1. Explain why topsoil is such a precious resource. Why is there cause to worry about topsoil degradation or loss?
2. What are the basic ingredients of soil?
3. What is the role of legumes in the maintenance of soil fertility?
4. What is the role of organic compounds in mineral exchange and tilth within soils?
5. List and describe the various ways that water is held in soil.
6. What soil textures are generally more amenable to agriculture?
7. Give some examples of common soil structures. What is the essential difference between soil texture and soil structure?
8. Explain why soils developed in different environments may have greatly variable depths.
9. What conditions do soil colors usually reflect?
10. What techniques are employed to "sweeten" acidic soils? To neutralize basic or alkaline soils?
11. List and describe the characteristics of a typical mature soil profile.
12. Compare and contrast eluviation and illuviation.
13. What are the factors that produce a particular soil type?
14. In the Russian-American Soil Classification System, what are the principal soil-forming processes? Characterize each.
15. Distinguish between zonal, intrazonal, and azonal soils.
16. What is the major distinction between pedocals and pedalfers?
17. You should have an essential understanding of the location and characteristics of the Great Soil Groups.
18. In the Comprehensive Soil Classification System, what characteristics are stressed in classifying soils?
19. You should be familiar with the essential characteristics of the CSCS soils discussed in this book.
20. What is the Green Revolution?
21. Briefly explain the relationship of soils, and native plants and animal associations.
22. Explain why Africa seems to experience repetitive famine.

Selected References

Butzer, K. W. 1976. *Geomorphology from the earth.* New York: Harper and Row, Publishers.

Gabler, R. E.; Sager, R. J.; Brazier, S. M.; and Wise, D. L. 1987. *Essentials of physical geography.* 3d ed. Philadelphia: Saunders College Publishing.

Gersmehl, P.; Kammrath, W.; and Gross, H. 1980. *Physical geography.* Philadelphia: Saunders College Publishing.

Hunt, C. B. 1972. *The geology of soils.* San Francisco: W. H. Freeman.

———. 1938. *Soil and man: Yearbook of agriculture.* Washington, D.C.: United States Department of Agriculture.

———. 1957. *Soil: Yearbook of agriculture.* Washington, D.C.: United States Department of Agriculture.

Strahler, A. N., and Strahler, A. H. 1984. *Elements of physical geography.* 3d ed. New York: John Wiley and Sons.

18

Biotic Resources

The biosphere is a place where atmosphere, hydrosphere, and lithosphere meet, as in this photograph of Lake Matheson, New Zealand.
New Zealand Tourist and Publicity Office.

> The snows have dispersed, now grass returns to the fields and leaves to the trees.
> Horace, *Odes, Book II*
>
> To him who in the love of Nature holds Communion with her visible forms, she speaks A various language; . . .
> William Cullen Bryant, *Thanatopsis*
>
> "Then, cleaving the grass, gazelles appear . . ."
> Sturge Moore, *The Gazelles*

*P*lants and animals are inextricably connected to other elements of the physical environment, especially climate and soil, although elevation, slope, exposure, drainage, and a host of other environmental factors are interlocked with life in producing the physical fabric of the earth. It's noteworthy that a number of the climatic regions are named for the natural vegetation dominating within a region, and, of course, the Köppen Climatic Classification was based on assumed vegetative relationships. The USDA soil classification system used climate and vegetation as principal generative factors in developing the classification of a particular soil type. All stands of natural vegetation are a consequence of evolutionary processes involving plant differentiation and adaptation over substantial periods of time. It also appears that there are constant subtle changes in plant associations, which may not be noticed by the casual or transient observer. The variety of plants and their interrelationships are surprising when a particular assemblage is examined in detail.

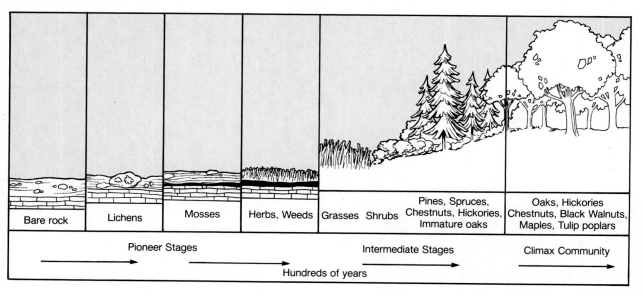

Bare rock	Lichens	Mosses	Herbs, Weeds	Grasses	Shrubs	Pines, Spruces, Chestnuts, Hickories, Immature oaks	Oaks, Hickories Chestnuts, Black Walnuts, Maples, Tulip poplars

Pioneer Stages — Intermediate Stages — Climax Community

Hundreds of years

The first unicellular plants and animals, which developed from inert materials about 3.6 billion years ago were the pioneers involved in helping to create an atmosphere suitable for the development of higher orders of life. Evolutionary development over time has yielded a multiplicity of life forms, which literally beggars the imagination. Those forms have influenced the environment of which they are a part just as they have responded to that environment. Environments have been created, which are salubrious for settlement, whereas others have proven to be hostile for people. There is a persistence of life in even the most austere of environments. Few places, except for vast ice caps in Antarctica and Greenland, are devoid of plant life.

Even in those regions, certain unicellular and algal forms find footholds in favorable microenvironments. Life appears in some very unlikely places which, at first glance, appear hostile to life forms. Life forms have been found in hot springs that are near the boiling point and on or at the margins of snowfields where temperatures are almost freezing. The tenacity, resilience, and adaptability of life is awesome to contemplate. Plant and animal associations can withstand substantial ecological stress, but they are by no means invulnerable. Hundreds of species fade into extinction each day—an extinction that is often induced or accelerated by the actions or attitudes of people. Many life forms in tropical environments, especially, fall into the abyss of extinction before they have ever been accurately classified.

Ecological Succession and Climax Associations

On any landscape amenable to plant growth, there are a group of initial plant colonists that establish themselves; in turn, the colonizers are succeeded by other plant associations in a process known as **ecological succession;** and, finally, an association develops and remains as the **climax vegetation** (i.e., the stable vegetation characteristic of a particular climatological and pedological milieu). Each succeeding vegetative form, in an intricate and intimate physical and biochemical dance, creates a more favorable environment for the plant and animal associations destined to succeed it. Each stage in the cycle produces a slightly different edaphic environment than the previous one as new associations establish themselves (figure 18.1). In a kind of ultimate irony, the very success of certain associations and species tends to create a habitat more suitable for their successors than for themselves. They succumb to new successors, which, in turn, may be succeeded over time by other species.

This state of presumed ultimate equilibrium has resulted from the subtle interactions of vegetation with soil and climate until finally an association finds conditions amenable to growth, while inhibiting the encroachment of other species. It's clear that in certain environments, there is a shift of climax vegetation over time in response to climatological changes, overgrazing, fire, and, especially, the influences of people. Many biologists dispute the stability of climax stands, averring that there is a kind of rolling sequence that results in temporary stability, followed by instability, succeeded by a new stability. This process may

Figure 18.2
The common cattail is a type of hygric vegetation.

Figure 18.4
Both grass and trees are mesic forms.

Figure 18.3
A type of xeric vegetation.

Figure 18.5
Probable initial range of the American bison (buffalo). The bison population was decimated by profligate hunting, and small herds exist now only in parks or preserves.

be repeated interminably. In broadest terms, vegetation evolves in three kinds of environment: **hygric** (wet), **xeric** (dry), and **mesic** (intermediate).

The existence of a climax stand assumes a substantial period of time of environmental conditions within a narrow range of variance. Since stasis (the same conditions) probably does not exist for any significant period of time in geological terms and because climate clearly changes on both a long- and short-term basis, it appears that biological features establish a kind of equilibrium with other environmental factors. As the environmental balance tips in one direction or another, there is a resultant biological shift to establish new equilibrium conditions.

While certain animal associations characterize particular regions, the range and mobility of many animal types makes their distribution patterns and regional associations somewhat less well fixed. Of course, many migratory species inhabit different edaphic and climatological regions in different seasons. Others readily adapt to an environment other than their native habitat as they are introduced accidentally or by design to a new region. Even domesticated species may go wild in certain areas and develop wild populations. In some instances, such introductions can have catastrophic effects on other animal populations and plant species. Nevertheless, as practicable, principal animal associations will be considered in the discussion of vegetation patterns that follow.

The distribution and character of the world's biotic resources exerts a strong influence on human settlement patterns and economic activities. In forested areas, man may rely on wood products for food, shelter, and foreign exchange. The native animal life may pose a threat to be exterminated or offer a promise of food or motive power to be used. Grasslands with their native herbivores are strongly suggestive of a grazing or agricultural economy. Usually, people respond positively to this biological suggestion. Rarely is the human species satisfied to leave existing associations unexploited, although in recent times, a spirit of conservation has been born in certain countries. In the main, human beings have proceeded in the process of development to exploit the environment, destroying thousands of plant and animal species in the process. The pace of plant and animal extermination has been so alarming that a national project was recently begun in the United States to save the seeds of thousands of endangered species of plants in a kind of germ plasma bank to be used, if needed, at some future time.

At present rates of exploitation of the tropical rainforest, for example, it is estimated that 10 to 20 percent of *all* living species will have been destroyed by the year 2000. Many species have been or will be destroyed before they have been adequately catalogued and classified. This dramatic reduction of species and diminution of the world's gene pool almost certainly will diminish the world's diversity and beauty. Our descendants are likely to have a biological environment much impoverished in comparison to what we or our ancestors enjoyed.

On the other hand, there are some attempts to save the remnants of certain animal species by captive breeding programs. Zoos may retain the last remnants of certain populations, especially large wild mammals, as the inexorable advance of human populations destroy habitat for wild creatures.

Even where some species have made a comeback through human protection, they may be vulnerable because of a limited gene pool. The elephant seal of the California and Mexican coast has made a dramatic recovery in numbers from fifty a few decades ago to more than one hundred thousand in 1988. On the other hand, biologists worry that the lack of genetic diversity may make them vulnerable to some disease or biological shock in the future.

Human Modification of the Biota

People began large-scale modification of the biota when domestication became commonplace. Subsequently, people hybridized large numbers of plants and animals to provide food. Although these hybrids have multiplied food production several times over, there is considerable concern that a concentration on a limited number of varieties makes production vulnerable to disease or insect pests. These changes resulting from hybridization have forever altered the biota. It is increasingly apparent that the success or failure of particular life forms may, in fact, influence as many other elements of the environment as the environment influences the life forms.

The possibilities for widespread genetic manipulation of plant and animal species are increasingly within human understanding, and the ultimate effects of such biological manipulations are unknown. It's certain that they will initiate broadened ecological and ethical arguments. Nevertheless, principal plant and animal associations remain, and their distribution and types give character to the physical environment, adding another element to the mosaic that is the cover of the blue-green planet, third from the sun, which is our home. How gene splicing and other modern microbiological manipulations may alter those patterns in the millennia ahead remains to be seen. Whether human beings will begin to obey an ethic of conservation—which has two basic tenets: (1) you shall maintain energy flow, and (2) you shall not sacrifice the abiding and eternal to the temporary or expedient—remains to be seen. Life is so precious—perhaps unique in the universe—that we need to exert every effort to protect its fragility. Halting conservation efforts have begun, and some encouraging signs, at least among scientists, suggest a concern for and an interest in environmental preservation and enhancement.

Different discrete biological associations, called **biomes,** will be analyzed in terms of their spatial distribution and environmental relationships. Those discussions follow.

Forests

Although different tree species have varying environmental requirements because of their inherent physiological differences, there are certain species or combinations of species that tend to be dominant in certain discrete geographical areas. The distribution of plant species, including trees, depends on a number of interconnected factors: (1) length of daylight and darkness (**photoperiod**); (2) temperature means and extremes; (3) length of growing season; (4) precipitation, including amount, distribution, and intensity; (5) winds; (6) soil; (7) slope; (8) exposure; (9) drainage; and (10) other plants and animal populations. These environmental factors combine in intimate and intricate ways to produce varieties of conditions favorable or hostile to particular plant species and associations.

Forests are limited primarily to tropical wet, tropical wet and dry, humid subtropical, Mediterranean, marine west coast, humid continental, and subarctic environments, although there may be some extensions of woodlands into drier or colder regions, especially in protected or better watered microenvironments within larger

Figure 18.6
Typical midlatitude forest along the margins of the Highland Rim of the Nashville Basin.

Figure 18.7
Colder conditions in high elevations of the San Juan Range in Colorado limit tree growth. Above that tree line only low-order vegetation can survive.
Colorado Tourism Board.

climatic regions. The climatic types listed meet tree requirements of adequate moisture, unfrozen subsoils during the growing season, and sufficiently long growing seasons to allow flowering and fruiting between periods of killing frosts or periodic drought. Other regions are too dry, too cold, or have permafrost too close to the surface to permit trees to grow and reproduce.

Figure 18.8
The edge of the selva in Mindanao in the Philippines.

Tropical Rainforests

Areas of low to intermediate elevation astride the equator in South and Central America, southeast Asia, and central Africa are clothed with tropical rainforest, also known as **selva.** About one-twelfth of the earth's surface and nearly one-half of the earth's remaining forested area is covered with tropical rainforest, but rapid cutting of such forested areas and subsequent degradation of the regions that have been cut are leading to an alarming reduction of selva regions. The annual loss of selva vegetation amounts to more than 44,000 square miles (almost 71,000 square kilometers), an area about the size of the state of Ohio. Indeed, except for a few protected sanctuaries and small inaccessible regions, the selva may disappear before the depredations of chainsaw and the squatter in this generation. If this does happen, the earth's biota will be irrevocably impoverished, and humankind will be markedly diminished. Not only that, but broad clearing will certainly have major impacts on hydrology, soils, and perhaps climate.

The typical selva is tristoried; that is, a main canopied area at intermediate levels is pierced by a few forest giants called **emergents,** which extend to great heights, and beneath the general canopy area a few species, which constitute a forest understory exist. Most species grow to intermediate levels; both the emergent and understory species exist in smaller numbers. Overall, the forest is characterized by tall trees averaging more than 100 feet (30 meters) in height with emergents often soaring almost 200 feet (60 meters) above the forest floor. Trees are closely spaced and interlaced with a multiplicity of climbing vines called **lianas** and numerous **epiphytes,** or air plants. The

Biotic Resources 315

Figure 18.9
Underbrush in jungle areas makes penetration difficult.

but there is the very real danger that, except for a few isolated or remote areas, the association will disappear from the planet early in the twenty-first century.

In tropical regions adjacent to coasts, mangrove trees thrive. Tidal flats, estuaries, and other areas of brackish water provide the ideal habitat for the stilt-root mangrove, which manifests itself in several species. These mangrove swamps also serve as rich habitat nurseries for a variety of aquatic and amphibian forms. Especially broad expanses of mangrove are to be found in the delta areas of the great rivers of southeast Asia. Mangroves often march seaward over the sediment trapped around their stilt roots. Significant accretions of territory are built up over time with this seaward march of mangroves in certain stretches of the continental shelf, especially in southeast Asia.

Tropical rainforests are composed of mixtures of scores of tree species, which include rich reserves of such valuable hardwoods as mahogany, rosewood, and teak, and such useful tree products as camphor, quinine, vegetable gums, and rubber. Although species variety characterizes all selva, there are differences in those tree aggregations among the several continents. The forests are, however, being rapidly exploited for lumber, firewood, and charcoal, and they are being cleared for agricultural and grazing land at an alarming rate.

It is difficult to convince peasant farmers with families to feed that the forest giants should not be sacrificed to a transitory agriculture. Those without fuel to cook their meager rations or to heat their rude hut turn deaf ears to

combined thick tree foliage, epiphytes, and woody lianas practically exclude sunlight from the forest floor. This is a dim, shadowy world, even at midday. There is surprisingly little undergrowth because of the restriction of light. Only shade-loving plants can compete effectively on the forest floor.

The forest is choked with underbrush in only a few specialized areas. Where light can penetrate to the forest floor as along a river or stream, on a steep slope, or at the edge of a forested area, that section becomes an almost impenetrable wall of vegetation called a **jungle.** Peoples from middle latitudes often confused the essential character of the selva because their approaches to the jungle margins of the rainforest obscured the nature of the forest some distance away from coast or stream. As a result, early writings about and impressions of the selva were erroneous. Scientists now have a much better understanding of the selva, but a great deal of additional work is still needed to unlock the forest's detailed secrets. The selva has existed for millions of years as a distinct association,

Figure 18.11
Buttresses are needed to support the forest giants of the selva.
Jerome Coling.

Figure 18.12
Savanna grass (foreground) is succeeding in an area of selva cleared for cultivation in Malaysia.

Dead trees, broken limbs, and leaf litter are quickly reduced by fungi, bacteria of decay, and insects. This rapid decomposition in a hot, humid environment accounts for the thin layer of leaf litter. The decayed and decomposed material is consumed rapidly by the plant roots, resulting in a soil with little organic matter.

Areas of tropical rainforest that have been cleared can support pasture or crops for very short periods unless the soils are heavily fertilized. Governments with large areas of selva and burgeoning populations, like Brazil, have apparently not learned this lesson as they extend settlement further into the reaches of the rainforest. Each new road bulldozed into the hinterland accelerates the destruction of plant and animal habitat. As settlers have followed these roads, they have literally laid waste adjacent areas of forest without thought for or appreciation of the ecological consequences.

As forests are cleared, the heavy tropical rain quickly leaches the small reservoir of residual fertility, and the symbiotic interconnection between soil and forest is broken. A kind of biological degeneration often occurs, which leads to a succession of scrubby growth and savanna in formerly forested regions. Indeed, significant expanses of savanna may, in fact, have resulted from rainforest areas that were irreversibly disturbed. Large areas of cogon grass (*Imperata cylindrica*) savannas in the Philippines are the result of such disturbances by people. Their persistence as a new vegetative association results, in part, because of repeated burnings either by accident or design.

the needs for conservation. Better prospective agricultural regions must be developed and cheap fuel sources must be made available, or the corporate lumbering giants augmented by the depredations of encroaching peasants will eliminate this great biological legacy. In the end analysis, population must be stabilized because people's needs and wants are causing so much pressure on the environment that people may be in imminent danger of destroying the home that sustains them.

The typical tall, straight selva tree is flanked by a series of buttresses around the base, which help to support the shallow root system. When a forest giant falls or is cut, there is fierce competition among seedlings to reach the light and achieve their place in the sun. This fierce competition and varying degrees of success of different species of seedlings accounts for the heterogeneity of mature tree species.

Selvas are reservoirs of an astonishing array of plant and animal life. Literally hundreds of new species of plants and animals are identified each year. As the tropical rainforest shrinks in area, numerous plants and animals face extinction. Many will probably become extinct before they have been scientifically identified and classified. The rate of extinction has skyrocketed in recent years. Fortunately, a few governments (notably Costa Rica, Panama, Brazil, and Zaire) have set aside certain areas as reserves, to preserve at least a portion of the great biological treasure resident in the rainforest. Unfortunately, in Brazil, especially, even these reserves are being reduced, often illegally, as roads are cut deeper into the interior of the country. It is difficult to convince the illiterate squatter not to plunder the forest, because he is usually surrounded by a brood of hungry children demanding food and shelter. At this time, it is not known how large an area of selva must be retained in order for it to sustain itself. Current data suggest that the area must be substantial if it is to retain its essential integrity.

The poacher, pandering to the blandishments of the unscrupulous, kills or captures rare animals. Certain of the great apes face extinction as the margins of their preserves are whittled away and their numbers reduced by gun or snare to supply skins, heads, or hands. The forest elephant may be in imminent danger because of the continuing trade in ivory. Habitat destruction will ensure the demise of other animals, large and small.

Rainforests are paradises for insects, which have blossomed into a mind-boggling plethora of form, function, size, shape, and color. In addition, there are a number of reptilian forms including significant numbers of poisonous and nonpoisonous snakes, alligators and crocodiles, and turtles and lizards. Arboreal forms include the primates, sloths, and birds of incredible variety and color. The forest floors are prowled by some predatory creatures, including several of the big cats.

Representatives of forest floor, selva streams, and arboreal realms include some very exotic creatures. All of these environmental niches are filled in each selva realm in significant part because the selva has existed for a protracted period of geologic time, but there are significant differences in species from one continent to the next.

Rainforests typically have been hostile areas for human settlement because of the enervating climate and endemic disease and parasites. Further, so long as man's muscle was augmented only by beasts of burden, his attack on the forest resulted in minimal destruction. Indeed, the forest was able to provide new vegetative growth at a rate that insured quick recovery from man's incursions. With the coming of mechanized equipment and advances in public health, however, the pressure on the forest has been absolutely enormous, and the vast green expanse is being steadily diminished as it is attacked on all sides by the saw, ax, bulldozer, fire, and plow.

Monsoon Forests

At the margin of the tropical rainforests in tropical areas with a distinct wet and dry season such as northeastern India, Burma, Thailand, Kampuchea, Laos, and Vietnam, tropical evergreen forests give way to varieties that drop their leaves during the dry season. Many species common to the selva are also found in the monsoon forest, but they respond to drought by a period of dormancy and leaf fall.

Trees are typically more widely spaced in the monsoon forest than in the selva, reflecting the seasonal competition for moisture. Greater light penetration to the forest floor results in the development of dense underbrush in a number of places. These dense thickets of underbrush, or jungles, are difficult to penetrate. Although most monsoon forest trees drop their leaves in the dry season, some are evergreen and do not. When it does occur, leaf fall does not happen simultaneously because different species tend to drop their leaves at different times.

Areas of monsoon forest are somewhat easier to clear for crop production or pasture since the dry season facilitates burning. Lumbering is somewhat easier, too, because there are larger areas with almost pure stands of trees or the number of species is less than in the selva. For example, the rich teak forests of southeast Asia are part of a monsoon forest association.. Bamboo thickets are sometimes interspersed with forest stands. Monsoon forests may result from clearing of selva as well as in response to climatological rhythms or pedological characteristics. In large areas of southeast Asia, especially India, the monsoon forests have been removed to accommodate the agricultural and settlement needs of people, except in remote areas of rugged terrain.

The native animal life is very similar to that encountered in the tropical rainforest area, and the number of large cats, which penetrate the margins from nearby savannas, may be somewhat larger. The monsoon forest, like the tropical rainforest, is feeling the pressure of the chainsaw, bulldozer, plow, and urbanization. Many areas of southeast Asia are characterized by an almost pervasive haze derived from the fires of the slash and burn agriculturalist. Indeed, the suffocating pressure of people on land has reduced the monsoon forest at a rate exceeding that in the selva. At the same time, the primates and great cats have been decimated. It is probably already too late to save any broad expanses of monsoon forest from massive human encroachment.

Humid Subtropical Forests

Humid subtropical forests are characterized by an abundance of coniferous trees, although broadleaf evergreen and deciduous species are also common. About ten species of pine dominate the humid subtropical environment of the southeastern part of the United States. Pines are especially dominant on poor, badly leached, and scanty soils of the coastal plains sections. Well distributed precipitation and a long growing season foster the growth of trees, although marginal soils tend to inhibit growth.

In large areas, the original forest cover has been removed for lumber or pulp, but thousands of square miles have been replanted in pines to provide a sustained forest yield. Pines are the species of choice for such planting, since they grow rapidly to marketable size. Many of these forest plantations are owned by lumber or paper companies that cultivate large tracts of forests to sustain the operation of a planing mill or pulping facility. In western Florida, for example, orderly rows of planted pine trees in every stage of development are everyday facts of life.

Figure 18.15
Humid subtropical forest in west Florida.

Figure 18.16
Trees in humid subtropical forests grow rapidly to pulping or timber size.

A lack of seasonal leaf fall, such as is characteristic of deciduous trees, dense undergrowth, and prevalence of climbing plants are typical of many humid subtropical forests. On cutover lands, jungle-like thickets sometimes succeed. Bamboo thickets, for example, are common in the lowlands of China, and the citizen of the American southeast can testify to the tenacity of greenbriar, wild honeysuckle, and yaupon. In some areas, kudzu introduced to stabilize road cuts has marched unchallenged into the adjacent forests or woodland strangling competing vegetation as it goes. Those green tendrils suffocate trees, grass, and almost any competing vegetation. It's fair to say that in most regions, kudzu has become a pest defying most efforts at eradication.

Mixed deciduous forests of excellent quality are found on certain sites where soil, slope, and climate are especially favorable. Oaks are common deciduous trees, along with gums, tupelo, magnolia, and hickory.

Humid subtropical forests are largely coincident with humid subtropical climates. These regions are in North America, Asia, South America, Africa, and Australia. Very significant shares of these forested lands have been cleared, especially in areas of Asia with high population densities.

Animal life within humid subtropical forests is abundant with numerous representations of reptilian, mammalian, and avian forms. Insects are also particularly numerous, and many of them prove to be troublesome for human occupance, not only because of unpleasant bites or stings, but also because the insects may serve as disease vectors for a number of uncomfortable, even fatal, diseases. The relict animal forms in Australia are particularly interesting because long isolation of that continent has produced a number of unique and somewhat more primitive forms of both animal and plant life.

Curiously, in spite of the steady encroachment of settlement and urban accouterments in the United States, a number of animal species, especially game animals, may be more numerous than when Europeans first arrived on these shores. For example, the whitetail deer has flourished in significant part because the principal wild predators have been decimated, and game departments have controlled hunting. Where these populations of animals have grown, the increase can be attributed to the destruction of many of the normal predators as well as to restocking and conservation efforts supported by hunters.

Humid Continental Mixed Forests

In humid continental climates, most original natural vegetation consisted of deciduous, broadleaf trees. After initial cutting, there has been a substantial addition of fast-growing conifers by people intent on producing a tree crop in a shorter period of time. At both the poleward and equatorward margin of these forests, there is a substantial interdigitation of conifers. At the poleward margins, there

Figure 18.17
Mixed forests growing in northern Japan. Note columnar jointing in basalt along valley walls.
Japan National Tourist Organization.

is a transition to species like spruce and birch. At the equatorward contact, the pines encroach on and mix with the deciduous forests.

The humid continental mixed forests usually require upwards of 25 to 30 inches of precipitation per year. Tree growth is limited in those areas where precipitation falls below this level, where the frequency of drought is increased, or where evaporation is speeded by persistent desiccating wind. On the drier margins of these regions, grasses become dominant.

Humid continental mixed forests are characterized by several different associations that have evolved in response to differences in the pedological and climatological environments. Oak, maple, beech, hickory, and elm are commonly found in various associations within the United States and abroad.

A number of devastating diseases have virtually eliminated certain species. Examples include the Dutch elm disease and the chestnut blight. These diseases have made the forest chestnut and the American elm rare in the United States, and they reflect the dangers in the accidental introduction of disease pathogen or insect vectors.

In addition, pollution in the form of acid rain and noxious gases has become a major concern in the east-central part of the United States. Certain sections of Germany, notably Bavaria, have seen the mysterious death of large numbers of trees. This same mysterious plague of tree deaths is now appearing in New England and certain portions of the southern Appalachians. In other regions, growth has obviously been slowed, probably because of the toxic effects of wind- and water-borne pollutants, although verifiable scientific data to support this contention are still unavailable.

Trees within this forest do not normally exceed 50 to 60 feet (15 to 20 meters) in height, but there are variations, depending upon the species involved and the location. Many of the hardwood varieties are especially valued for construction or furniture-making, and certain types of wood have become rare and very expensive. It's no accident that much furniture is constructed of particle board made from wood wastes and covered with a thin veneer of quality wood for aesthetic purposes. This trend is likely to accelerate as valuable species of trees become increasingly scarce.

The value of the wood has contributed to the reduction of these forest areas, but clearing of the land for agriculture and urban accouterments has been most significant in forest reduction. Forests remain principally on those hilly and rocky areas that are less suited to agricultural purposes, pastoral activities, or urban pursuits. There is, nevertheless, a steady diminution of the areas occupied by these forests. Even in areas where sustained yield forestry is attempted, slowed rates of growth result in the lapse of substantial periods of time before an area can be logged again.

Animal life is largely transitional between the subtropics and subarctic. Overall, there has been a reduction in animal population as the human population has expanded, although a few species in the United States, notably the whitetail deer, moose, certain species of bear, and wild turkey have made a comeback because of the conservation efforts of a concerned citizenry. There seems to be a recognition at last that without other life-forms, the world would be a lesser place.

Marine West Coast Forests

The marine west coast climatic region of North America supports magnificent coniferous forests. In the Pacific Northwest, trees soar well above 100 feet (30 meters) and often have waist-high trunk diameters in excess of 6 feet (2 meters). Common tree species include spruce, hemlock, redwood, Douglas fir, and Sequoia. Among the Sequoias and redwoods of northern California are numerous examples of trees more than 1,000 years old with a few specimens exceeding 2,000 years in age. Only the gnarled bristlecone pines, which occupy high windswept locales in western American mountains are older, among the world's living things. It's awe-inspiring to contemplate that a living thing existed and continues to exist from the time of Christ or before.

Most other marine west coast regions have less striking forests, although there are magnificent stands in certain portions of northwestern Europe, New Zealand, and southern Chile. All of the areas listed continue to be major sources of timber for a wood-hungry world. Where marine west coast climates have precipitation amounts augmented by mountains backing the coast, a kind of middle latitude rainforest may develop.

Figure 18.18
Douglas fir forest in the marine west coast environment of Oregon.
Daniel Ehrlich.

Figure 18.19
Brush and forestland in New Zealand.
New Zealand Tourist and Publicity Office.

Deciduous trees, which are less prepossessing in size, are significant components of the forest in northwestern Europe. Continuous heavy settlement in Europe over a protracted period has reduced the forested area, although careful management of forest preserves in countries such as Germany has retained pieces of the forest environment much as it was. Many of the forests of Germany are note-worthy for the absence of underbrush and wood debris carefully removed by foresters and gleaners. Considerable concern has been expressed in recent years over the fact that numerous forests are showing unmistakable signs of slow growth and decline, apparently from the effects of pollution attendant to expanding industrialization.

Remote sections of southern Chile, on the other hand, remain almost untouched, and the grandeur of their virgin state is a sight to behold. Sparse population, remote location, and difficult terrain have combined to preserve forested areas, which in other places would have fallen to the insatiable appetite of civilization. Even here, logging expands, and the quality and areal extent of such forests seem sure to decline.

A considerable variety of animal life can still be found in the least accessible areas of marine west coast forests of the world, although the encroachment of human settlements has resulted in significant declines in other areas. Only concentrated conservation efforts can retain or restore animal populations to levels approaching those that existed in the past. Game reserves for royalty or rich citizens provide some game and forest sanctuaries.

Taiga Forests

The **taiga,** or subpolar coniferous forest, is essentially coincident with the Dfc(d) and Dwc(d) environments. Large sections of Alaska, Canada, Scandinavia, and the U.S.S.R. are cloaked in this forest. The taiga is primarily composed of spruce, fir, certain pines, and other conifers, although birch, poplar, willow, and ash are interspersed with the conifers. Where permafrost (permanently frozen subsoil) approaches the surface, the root development of trees is restricted until, at last, trees can no longer sustain themselves near the border of the tundra climate. Tree size diminishes, and tree spacing increases near the poleward margins of the taiga in response to cold and the lack of available moisture. A short growing season and relatively small amounts of precipitation reduce the rate of growth of taiga trees markedly. Once cutting has occurred, protracted periods must elapse before trees again reach marketable size.

Although exploitation is made difficult by severe cold at one season and insect pests and difficulties in dealing with thawing permafrost at others, there has been an increased exploitation of taiga forests for wood pulp and timber. This is true especially in the Soviet Union where timber cutting produces useful raw materials and valuable foreign exchange while helping to open up an area for settlement. Basically, the eastern and northern portions of their country is the frontier for further development. Large areas are covered by only a few species of trees, and this characteristic facilitates logging.

Figure 18.20
Coniferous forest in western Canada.
Department of Regional Industrial Expansion photo, Government of Canada.

The taiga remains the home for a fairly abundant animal population largely because it is remote and sparsely settled. Herbivorous animals such as deer, moose, elk, and caribou are at least sometime inhabitants of the taiga. Carnivorous animals such as wolves and bears thin the herds by dispatching the old, the weak, and the infirm.

Smaller predators like otter, mink, ermine, and fox feed on rodents and migratory birds, and they, in turn, are taken by man because of their fur and by larger predators as a source of food. The trapping of fur-bearing animals remains a significant occupation for a few within this harsh and forbidding land. Fur farming has developed in such regions, since cold climates continue to insure the production of superior pelts.

The availability of raw materials, notably petroleum, natural gas, and gold have led to exploitation in the region. Mineral extraction and lumbering, along with roads and settlements, are reducing the area of taiga, and the very slow rate of plant growth is an ominous sign for reforestation within a reasonable span of time. The selva of the humid tropics and the taiga of the subarctic are the

Figure 18.21
A relict stand of forest trees in an area surrounded by grasslands in western Oklahoma.

Figure 18.22
Open oak forest and grassland in California.
Daniel Ehrlich.

Figure 18.23
The Mackenzie River Delta in Canada supports a few low-order trees and various shrub forms.
Department of Regional Industrial Expansion photo, Government of Canada.

largest forest reserve areas of the world, and the rate of cutting in both regions now exceeds regrowth. It is reasonable to assume that these forest reserves will be markedly reduced during the lifetimes of most readers of this book.

Forests dominated a high-order vegetation for millennia, and it is only relatively recently, within the context of geological time, that grasses have become major colonizers. Indeed, grass has filled an important niche in cloaking our terrestrial planet in green, because substantial areas that support grass will not support tree growth.

The frayed boundaries of forests with other vegetation communities may occasionally show disjuncts with islands of forest surrounded by other plant communities. These may occur as small relict islands in especially suitable microclimatological or microedaphic environments. For example, certain areas within grasslands protected from the omnipresent wind or in better watered regions may be tree covered. The Black Hills stand as a tree-covered island surrounded by the sea of grass of the Great Plains. Somewhat greater precipitation in the higher elevations fosters the growth of trees, whereas the surrounding subhumid to semiarid areas will support only grass or scrub growth.

Islands of birch or willow may exist along streams, in pockets, or in areas where the permaforest is deeply buried within the tundra. Elsewhere, pathetic crawling vine-like forms of trees may be the only representatives of aboreal forms within the tundra.

Shifts of vegetation patterns have clearly occurred during periods of climatological change. For example, boreal forms (taiga) existed in the mid-South during Pleistocene glaciation. Except for relict stands, these associations marched north as the ice retreated. There are still small areas deep within the humid subtropical regime, which are dominated by these boreal forms. During the Pleistocene, the species of animals that now exist only in far northern latitudes were driven south by the ice front.

Grasslands

Grassland is usually dominant where (1) annual precipitation is low, (2) precipitation is unevenly or erratically distributed during the growing season, and (3) evaporation rates are high. Grasses may also exist in response to physiological aridity created by permeable soil. Tropical grasslands usually prevail in areas that have a distinct wet and dry season with high evaporation rates. Increasingly, however, tropical grasslands are expanding as selva and monsoon forests are cleared, and grasses succeed in areas formerly covered with forests. Some of these grasses with rhizomes and deep tap roots are exceptionally difficult to eradicate once established in formerly forested regions.

Middle latitude grasslands are found in subhumid to semiarid areas principally, although certain prairie grasses developed in humid environments. Perhaps these humid land prairies resulted from repeated burnings or because of soil peculiarities. The effects of fire, either from natural causes such as lightning or from the purposeful fires of native peoples, cannot be discounted. A fair body of scientific opinion holds that fire is an essential component for the continued existence of certain grassland associations.

Soil types and soil moisture are significant factors in the development of grassland associations. Grassland soils typically test basic, although neutrality or a slightly acidic pH characterize some. For the most part, grasses evolve in regions that are climatologically or physiologically too dry for trees, or where the soils are too shallow to support tree growth.

Tropical Grasslands

Tropical grasslands predominate in the wet and dry tropics (Aw) and certain portions of tropical monsoon (Am) environments. These grasses are characterized by tall, coarse grass, if left undisturbed. Indeed, such grasses may reach up to 6 feet (almost 2 meters) in height if left ungrazed. Generally, however, they are grazed down to much shorter height. These grasses are sere and brown during the dry season and subject to frequent fires, either naturally caused or purposely set, whereas they sprout clean and green with the coming of the rains of the wet season. Native people often deliberately set fires to remove the dry plant remains to make it easier for new shoots to grow to effective grazing heights with the beginning of the wet season. Fire may be necessary for the continuing dominance of savanna grasses.

Two principal types of grassland associations should be noted. At the drier margins, the grasses are shorter with widely scattered scrub brush which may grade into bush steppes. At the wetter margins, the grass is taller with interspersed trees occurring at more frequent intervals until the savanna grades into adjacent monsoon forest or tropical rainforest.

Figure 18.24
Park savanna grassland in southwestern Puerto Rico.

Few areas of savanna grassland are devoid of trees or shrubs. Areas of grassland and scattered trees, frequently flat-topped acacias, which have been pruned by browsing giraffes, are called **park savannas.** Where trees extend along banks of streams in sinuous extensions of adjacent forests into grasslands they are known as **gallery forests** or **galeria.**

Savannas are the home of big game, especially in Africa. Large numbers of herbivores, such as the zebra, gnu, deer, antelope, gazelle, and buffalo, take advantage of the abundant grass. They are stalked relentlessly by large carnivores, such as lions and other big cats, wild dogs, and hyenas. Other large herbivores, such as the rhinoceros, elephant, and hippopotamus, are largely immune from attack as adults because of their large size. The very young, the very old, the weak, or the sick all fall prey to the efficient carnivores. Hordes of scavengers wait in the wings for scraps from the carnivore's table. These scavengers, such as jackals, hyenas, and vultures, serve an important function, in that they clean the landscape of refuse. It's now clear, too, that hyenas and jackals acquire a significant share of their food by hunting. As a matter of fact, the lordly lion probably acquires a greater percentage of its food through scavenging than the jackal or hyena. The safari ants remove the last elements of organic material from the bones picked almost clean by larger scavengers. Birds and reptilian forms add their variety to the chorus of life, which is the savanna.

The savanna, especially in Africa, throbs with life and exemplifies the cycle of life and death, which are so intimately and intricately intertwined. The panorama of migrating herds and predatory hangers-on is a sight vastly appealing to more and more of the world's urbanites, and tourists are becoming almost as numerous as herds of wildebeest. Although herd animals have been reduced somewhat in numbers, their annual migrations continue to be one of earth's great spectacles.

In some areas where animals have been protected by governments, some of the large herbivores, notably elephants, have destroyed trees in park savannas to acquire the necessary browse. Although declining herds are of concern in most areas, in a few, the overabundance of animals may be a threat to habitat.

Prairie Grasslands

Prairie grasslands developed largely within the central part of the United States in portions of Illinois, Iowa, Kansas, Nebraska, and Oklahoma, and in a few other areas of the world like the Pampas of Argentina. Isolated patches of prairie developed in areas otherwise surrounded by forest, but the numbers and extent of such relict forms have been few and small. Essentially, they developed along the drier margins of the humid subtropical or humid continental climatic regimes. Certain prairies, such as those of East Texas, apparently came into existence because of the presence of the vertisols.

Prairies are characterized by tall, deep-rooted grasses, which grow naturally to heights of 3 to 6 feet (about 1 to 2 meters) over soils with abundant organic matter and adequate available minerals. There is some speculation that these grasses may have been perpetuated by repeated burnings initiated by aboriginal people, but there is no definitive proof of such conjecture.

Few stands of prairie grass remain, since most have long since been put to the plow or used for intensive grazing. Certain sections of virgin prairie have recently been set aside as state reserves in Kansas, and there are efforts in Oklahoma to protect some prairie areas on federal land. The grasses, in combination with various forbs and flowering plants, are a beautiful sight at the height of their growth and are valuable protectors of the soil that sustains them.

It is appropriate to protect some sites to be enjoyed by those who will follow us. These regions were avoided by early settlers because of a misconception that forested soils were more fertile, and because wooden plows could not effectively turn the thick turf. Subsequently, however, these regions have become the great granaries of the world. The iron plow, followed by steel plow, broke the sod, and abundant crops insured that these regions would be heavily used for agriculture. The technological advances that permitted their cultivation and their inherent fertility were the seeds of their destruction.

Figure 18.25
Prairie grasses and scrub oak on loess hills in Iowa.
Wallace E. Akin.

Herbivores, such as the bison, once roamed these grasslands in vast herds, and the pronghorn (antelope) were omnipresent in the biological landscape. Few large predators thinned their numbers, except for the wolf. Game birds such as the pheasant, quail, and dove have succeeded in the domesticated landscape on what was once the prairie. The prairie chicken still drums its mating song, but numbers have been depleted except in a few scattered locations. The bison has recovered from near extinction in protected reserves and parks. The pronghorn has made a modest recovery, responding to conservation efforts. Rodents, such as the jackrabbit, prairie dog, and ground squirrel are present in significant numbers, although they are erratically distributed. Reptiles, including poisonous snakes, prey on the rodent and bird populations. The herds of millions of wild herbivores will never again return, however, unless man is removed from the scene by some cataclysmic event like nuclear war.

Steppe Grasslands

The semiarid (BS) environments of the world have supported steppe grasslands or shrubs. The middle latitude steppes are more commonly grass covered. Steppe grasses are shorter than those of savannas or prairies, and different species dominate in the plant association. Toward

Figure 18.27
Thorny, scrub vegetation in southwestern Puerto Rico.

the humid margin, these short bunch grasses provide a continuous cover. At the drier edges, as steppes grade into deserts, the grasses become less continuous (i.e., there is more bare earth between grass clumps). Where overgrazing has occurred, more xeric (xerophytic) forms have succeeded at the expense of the grass cover.

Steppe grasslands have traditionally supported vast herds of herbivores, such as the bison and antelope, which once blackened the plains of the American west with their sheer numbers. Carnivores like wolves, cougars, and coyotes could barely dent their vast numbers. Buffalo hunters, however, decimated the herds, and only pathetically small numbers remain in protected enclaves. The fragile steppe grasslands now support domesticated animals or have been put to the plow for the production of small grains. Protracted drought threatens to reintroduce the dust storms of the 1930s, perhaps compounding people's misuse of the steppe grasses with the natural disaster of long periods of below-normal rainfall. In numerous areas, grasslands have been succeeded by scrub vegetation like sage or mesquite as overgrazing or cultivation has degraded the biological environment.

Shrub and Scrub Vegetation

Woody scrub plants have evolved or succeeded in a number of climatological niches including certain tropical steppes, overused middle latitude steppes, and Mediterranean regions. Within the Mediterranean areas, for example, an open, low-growing forest called **schlerophyll** developed.

Widely spaced trees, usually not more than 15 to 20 feet (about 5 to 6 meters) high, were able to survive with relatively small amounts of precipitation and the almost completely rainless summers. Certain of the species such as the cork oak, olive, carob, and others proved to have use for man, and they were spared and even propagated as the rest of the woodland was cut for timber or firewood. Because of long human occupance of the areas around the Mediterranean Sea, few regions exhibit the characteristics of the original schlerophyll forest cover. As a matter of fact, the total cover has been much reduced, and the species associations have changed, reflecting the preferential selection of certain types having important use. Significant stands of the original schlerophyll forest are missing from most areas with a Mediterranean climate today.

Other plants of lesser height grow on less favored portions of the Mediterranean regime. These scattered distributions of drought-resistant, dense, shrublike growth carry a variety of regional names. In the United States, the association is known as **chaparral,** in France as *maquis,* in Italy as *macchia,* and in Australia as *mallee.* These associations are found typically where maximum temperatures are less than 100° F (38° C), and average annual precipitation is between 15 and 25 inches (between 38 and 64 centimeters). This shrub-form vegetation is frequently equipped with thick bark and small, stiff leathery leaves, which are sometimes densely covered with fine hairs that help to limit moisture loss through evapotranspiration. Some bushes have thorns that tend to discourage browsing animals. Although these are not true xerophytes, many of the forms exhibit certain of the drought-tolerant characteristics of the true xerophyte.

Most of these plants are of little direct use to man, but they help to hold the soil in place and to protect it from the ravages of erosion. Southern California, especially, has seen a number of disastrous fires that have caused property damage and loss of life. These denuded areas have almost inevitably succumbed to the ravages of mudslides in the succeeding rainy season. Natural and accidental fires are facilitated in the very dry summers, and the vegetation, much of it with high percentages of resins and oils, burns very readily.

In a few such areas, vegetation is surprisingly dense, reflecting the tenacity of life forms as well as the ability of many resident species to tap subsurface water supplies. Animal life of considerable variety characterized the regions initially, but rather close settlement in most areas of Mediterranean climate has drastically reduced indigenous populations. Occasionally, a cougar may wander into a settled place or a coyote may raise his voice in song in a busy suburb. Birds continue to find favorable habitat in existing surroundings, but their numbers thin before the inexorable march of human settlement.

Desert Vegetation

Deserts are hostile environments for most plant and animal species, since evaporation exceeds precipitation, and available supplies of moisture are greatly restricted. Nevertheless, certain deserts support a surprising array of plant and animal forms, reflecting the adaptability and tenacity of life forms, even in very austere environments.

Although mesic forms may be found along stream courses or in other well-watered areas, most desert plants are **xerophytes** (i.e., plants that are able to subsist in environments with little available moisture). Xerophytic plants have developed a series of evolutionary strategies to subsist in a moisture-deficient environment. Essentially, all of them have developed characteristics to resist moisture loss through evapotranspiration.

Some, like the **mesquite,** have developed deep tap roots to reach for subsurface moisture. In fact, mesquite may send tap roots 30 feet (about 10 meters) or more into the subsoil to search out subterranean water. In many areas of Texas and the Southwest, vast overgrazed areas have been invaded by mesquite. It has proven to be singularly difficult to eradicate in spite of an arsenal of physical, biological, and chemical weapons used against it. Perhaps the fact that it has become popular to burn mesquite in cooking fires to flavor meat and fish may lead to a commercial exploitation of mesquite, which will help to eradicate it.

Figure 18.28
Dry environments, even with thin, rocky soils usually have some vegetative cover.

Other plants, like the creosote bush, have developed a root system that extends out laterally a great distance from the center of the plant to take advantage of the moisture available over a considerable area. Because of this lateral root spread, the creosote bush is a very tough biological competitor, and other plants have a difficult time establishing themselves in the same area.

Certain other plants have developed small, shiny, waxy leaves with few stomata (the plant equivalent to the pores of animals) to resist moisture loss. Others have developed silvery undersides, apparently to increase reflectivity of the searing rays of the sun, and, therefore, to reduce leaf temperatures and evapotranspiration rates. It is not always possible or profitable to know the specific purposes of plant characteristics as related to climatic adaptation. The persistence of various characteristics, however, provides useful inferences about the reasons for such adaptations.

Cacti typically are thorny, apparently to discourage browsing animals, although certain browsers are able to penetrate the spines. Cacti also often have fleshy bodies, stems, or leaves, which swell markedly as the moisture from a rare storm is absorbed. These sponge-like portions of the cactus may shrink markedly during protracted dry periods, only to bulge again after rains. Plants with thick,

Figure 18.29
The prickly pear (left foreground) and saguaro (background) are two of the many varieties of cacti that grow in desert environments.
Arizona Office of Tourism.

Similarly, a few desert playa lakes come to life with myriads of brine shrimp after a lake is temporarily filled, and the eggs hidden for years in the dry lake bed swell to life when water is again available. When this happens, predatory birds may be attracted from great distances for an unexpected feast. Water is the life giver, and it must be husbanded when in scarce supply. Life almost always responds to its presence.

Xerophytic plants may be spaced with amazing evenness, depending on the availability of moisture. They are widely spaced at the drier margins of the desert, and they are situated closer together at the more humid margins. In fact, a salient characteristic of most xerophytic associations is the large expanses of bare ground, which exist between adjacent plants. Several desert regions are so dry as to exhibit no high-order vegetation at all. This is especially true in certain areas of the Atacama and in the interior of the Sahara.

Animal life is surprisingly varied in most desert locations, but members of the reptile, rodent, insect, and bird families seem to subsist best. Many species are nocturnal, avoiding the searing heat that so frequently characterizes the desert day. They sally forth at night to acquire food. Since there are many levels of predators, some of the would-be eaters are themselves eaten. Small animals are more numerous in the desert, although larger representatives like the camel, gazelle, donkey, and certain species of sheep and goats live there, too. In North Africa and the Middle East particularly, some of the wild sheep and gazelles have been hunted to extinction. The lion, which formerly prowled those regions, has long since disappeared. The herbivores have adapted to scanty browse, and many have penetrated plant defenses by their ability to eat and digest thorny forage. The camel can eat thorny material, which would literally destroy a horse or a cow. Camels eat with relish the camel thorn, which would severely lacerate lesser beasts.

Several desert animals, such as the fat-tailed sheep and camels, have developed fatty food storage in their hump, tail, and rump to carry them through periods when forage is scarce or nonexistent. The camel, although native to Africa and Asia in historic times, was introduced to Australia, where it now thrives in the outback. In a curious irony, Australia now exports camels to the Middle East where they are used for their usual riding and draft purposes as well as for meat. Others, like the kangaroo rat, have developed the ability to go without drinking by releasing moisture from foods in the metabolic process.

Most desert animals have developed mechanisms to enable them to cope with minimum water and scanty food supplies, but in many ways the camel exemplifies evolutionary adjustments to a hostile environment. The camel can store significant quantities of water in its stomach for protracted periods; the hump is a storehouse of fat to be used in lean periods; the lips and mouth parts can withstand the thorns and spikes encountered in browsing; nostrils are equipped with an effective hair-filtering system

fleshy bodies, leaves, and stems are called **succulents.** Several varieties of cactus are indigenous to specific desert locales. The giant saguaro cactus, for example, is native to the Sonoran Desert and is especially prevalent in southern Arizona. The cholla cactus, like the ocotillo, also exhibits spotty distributions. The prickly pear is much more widespread in terms of areal extent. In fact, varieties of prickly pear are found in virtually all of the world's deserts as well as in a number of better watered areas.

Still other xerophytic plants have dramatically shortened life cycles. Seeds may sprout, grow to maturity, flower, and produce new seeds in response to a single shower. Seeds may then remain dormant for long periods before again springing to life with the next life-giving rain during the growing season. It has been demonstrated that such seeds may remain viable over scores of years. A few seeds may retain this viability for hundreds of years.

Figure 18.30
The desert nomads of Iran have learned how to use the camel in transporting goods.

to keep out sand and dust; heavy lashes protect eyes from dust-laden air; and large, padded feet enable the camel to walk efficiently in loose sand or soil. Camels may be driven long distances to market, and they continue to be the principal mode of transportation for nomads who ply the deserts of North Africa and the Middle East.

Animals and plants of desert regions hang tenaciously to life in a very hostile environment. At once, deserts exhibit the tenacity of life and fragility of that life if a delicately-balanced ecosystem is upset by external forces. Vegetation is slow to return where it has been overgrazed or overused for fuel. Deserts and desert vegetation are expanding because of overgrazing and destruction of steppe vegetation by man.

Xerophytic plants may also be found in humid areas where they have been introduced by accident or design. Occasionally, they may become great pests when they encroach on more mesic forms. They also succeed naturally in certain pockets of physiological aridity within otherwise humid regions.

Figure 18.31
Desert vegetation in southern Utah.

Tundra Vegetation

The most important factors affecting life in tundra regions are short growing seasons, cool temperatures, limited precipitation, and permafrost. Each summer the topsoil thaws, and vegetation grows, flowers, and produces seed in another hostile environment. Because the region is underlain with permafrost and the frost-free season is short, the soil does not thaw to any great depth; hence, roots cannot penetrate far into the soil. Trees are effectively eliminated except in special circumstances, and sedges, shrubby brush, and lichens are typical forms, although willows and birches may appear along streams. When trees do appear, they are short and stunted, and they may actually creep along the ground like vines.

Tundra vegetation is appropriately adapted for the short growing season. It is generally stunted, low-growing, and compact. Most tundra plants have the ability to sprout, grow, flower, and produce seeds in a few weeks. During the short growing season, flowering vegetation may produce a veritable carpet of loveliness in this ordinarily bleak region. Few human eyes are available to feast on the beauty, but insects and grazing animals appreciate the nectar or the browse.

The evolutionary struggle is evident in the relative paucity of species able to adjust to the severe environment. As in the desert environment, this is a true test of survival of the fittest for both plants and animals.

In a demanding environment such as this, there is a surprising abundance of animal life, especially in the summer season when migratory waterfowl add their teeming millions to the permanent population. The birds are able to subsist quite well on the teeming populations of mosquitoes and flies of various types.

Predatory birds as well as carnivorous mammals are on hand to rob nests of eggs and young. The unwary, sick, or injured adult may also fall prey to the vigilant fox, wolf, or weasel. Rodents are near the bottom of a food chain, which culminates in the Arctic fox, bear, and wolf.

Large grazing animals, such as the caribou and musk ox, are important to humans in those areas. Some caribou migrate in large numbers to the relative protection of the taiga in the winter season. The barren ground caribou, however, is a creature of the tundra, although it, too, may migrate over considerable distances in search of food.

Marine life in adjacent seas yields harvests of fish and mammals like the whale, seal, and walrus. The polar bear stalks the margins of ice in search of seals, especially. Mature male polar bears are the world's largest carnivores. They may migrate long distances in search of food; they prove to be excellent swimmers; and they are certainly very efficient predators.

Figure 18.32
Tuk pingo in the Northwest Territory of Canada. Department of Information, Government of the Northwest Territories.

Any notion that there is a uniformity in animal distribution in tundra areas should quickly be discarded. There are regions of relative animal abundance and other regions that are almost devoid of life. The availability of food is always a determining factor in clustering of animal populations.

Fish in Antarctic waters have proven to be of considerable interest to biologists. Some of them apparently have a kind of built-in antifreeze which permits them to function in very cold waters. How such materials are produced and used is a question of considerable scientific interest.

The nearby ice-cap environment has no high-order vegetation, but certain creatures, such as polar bears in the Northern Hemisphere and various species of penguins in the Southern Hemisphere, live on the ice-cap margins adjacent to the sea, which supports them. The surrounding seas support varieties of fish, walrus, seals, and whales. As a matter of fact, remote waters, like those of the Antarctic, have become the last significant redoubt of whalers.

The peculiar nature of permafrost often creates, as the result of repeated freezing and thawing of areas above it, areas of **polygonal ground,** which creates a curious surface pattern. From the air, especially, the pattern of polygons is pronounced, since the edges may often be marked

by water. Similarly, ice heaving through expansion from beneath the surface creates mounds known as **pingoes.** On a small scale, these pingoes are like miniature domes, even to the hogback-like features at their margins.

Vegetation, animal life, and peculiar localized landforms represent other elements in the variegated pattern, which is our physical earth. Movements in the earth's crust, the positions of continents and seas, the continuing effects of aggradational and degradational forces, the various aspects of weather and climate, along with the biological cloak of plant and animal species, produce an elemental home for mankind.

Man has left an indelible imprint in the production of a material culture. In addition, however, *Homo sapiens* has modified the physical environment in many significant ways. More and more people are making pervasive and long-lasting changes in our physical world. A review of people's modification of the physical environment is set forth in the next chapter. Ultimately, we are actors in the drama of life and creators of part of the scenery as well as spectators observing the most interesting play of all.

The end of the play has not yet been written, and the conclusion of the drama has several possible scenarios. *Homo sapiens'* run on earth may be long or short, depending in significant part on whether the species is really sapient (wise).

Figure 18.34
Generalized natural vegetation map of the world.

World Natural Vegetation

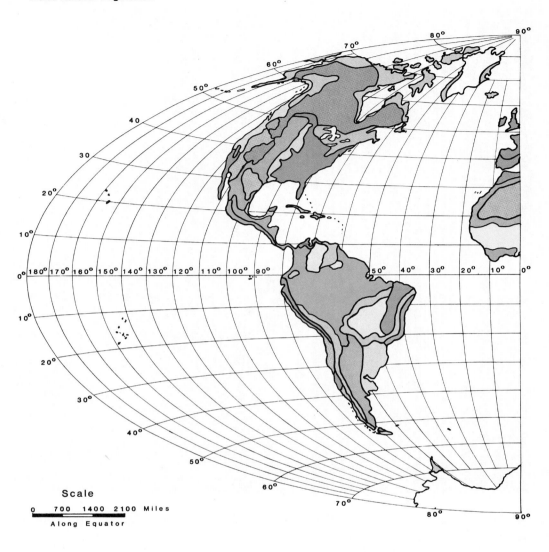

Scale

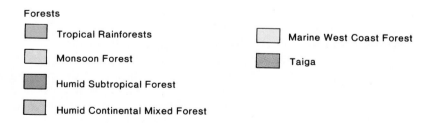

```
0   700  1400  2100  Miles
```
Along Equator

Forests

Tropical Rainforests		Marine West Coast Forest	
Monsoon Forest		Taiga	
Humid Subtropical Forest			
Humid Continental Mixed Forest			

Aitoff's Equal Area Projection

Grasslands		Other Forms	
	Tropical Grasslands		Shrub and Scrub
	Prairie		Desert
	Steppe		Tundra

Study Questions

1. Approximately when did life develop on earth?
2. Briefly explain ecological succession. Climax vegetation.
3. From the standpoint of soil moisture, what are the principal types of plant environments?
4. What does the term *biota* refer to?
5. What interconnected and interrelated factors are principally responsible for the distribution of plant species?
6. Characterize the selva. What is the jungle variant of the selva like in terms of forest physiognomy?
7. Why do rainforests often degenerate into a kind of man-made savanna after they are cut?
8. In what ways are monsoon forests distinguished from selvas?
9. Characterize a forest that would be typical of the American Midwest.
10. Why do several marine west coast environments remain as significant reserves of marketable timber?
11. Characterize the tundra. Briefly contrast it with the selva.
12. What conditions are conducive to the dominance of grasslands?
13. What is a gallery (galeria) forest?
14. Describe the salient characteristics of a park savanna.
15. What are the essential characteristics of prairie grasslands? Why were such grasslands largely ignored by early settlers in the American Midwest?
16. Where do schlerophyll forests tend to be dominant?
17. List and describe some of the characteristics that xerophytic plants have developed to resist moisture loss through evapotranspiration?
18. Why are tundras so vibrantly alive during the summer season?

Selected References

Cain, S. A. 1971. *Foundations of plant geography.* New York: Hafner.
Gabler, R. E.; Sager, R.J.; Brazier, S. M.; and Wise, D. L. 1987. *Essentials of physical geography.* 3d ed. Philadelphia: Saunders College Publishing.
——. 1948. *Grass: Yearbook of agriculture.* Washington: United States Department of Agriculture.
Polunin, N. 1960. *Introduction to plant geography.* New York: McGraw-Hill Book Company.
——. 1949. *Trees: Yearbook of agriculture.* United States Department of Agriculture.

19

People as Modifiers of the Physical Environment

Yemeni tribesmen have modified the austere environment of their mountainous homeland by devegetating the hillsides and terracing the slopes.
Photo by Lynn Abercrombie. © 1985 National Geographic Society.

All nature wears one universal grin.
Henry Fielding, *Tom Thumb the Great*

Great things are done when men and mountains meet; this is not done by jostling in the street.
William Blake, *MS/Notebooks*

S ince people became thinking creatures, the species has modified the physical environment in increasingly dramatic and pervasive ways, while, at the same time, creating a cultural realm of great complexity. The rate of change continues to accelerate, and the curve of change mirrors the exponential growth rate of human population. While other living creatures significantly modify their environmental milieu; few, if any, have the capacity to effect irreversible changes as people do.

The elephants of Tsavo may uproot trees and modify the patterns of vegetation, but their numbers are few and their impact is localized. People, on the other hand, inhabit virtually every corner of the globe and the population is rapidly expanding. People's capacity for mischief, usually unintended, is very great indeed, and that capacity is expanding apace, carried on the wings of technology and an expanding array of human wants.

Until relatively recent times, such influences have not been systematically studied. Since the impact of people is expanding swiftly, however, it is singularly appropriate to consider the effect of people in the physical environment. People affect all aspects of the physical world. In some cases, these influences may be subtle, and in others, dramatic. In a few, the influences are transitory, whereas in others, they may be permanent. Indeed, human potential for doing mischief to the environment may threaten the survival of the human species just as past depredations have resulted in the destruction of thousands of plant and animal species. In other instances, of course, people's modifications have created a more suitable environment for their purposes.

Although natural conditions have caused the extinction of thousands of species since life began, it's clear that, among animals, people have caused the greatest damage in the destruction of other forms of life. Only relatively recently have people, in certain limited circumstances, become a conservator of the biota. In a few instances, seed banks and captive breeding of animals may be the last bastion against extinction.

Obviously, numerous factors have contributed to the dimension and scope of environmental modification. A small number of people with limited needs and primitive tools typically had a minimal effect on the environment. An exception may have been in the use of fire. Either by accident or design, fire may have had significant impact on the biota, even when people were few in number. As indicated earlier, it has been suggested that certain grassland areas, especially the areas of tall grasses, may have been created or perpetuated by repetitive burning.

For a hunting species like early people, fire was a powerful tool—both to drive prospective quarry into places where such creatures might likely be killed, or to remove dead plant matter to foster the growth of new shoots of vegetation in the likely prospect that animals might be attracted to such locales. Although primitive people almost certainly didn't understand that burnt plant residue provided some fertilization and the removal of plant debris through burning made it easier for new growth to be projected, they did know that the herbivores were attracted to the new grass. Evidence demonstrates that primitive people used fire consciously to modify or enhance the environment.

As people became more conscious of their ability to alter the environment in ways to suit their purposes, they began to effect greater and greater modifications. Awareness of an ability to modify and tools to effect changes were twin factors in an accelerated alteration of the physical world. Change begets and usually accelerates change.

Figure 19.1
Monsoon forest in Thailand being cleared, in part, by burning.
Joseph Castelli.

New tools, new perceptions, and the rapid expansion of population in this century, along with the development of sophisticated technological equipment, and the creation of wants—in addition to the expansion of needs—have all contributed to a rapidly expanding period of environmental modification. For example, the population had doubled since World War II, and a continued growth of more than 1.5 percent per year in 1988 suggests, if growth continues at that rate, a doubling of the current population of more than 5.0 billion will occur in forty years. Providing for the needs, satisfying the desires, and handling the waste products for such a large number of people suggests an acceleration of environmental modification heretofore unprecedented. Indeed, daily news accounts are replete with examples of harmful changes in the environment. Surface and groundwaters are polluted, the atmosphere is choked with the effluent issuing from smokestacks, vegetation is killed by the release of toxic materials from factory and farm, and new plagues of pests are visited on us as their predators are destroyed by design or mishap. Wherever people have upset a previous balance of plant and animal forms, typically the result has been unfavorable from a human standpoint. There has been a much greater tendency to degrade the natural environment than to enhance it. Few consciously set out to destroy an environment, but the prospects for economic gain usually outweigh a respect for the physical world.

Every environmental sphere has been modified by people's actions. No corner of the earth has escaped human attention. Some specific examples of human modification of the physical world follow.

Figure 19.2

The developing world is increasing in population at a much more rapid rate than the developed world. The environmental impact caused by such growth is a cause for concern within such regions and outside of them.
Data from Carl Haub and Mary Mederios Kent, World Population Data Sheet *1987 (Washington, D.C.: Population Reference Bureau, Inc., 1987).*

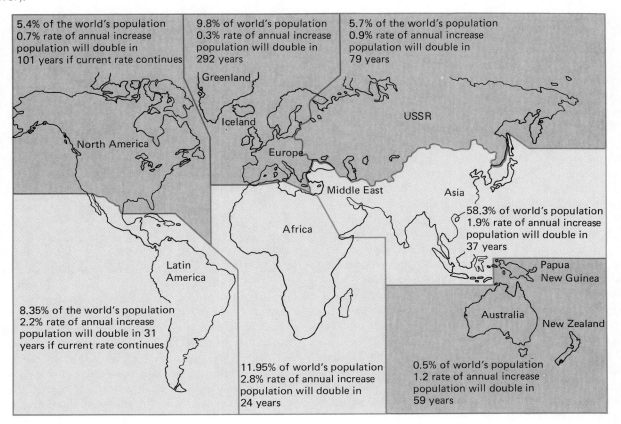

People As Modifiers of Landforms

Geomorphological features are held in a system of dynamic equilibrium so long as "natural" forces interact. From a conceptual standpoint, for example, as has been seen earlier, the entire modern precept of landforms created by erosion is one of erosional and depositional forces being in a state of dynamic balance. Obviously, a cataclysmic event such as a major volcanic eruption, earthquake, or major storm may interdict and interrupt the equilibrium of a specific landscape, but, overall, the forces of nature tend to develop a stability of active processes. Even with the interdiction of catastrophic events, nature quickly reestablishes an equilibrium, which modulates the disruption. An interrupted old process may be renewed towards equilibrium, or a new process may be begun towards the establishment of a new equilibrium. In a sense, these modifications of equilibrium conditions and the development of new equilibrium conditions are analogous to the notion of punctuated evolution now enunciated by some biologists. Basically, the notion that the slow processes of natural selection are accelerated by some significant event, whereupon changes can and do occur very rapidly, is not dissimilar to the idea that abrupt environmental changes may be met by equally abrupt changes in equilibrium conditions in the geomorphological landscape.

Man and his mores, however, tend to introduce an element of instability into the processes of nature. The acceleration of that element of instability threatens to reduce certain elements of order in the landscape to chaos in the environment. The sophistication of tools and the intensity of human efforts in the use of tools seems certain to accelerate landform change. A human being with a digging stick planting a yam didn't change much, a man on a bull-dozer might cause significant change. Human interdiction is characterized by persistence and repetitiveness, which affects the ability of natural forces to establish equilibrium conditions. Further, the frequency of changes induced by human actions makes it difficult for equilibrium conditions to be established before a new element of disequilibrium is introduced. Perhaps chaotic changes introduced by human intervention will be analyzed and amalgamated into a comprehensive system of landform analysis through chaos studies. Or perhaps scientists will despair at developing such an analysis because the number of human variables will exceed even the ability of the fastest and most sophisticated computer to compass them. Some examples of human intervention are quite obvious, however, and their impact, which will be examined in the following sections, can be measured and analyzed.

Grazing, Cultivation, and Burning

The first significant human geomorphological influences seem to have occurred through the process of devegetation, either through overgrazing, burning, or the clearing of land for cultivation. Obviously, such actions had biological consequences, and those will be discussed subsequently, but these activities also modified landforms. As slopes are devegetated, runoff is accelerated with an accompanying increase in the rate of erosion and the sediment burden of proximate streams. Devegetation may be as simple as the girdling of trees with primitive tools in slash and burn agriculture, or it may be as sophisticated as the defoliation of vegetation by the use of chemicals such as Agent Orange. The effects of such devegetation are similar no matter how it may be accomplished. Accelerated runoff results in reduced water infiltration, which, in turn, slows revegetation. Heavier loads of eroded material increase the rate at which deposition occurs downstream with predictable effects on stream and valley profiles, and stream patterns. Natural flooding is accelerated to floods of greater frequency, extent, and duration by the removal of vegetative cover. The dimensions of flood damage are rising in no small measure due to the stripping of vegetation from watersheds.

Tillage has exposed bare soil to the effects of wind erosion as well as water erosion. In areas where the wind has considerable velocity and fetch, the removal of topsoil may be significant and the deposition of these materials in adjacent regions may modify preexisting slopes. The dark dust-laden skies of the 1930s in the United States came as a direct result of tillage of the land combined with the protracted drought of that decade. It's obvious, even with modern cultivation techniques, that those same regions are vulnerable to severe wind erosion in a time of prolonged drought. In fairness, it must be said, however,

Figure 19.3
This lumbering operation in Mindanao is having a devastating effect on the selva. Normally, after logging, squatters take up occupancy, and their subsistence pursuits usually accelerate forest destruction, preclude revegetation, and speed up erosion.

that the drought of the 1950s which was of similar dimensions to that of the 1930s did not yield as large a harvest of soil destruction and dust as the 1930s, although the author's sojourn in Oklahoma in the 1950s was marred by dust-laden skies on more than one occasion. Fortunately, agricultural practices had improved so as to reduce the effects of wind erosion. Similarly, the drought of the 1970s did not see as much wind damage as the period of forty years earlier. If a postulated drought comes again to the Great Plains in the 1990s, will the dust of exposed fields swirl again? Will the sun shine through an ever-present brown haze? The threat is still there, and dry periods of greater length or severity may again lift quantities of topsoil from exposed fields to darken the skies of the Great Plains. Those dust-laden skies, not only create problems for local inhabitants' health and sanitation, but they also signify a loss of topsoil, which, in turn, means less efficient and effective agriculture in the region.

Similarly, agricultural efforts may produce direct effects in the modification of slope. In hilly terrain, the construction of terraces by human beings produced distinctive slope-like features that are as real and as persistent as wave-cut or wave-built terraces. Such features are salient aspects of the landscape in certain regions, especially in east and south Asia where age-old cultures and high population densities have produced an indelible imprint on the land over centuries. The rice terraces of Banaue and Bontoc in the Philippines or on volcanic slopes in Java or

Figure 19.4
Rice terraces in the Banaue-Bontoc area of the Philippines have markedly changed the character of slope and drainage in these mountains in central Luzon.

Figure 19.5
Used and abandoned terraces on steep mountain slopes have markedly changed the land's profile.

Figure 19.6
Terraces on modestly sloping land, although not so dramatic as those on the mountains, influence slope profiles and drainage characteristics.

Bali are significant features of the landscape and have affected the erosional stability of adjacent mountains, not only through modification of preexisting slopes, but in changes of the patterns, volume, and velocity of streams.

The risers of terraces are significantly steeper than preexistent slopes, whereas the terraces are almost flat and have slopes much less than the mountainside into which they were cut. Streams are led off in a circuitous route to provide irrigation water, and initial relatively straight courses are drastically altered. Even the relatively modest terrace systems on gentler slopes in the United States have modified slopes and changed drainage patterns. Rational agricultural practices in sloping terrain suggest the use of terraces, and such terraces become essentially fixed features of slopes. Nevertheless, the introduction of terraces modifies rates of runoff and establishes an element of disequilibrium in the geomorphological process. In the Philippines, the risers are about 6 feet (2 meters) high, and the flat cultivated section may be 15 to 30 feet (about 4 to 8 meters) wide. In Java, the dimensions of the risers and the cultivated sections tend to be somewhat greater. The patterns in Bali are more nearly comparable to those in the Philippines. Although terraces do not interrupt the overall concave or convex profile of the mountain slopes, they do produce irregularities on it.

Of course, there are other terraced areas in mountains that are cultivated. They are observed in almost all continents, except for Antarctica, but they are especially noteworthy in Asia.

Terracing in areas with less steep slopes is especially prevalent in the U.S. They are less prepossessing features of the landscape, but they do exert an influence, especially in slowing the runoff and reducing the rates of erosion. Terracing, even on modest slopes, is a sound agricultural practice that slows runoff; hence, reducing erosion while maximizing the percolation of water into the soil. Such terraced areas usually replenish the water table more effectively than sloping terrain that has not been terraced.

The development of irrigation reservoirs, canals, and drainage systems to support agriculture has created new features, and the indirect effects on landforms and other facets of the physical environment are significant as well.

Irrigation reservoirs often have many other purposes, including flood control and the production of hydroelectric power. These enormous lakes are, however, as real as those created by ice scour or glacial deposition. The canals may be as significant as natural streams. The drainage ditches may produce an entirely new surface subject to subaerial erosion and deposition. Because people tend to make canals or drainage ditches straight, rather than comparable to the natural sinuosity of a stream, water tends to run off more rapidly, and that quick removal of surface water tends to depress the water table. Artificial reservoirs may increase water infiltration into adjacent areas and, therefore, have the effect of raising the level of the water table in proximate regions. Straightening of existing stream courses also tends to increase stream velocity, which changes the character of a stream in the erosion-deposition ratio. Most straightening of water courses results in permanent alteration of preexisting patterns. In at least one area in central Florida, work began in 1988 to divert waters from a straightened course of a stream back into the sinuous pattern of the Kissimmee River, which existed before straightening was undertaken. Virtually everyone now agrees that the twisting course of nature is better for wildlife and the water table than the straight course, which was excavated. Water is led off so rapidly in a few instances that a permanent stream may become intermittent.

The modification of landforms to support or facilitate agriculture has touched virtually all sections of the globe except for very high latitudes, the most arid reaches of the earth, and a few extremely rugged or remote places. Examples of past disruptions are still to be seen in the silted up ruins of abandoned irrigation facilities in deserts. These abandoned irrigation facilities are especially prevalent in the Middle East and North Africa, perhaps reflecting the long period of time that those regions have been settled.

The construction of giant irrigation facilities like the Aswan High Dam, which created Lake Nasser in Egypt has had a variety of expected and unexpected consequences. It was built to insure an adequate supply of irrigation water for the Nile Valley and delta. By reducing or eliminating the annual floods along the Nile, however, the natural enrichment of the soils has ceased, and farmers incur new costs in purchasing commercial fertilizers.

Siltation of the reservoir was anticipated, but the rate has been greater than expected. In addition, the settling of silt in the lake has caused clearer water to be released downstream and accelerated the erosive effects of the water, since clear water has greater capacity for carrying materials in solution or suspension than water which is saturated with material. In short, the greater reliability of water availability has been offset, in part at least, by other costs, both physical and economic.

The cutting of canals has created water courses where they didn't exist. The Panama Canal connects the Caribbean and Pacific; the Suez Canal connects the Mediterranean and Red Sea. East–west canals connecting

Figure 19.7
Reservoirs have changed hydrological patterns and microclimates.

north–south flowing rivers in northwestern Europe (discussed in the section of this book on glaciation) have made a latticework hydrological pattern in that region. Even unfinished canals, like the cross-Florida Barge Canal in Florida, which was halted because of feared environmental damage, has created a new waterway used by fishermen and boaters. Other canals have had biological impact, and some of those will be discussed in a subsequent section of this chapter.

The Extraction of Minerals

Other examples of dramatic localized influence on landforms include the extraction of minerals from, or beneath, the earth's surface. Underground mining frequently affects the groundwater level and quality, and those underground tunnels that have been improperly shored up may subside, affecting the surface, occasionally, with karst-like sinks, and frequently with modified surface slope and drainage patterns. Earth movements in mined-out areas create small tremors not unlike minor earthquakes. The author, who grew up in a coal mining area of southern Illinois, was wakened many nights by slight tremors associated with the "squeeze," or subsidence in mined-out areas. These are minor earthquakes as real as those occurring along fault lines. Surface damage can result. In fact, many of the hills in that small town, which gave thrilling rides to a small boy on a winter sled, have been reduced in dimension by subsidence to mere nothings offering no challenge or opportunities to today's youth, and a school had to be abandoned because of such surface subsidence.

Subsidence in many areas of the Appalachian and Eastern Interior Coal Provinces of the United States has left swampy areas as a reminder of such movements. Lifeless branches sticking out of water and muck lie where magnificent woodlands existed only a half-century ago.

Figure 19.8
Areas affected by coal mining through 1977.
From 1983 Yearbook of Agriculture, *U.S. Government Printing Office, Washington, D.C., 1983.*

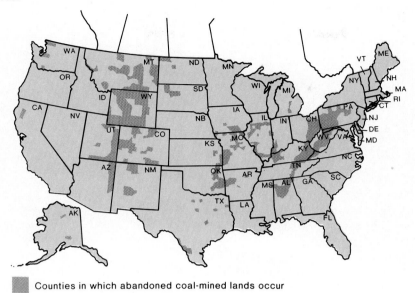

Counties in which abandoned coal-mined lands occur

The yawning pits in the tri-state lead and zinc district of northeastern Oklahoma, southeastern Kansas, and southwestern Missouri are as obvious as uvalas in an area of karst. Within the city limits of Picher, Oklahoma, for example, man-made sinks have been roped off to protect the unwary. Refuse fills holes, which were city streets a mere human lifetime ago. Abandoned rock quarries, some filled with water, are ubiquitous examples of a kind of man-made karst. Young people splash and play in the waters of an abandoned quarry, which was a hill of limestone, granite, or marble furnishing employment to their fathers. Virtually every state has examples of such abandoned rock quarries, or gravel, sand, or clay pits.

The galleries of abandoned coal mines in Illinois, salt mines in Kansas, metal mines in Montana, and potash mines in Germany are man-made analogues to natural caves everywhere. Again, these features are very widespread. Since many countries are now producing radioactive wastes either in weapons manufacturing or in the generation of nuclear power, they are looking at such man-made caves as possible sites for the long-term storage of nuclear wastes. Obviously, any such storage will have to take place in deep mines in geologically stable areas where there is no possibility of contaminating the groundwater supply.

Surface mining often produces a dramatic impact on the preexisting landforms. The enormous pits from extracting the iron ore of Minnesota, the copper of Arizona, the granite of Vermont, the coal of Montana, or the phosphate of Florida are large holes in the ground, which are certainly more impressive than a kettle hole in the glaciated regions of Wisconsin. Indeed, many of them are very significant surface features. The waste extracted and piled

outside underground pits in the Witwatersrand of South Africa, or the waste thrown up in ridges in a coal strip mine area of Illinois, are more impressive than countless hillocks produced by natural forces. These landform creations of man continue swiftly as sophisticated tools and equipment are used to dig out minerals buried at more inaccessible depths.

The area affected by surface mining and quarrying activity in the United States amounts to more than 5,000,000 acres, an area substantially larger than Connecticut and Rhode Island combined. All of this area is marked by some human geomorphological modification of the preexisting landscape. The area expands each year as more remote locations are tapped for fuel or minerals, and as lower quality ores are commercially exploited to replace minerals consumed in the insatiable maw of civilization.

Although most states now have laws designed to ensure that slopes of stripped areas will be modified to reduce some of their badland-like character, the hummocky terrain left behind is usually in distinct contrast to the preexisting terrain. Revegetation usually occurs using different plants for replanting from those that are native to the area before stripping occurred. New surface, new slopes, and new exposed materials all become elements of disequilibrium in the geomorphological cycle. Also, it should be recognized that large areas were stripped before restrictive rehabilitative legislation was passed. As a result, some abandoned strip-mined areas in the Appalachian and Eastern Interior Coal Provinces, especially, are like mini-badlands. Many quarries have been abandoned after the last piece of dimension stone has been removed only to fill with surface runoff after abandonment.

People as Modifiers of the Physical Environment 341

Figure 19.9
Coal stripping in Oklahoma is producing a miniature ridge and valley terrain. Subsequently the land will be leveled and reseeded.
Kenneth S. Johnson, Oklahoma Geological Survey.

Figure 19.10
Strip pit lakes left after land leveling and reseeding.
Kenneth S. Johnson, Oklahoma Geological Survey.

Expansion of surface coal mining in many western states in response to demand for low sulphur coal will modify the existing landform, edaphic, and vegetative environment. Although legislation to rehabilitate stripped areas will limit the deleterious environmental impact of mining, both subtle and occasionally dramatic environmental changes will result.

Clearly, the slopes developed and the size of the hillocks depend on the depth stripped, the nature of the spoil, and the stripping equipment used. Hillocks with a local relief of 25 to 50 feet (8 to 15 meters) are commonplace in stripped coal areas of the Eastern Interior and Western Interior Coal Provinces, and this height may be substantially greater in coal fields of the West. These features are certainly as significant as many glacially deposited or glaciofluvial features. Slopes as great as 45° may be encountered, although they are usually less than that. Before grading, erosion and subsequent deposition from these exposed slopes may be great. As in other cases, the amount of erosion and subsequent deposition will depend on the slope, type of exposed soil, and amount and intensity of runoff. In coal stripped areas, especially, sulphur may acidify runoff causing damage to surrounding natural vegetation or cropped lands.

In addition to the obvious pits and ridges produced by mining activities, frequently water may fill the pits producing permanent lakes or swampy areas. The initial hydrological pattern, both on the surface and below ground, is certain to be altered. Natural water courses peripheral to stripped areas may receive an extra burden of sediment until revegetation and grading occur. This produces a pattern of disequilibrium in adjacent streams, and the quantitative effects of different loads almost certainly

will have ramifications on higher orders of streams, which are a part of the system. In some instances, the sediment burden may be so great as to cause the preexisting pattern to become deranged. Water quality is also likely to degenerate. In areas stripped for coal, these strip pit lakes may be quite deep and from a quarter to a half mile (0.3 to 0.6 kilometers) in length. The water surfaces, when augmented by tens of thousands of man-made lakes and ponds, change the hydrological patterns and increase relative humidity. Effects on larger climatological patterns cannot accurately be measured. Where man-made lakes and ponds are abundant, the water table is probably raised, although enough premining and postmining data are often not available to make a definitive judgment.

Not infrequently, especially in areas mined for coal, waters become contaminated by the sulphur that is present in many coal deposits. This exposure acidifies the water with resulting undesirable ramifications for the biota. This problem is especially apparent in the high-sulphur coals of the Appalachian and Eastern Interior coal provinces.

Acidified waters pumped to the surface from underground operations to prevent flooding can sometimes affect the surface in disastrous ways. As a boy in southern Illinois, the author observed such influences as subsurface pumps brought millions of gallons of acidified water to the surface to be dumped into adjacent creeks. Some of those creeks have not regenerated themselves a half-century after they were exposed to such toxic effluents.

Basically, the adjacent streams become more turbid, and their pH is lowered. Water quality is inevitably reduced.

Mining of all kinds has had some geomorphological and hydrological influences. Broadened areal extent of mining over time will cause these influences to have wider impact. The biological impact of such activities are numerous, and some of them will be noted in a subsequent section of this chapter.

Figure 19.11
The high sulphur content of coal waste limits revegetation.

Construction Activities

Reproductive industries represented by agriculture and grazing, and the extractive industries represented by mining have had numerous geomorphological effects, but the same can be said for the facilitative industries represented by transportation and communication. Basically, the construction of all such facilities is designed to move people and goods faster and easier; hence, there is an effort to reduce slopes and grades. Rivers must be bridged, swamps must be drained or filled, hills and mountains must be breached, and drainage and integrity of the surface must be maintained.

A gap cut through a hill or mountain may be as impressive in scale as a notch cut by a stream. A man-made pass is certainly as impressive as a col eroded by mountain glaciers over eons, or a wind gap eroded by a stream of long ago. A railroad embankment snaking across a low area may be a more obvious feature of the physical landscape than a sinuous esker in Finland or Minnesota. Certainly it is more impressive than a puny crevasse fill in Wisconsin. Of course, these gaps, cuts, and fills are omnipresent features of the landscape in developed societies.

The borrowing of fill to produce the interstate highway system has formed hundreds of lakes across the country, which are certainly more impressive than a filled kettle hole in Michigan or a flooded cenote (sinkhole) in Mexico. The impact of such man-made ponds and lakes cannot realistically be measured, but there is an effect on the hydrological, biological, and geomorphological landscape.

In addition to the obvious effects of such cutting, filling, leveling, and surfacing of transportation features, there are less obvious impacts. The rapid runoff from a six-land expressway in humid subtropical Miami may place a strain on natural and man-made drainage systems. Flash floods become commonplace in humid environments. This situation probably did not exist in such regions before widespread vegetative denudation and paving.

Figure 19.12
A pass through the mountains was produced for this railroad right-of-way. The earth-moving bulldozer acts like an agent of erosion.
Caterpiller Tractor Company and the Association of American Railroads.

Figure 19.13
The increased runoff from an urban area is depicted by this graph comparing normal runoff with the increase in runoff attributable to urbanization.
From Charles C. Plummer and David McGeary, Physical Geology, *4th ed. Copyright © 1988 Wm. C. Brown Publishers, Dubuque, Iowa. All Rights Reserved. Reprinted by permission.*

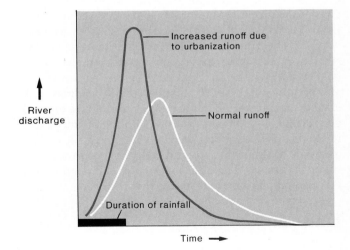

Certainly, the construction of an extensive highway system has exacerbated the problem. The additional runoff in a subhumid region may accelerate erosion on an adjacent surface. The increase of runoff caused by paving in urban areas and along highways tends to make flooding more frequent and to augment the area affected. Speeding of runoff probably has the effect of reducing water infiltration and of lowering the adjacent water table. In far too many instances, inadequate design studies and construction strategies have created flooding problems where none existed before.

Subsidence

In other locales, however, through other kinds of human intervention, the lowering of the water table has had other important hydrological and geomorphological effects. This drawdown of subsurface waters is by no means restricted to subhumid or semiarid regions. Indeed, some very dramatic impacts have occurred in humid and perhumid regions. A few examples will suffice to illustrate the point.

In the Great Plains of the United States, the depletion of water from the Ogallala Formation far beneath the surface has resulted in minor surface subsidence. In central Florida, rapid use of subterranean waters by a rapidly expanding population has led to the creation of surface sinks in natural karst areas, which have swallowed parking lots and automobiles in Winter Park and portions of superhighways near Gainesville. The rapidly expanding population of Florida and the increasing demands for water seem certain to increase subsurface drawdown, and it is likely that stories of collapse sinks in populated karst regions will become more commonplace. This is not a trivial inconvenience because a number of the areas in danger are urbanized, and the threat to life and property is real. It may become necessary to reinject waste water into some of the aquifers to reduce the number and frequency of collapse sinks, which develop.

Several of the great cities of the world are sinking because of the drawdown of water in the subsurface aquifer. Mexico City inexorably sinks into the unconsolidated sediments of the old lacustrine plain on which it was built. Venice continues to sink into the waters of the Adriatic. Houston is experiencing serious subsidence, not only because of the municipal use of subsurface water, but also because of the steady pumping of hydrocarbons from underground reservoirs. Subsidence, especially in areas of unconsolidated sediments, regions with soluble rocks, and in sections that have been mined out through subsurface means, is becoming more commonplace in occurrence and serious in impact. There are many cities not mentioned in this paragraph that are at risk.

Reclamation Activities

Obviously, there are many other examples of geomorphological modifications by human beings. The dredging of harbors, the filling of swamps, the straightening of rivers, and the reclamation of land from the sea, among many others, are having pronounced effects on the landforms of the world. In a sense, the Netherlands' prospects for the future continue to hang on the reclamation of land from the Ijsselmeer (Zuider Zee). Perhaps as much as a third of the national territory of the Netherlands is land that has been reclaimed from the sea (figure 19.14). New

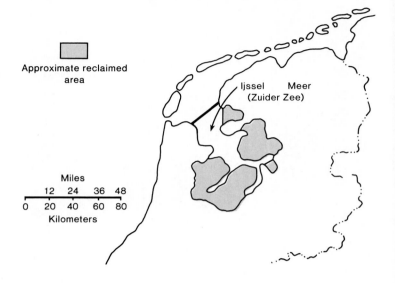

Figure 19.14
Approximate area reclaimed from the Zuider Zee.

polders continue to add to the national territory of this populous land. Eventually, the coastline of the Netherlands will become more regular as the result of the land reclamation.

In April 1987, the citizens of Utah began a project to ameliorate the effects of a rapidly expanding Great Salt Lake. A wetter period, especially in the 1980s, has greatly increased the depth and extent of the lake. This has encroached on settlement features while extending the effects of wave erosion and deposition far beyond past boundaries. A canal has been built from the lake into the western desert, and water is being pumped from the lake to lower the water level. The new playa to be created is expected to evaporate water rapidly, and the Great Salt Lake is expected to return more nearly to past levels. Will relative humidities be raised by expanding the water surface? Will climatological changes result?

There are no firm answers to these questions, but it's clear that a new lacustrine plain will be produced, and erosion will occur along the canal issuing from Great Salt Lake. Lake shore erosional and depositional effects will occur at several receding water levels. If the current pluvial period in Utah is transitory, the newly flooded area will dry up and become like its periglacial counterparts, which dot the Great Basin country.

Half a world away, other people in densely settled lands are also modifying their physical world. The pressure of people on land and water resources in Hong Kong has resulted in a change of shoreline configuration as well as hydrological patterns (figure 19.15). For example, the main runway of Kai Tak Airport is composed of fill dumped into the surrounding bay. Several bays or inlets of the sea have been dammed, the salt water has been pumped out, and they have been filled with rainwater to augment the existing water supply. The effect has been to

Figure 19.15
Approximate area reclaimed from the sea in Hong Kong.

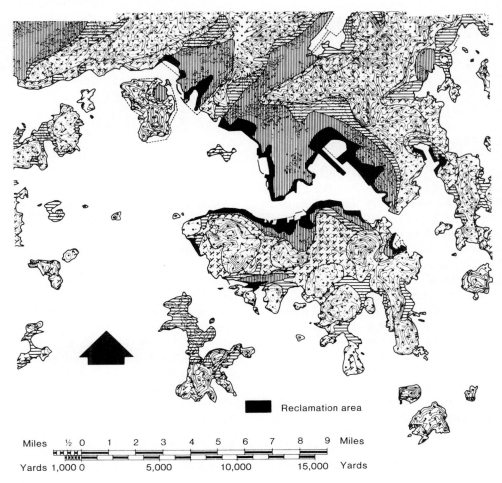

Reclamation area

Miles ½ 0 1 2 3 4 5 6 7 8 9 Miles

Yards 1,000 0   5,000   10,000   15,000 Yards

create a more regular shoreline and to increase the surface area of fresh water. These fresh water lakes now exist in what were once numerous arms of the sea. People have had a significant role in changing the shoreline configuration and hydrological patterns of Hong Kong.

In a book of this type, it is impossible to examine all of people's landform impacts, just as it is impossible to examine all aspects of natural phenomena. Suffice it to say that geomorphological impacts are persistent, pervasive, and rapidly expanding. The examples cited are designed to enumerate at the macrolevel some of the obvious modifications. Any reasonably perceptive observer can certainly cite many examples of modifications of landforms at a variety of levels in areas where he or she lives. Since one of the objectives of this book is to make readers more conscious of their environment, all are enjoined to note examples of large or small modifications of landscape resulting from human intervention. In any environment where a reader may live, there are certain to be microexamples of geomorphological change. Other aspects of the physical world have also been changed by human intervention. The results of some of those activities will be examined in the sections that follow.

People's Modification of Soils

The very act of clearing and tilling the soil modifies it: the soil biota is modified as soil is exposed, there is a tendency for soil moisture to be reduced, and there is some compacting by the movement of animals or machines over it. Soils that have shallow A-horizons may experience a mixing of subsoil at or near the surface by the act of plowing. Exposure of bare earth accelerates weathering, erosion, and organic breakdown, and results in a reduction of the thickness of the topsoil.

Tillage

Plowing to a constant depth may produce an impermeable zone in soils with a high-clay content. Such a **plowpan** may be increasingly compacted and result in a zone that resists water percolation and inhibits capillary action. These man-induced plowpan soils are very similar to naturally created **planosols,** which have developed in areas where eluviated clay from the A-horizon has been deposited in an almost impervious layer several inches beneath the surface. These natural claypans are quite common in regions where soils formed on glacial deposits.

A vegetable garden in the author's boyhood home in southern Illinois suffered through summer droughts, because the claypan, 10 to 12 inches (about 20 to 30 centimeters) beneath the surface, effectively limited the plants to the moisture contained in the top 10 to 12 inches of soil. Interestingly, however, years of intensive cultivation and the addition of enormous quantities of animal manure and organic material of all kinds over the years have so modified the soil that the claypan is no more. It has effectively been eliminated by human intervention. The impervious layer was as resistant to water percolation as caliche, which has been deposited in desert regions. In short, these physiologically dry soils have been altered by human intervention so they no longer are physiologically dry. It's worthwhile to recognize that these changes occurred within the span of a human lifetime, whereas natural development of soil characteristics might require hundreds or thousands of years. In terms of use to people, the soil is better now than it was fifty years ago.

Capillary action in subhumid to semiarid areas may draw water upward through tilled soil, and over time this may result in increased salinization. Undesirable mineral content may, however, be matched by loss of available minerals either because of existing tillage techniques or because of cropping patterns. Because organic material loss tends to be accelerated by clean tillage practices, subaerial erosion is accelerated and water retention is reduced.

Tillage practices in many areas of the United States, particularly, have changed in recent years. Clean tilling is declining, and tillage practices, which preserve organic litter, are being employed more broadly. This will have the effect of increasing water infiltration and retention, and raising the water table. The amount and frequency of tillage have declined as much to reduce fuel costs as to preserve the soil, but soil has been conserved in the process. Greater vegetation cover will also reduce surface erosion.

Although cultivation may impoverish some soils, its clear that soils may be enhanced by the addition of fertilizer, organic materials, or soil conditioners. Soils that have sustained cultivation over protracted periods are doubtless significantly different than the virgin soils from which they evolved. The subtle biochemical changes in such soils are often matched by a change in color, texture, biota, or structure of a given soil type. The rice-producing soils of China, for example, have been so modified by constant flooding and the addition of organic fertilizer over the millennia that they probably bear little relationship to the indigenous soil. Most areas that have sustained cultivation for protracted periods have, for all intents and purposes, produced a new soil.

Soils that have been enhanced by additions of organic matter typically exhibit an abundance of micro- and macro-organisms. Soils that have had large additions of chemical fertilizer without accompanying organic enrichment tend to show a decline in living organisms. Similarly, soils high in organic matter have improved rates of base exchange, water retention, and better tilth. Several high-intensity agricultural regions are beginning to exhibit diminishing returns with the continued addition of mineral fertilizers. As indicated earlier, soils are fragile resources vital to human survival, and they must be treated with care and respect if they are to continue to produce reliably.

In strip-mined areas, topsoils are usually buried by subsoil and bedrock. In a sense, that new surface awaits the development of new soil horizons, similar to the evolutionary process, which occurs on any newly exposed surface. The original soil in such circumstances is known as a **paleosol.** Paleosols may be buried by natural processes of deposition as we have seen, or they may be buried as a result of human intervention. Some areas in Europe have such rigid mining laws that the original topsoil must be replaced as topsoil when mining is completed. American laws are less rigorous, although most areas now require returning the land to something approximating the land profile prior to mining, and similarly, most such areas must be revegetated with indigenous plants.

In summary, it is safe to say that people's overall impact on soil has been to reduce the average thickness of the topsoil, to reduce inherent fertility, to diminish water retention, and to place in jeopardy the continuing prospect of people being able to meet the food and fiber needs of the species. The few areas where soil tilth or fertility have been enhanced by human activity are more than counterbalanced by the degradation of soil in other regions. Ignorance of sound agricultural practices in many developing regions, absence of capital to effect agricultural improvements, and the demands of short-term survival versus long-term benefits militate against sound conservation practices. The continuing negative effects of people on the soil must be reversed if the planet is to support the added billions of people destined to arrive in the next few years.

There appears to be an increasing awareness that the less cultivation necessary for the production of a crop, the better. It's ironic that diminished cultivation activities probably occurred primarily to save labor and energy before there was a recognition that such practices also conserved the soil. It is certainly true, however, that cultivation techniques in the United States in the 1980s are substantially more protective of the soil than were cultivation techniques of five decades ago. Nevertheless, soil erosion, fostered by cultivation, remains a serious threat to the continuing survival of people. The heavy sediment burden of the Mississippi River and its tributaries testifies, in part, to the vast cultivated areas in America's heartland. The turbid rivers carry our greatest wealth, our topsoil, to the Gulf of Mexico and surrounding oceans. The expanding delta, the changing course of the river, and the constant siltation makes dredging necessary and navigation hazardous.

Figure 19.16

These volcanically derived soils in the Philippines have been significantly modified by rice agriculture. Repeated waterlogging has modified mineral content and soil texture.

Some of the same types of human intervention, which results in modification of soil or landforms may also affect the climate. These influences on climate will be discussed in the section that follows.

People's Modification of Climate

There is an interesting interweaving and interlacing of the physical environment in such a way that the direct impact on one facet of the environment is likely to produce accompanying indirect effects on other parts of it. There is, essentially, a ripple effect throughout the environment when one aspect is disturbed or modified. The devegetation of vast areas, especially in the tropics, clearly modifies the microclimates. Shade temperatures in large cleared areas of the tropics may be several degrees warmer than in adjacent areas of the rainforest. Relative humidities may be somewhat less in such cleared areas, and temperature ranges may be greater. Many climatologists fear that rapid destruction of the rainforest may, in fact, cause serious large-scale climatic dislocations. The character and dimensions of such dislocations, if any, cannot be determined. Unfortunately, it is likely that devegetation will result in expansion of dry areas as well as to diminish climatological reliability. Urban areas may exhibit temperature increases as well.

Urban Areas

Large cities have been shown to be heat islands of considerable magnitude. Large expanses of reflective surfaces bounce heat from one area to another, and conductivity is increased. Measurements often show large cities to have temperatures several degrees higher than the surrounding countryside. Infrared sensors on satellites have persistently demonstrated that urban areas are heat islands. Because Western man relies increasingly on air conditioning in summer, the higher temperatures of cities outside are magnified further by the work of such units. The overall effects of these heat islands on atmospheric circulation, especially convection, are imperfectly understood, but it's clear that there are some influences. It's equally certain that cities are less comfortable to human beings, especially in the summer.

Cooling ponds associated with power generation facilities or certain manufacturing establishments often set up convectional systems replete with transient cumulus clouds. The power plant cooling pond, a stone's throw away from the office of the author of this text, produces a small cumulus cloud on most mornings in the cooler fall and winter seasons, although they may occur at any time. Occasionally, these cumulus clouds extend upward to considerable heights. An otherwise cloudless day may have such a man-induced towering cumulus.

The Greenhouse Effect

The burning of hydrocarbons may be adding to the carbon dioxide content of the atmosphere, which, in turn, accentuates the "greenhouse effect," which some climatologists assert is causing the earth to undergo a warming trend with possible enormous climatological ramifications. If such a warming trend continues, rising sea levels will threaten coastal margins with all sorts of prospective calamitous results. Although the overall tendency appears to be a gradual warming of the earth's atmosphere because of the increasing CO_2 burden, there are minor perturbations up and down of temperatures, which illustrate that this is not a well-understood phenomenon. The overall tendency towards warming is so pronounced, however, that in the spring of 1988, climatologists were already predicting that hurricanes a century from now will probably be up to 100 percent more violent than they are now. This postulation derives from the projected higher energy budget of warmed ocean waters at that time. If warming persists resulting in substantial reductions of ice caps, the rising seas will change the patterns of the world's coastlines.

Pollution

Vapor trails from high flying aircraft and particulate matter from industrial and agricultural activity seem to be increasing average cloudiness over humid parts of the earth. Such affects may ameliorate temperatures and modify precipitation patterns. The cloud cover and particulate burden of the atmosphere have certainly reduced visibility over the earth. The author served as an aerial navigator on military aircraft in World War II. Frequently, in training and in combat circumstances, it was possible to report "ceiling and visibility unlimited" (CAVU). Such conditions would certainly be rare in today's flying. One is struck by how quickly the clear, blue skies, which often succeed the passage of a cold front, are made hazy by the effects of pollution emanating from the surface as well as from high-altitude aircraft. The crisscrossed vapor trails of aircraft approaching big city airports begin to produce a high-level veil of clouds almost as soon as a cold front has cleared an area. The haze that stains the atmosphere is becoming thicker and more persistent. The careful observer may detect his or her approach to a city from far away by the blanket of pollution, which thickens around urban areas.

William L. Seaver of Virginia Polytechnic Institute and James E. Lee of the MITRE Corporation studied sky cover including clouds, smoke, and haze for two periods, 1900–1939 and 1950–1982. They discovered that sky cover was about 18 percent greater in the latter period. They also examined the record of cloudless days for forty-five American cities in the same time frames. They discovered that only one (Fort Worth) got slightly sunnier, whereas

Figure 19.17
Projections of temperature rise associated with the "greenhouse effect."
Electric Power Research Institute Journal.

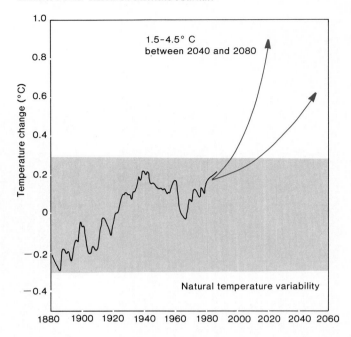

all the rest became cloudier. For example, St. Louis declined from 7.2 to 4.7 sunny days, Los Angeles from 10.0 to 7.6 days, and Washington from 5.3 to 4.4 days.

Although the reasons for these changes are not completely understood, it is reasonable to infer that increased human activity, especially in the burning of the hydrocarbon fuels, has added to pollution and diminished the clarity of the atmosphere. As population increases worldwide, it is reasonable to expect that this tendency will increase as more fuel is consumed and the particulate burden of the atmosphere is increased by agricultural activities, grazing, and devegetation of many forested areas.

As more surface water is available for evaporation over land areas because of the construction of reservoirs, relative humidities increase. Conversely, as areas are devegetated, relative humidities tend to decline, and the encroachment of deserts into semiarid areas is often attributed to physiological and climatological effects of overgrazing and overcutting. The average relative humidities, at least at local levels, may be changed significantly by human intervention.

The Ozone Problem

The use of fluorocarbons in spray cans has influenced the shielding effect of the ozone layer. Reduction in such use seems to have reduced the danger of increased ultraviolet radiation exposure, although a vast unexplained hole in that ozone layer over Antarctica seems to be expanding.

Figure 19.18
An 80-meter (more than 260-foot) rise in sea level, which would occur if existing ice sheets melted, would flood many heavily populated areas of the United States.
From Carla W. Montgomery, Physical Geology. *Copyright © 1987 Wm. C. Brown Publishers, Dubuque, Iowa. All Rights Reserved. Reprinted by permission.*

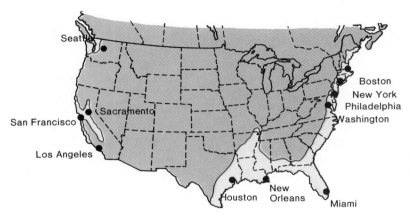

Scientists are concerned about this Antarctic hole, and investigations as to the magnitude and direction of this phenomenon are proceeding swiftly. If the ozone layer is sustantially weakened or eliminated, the deleterious effects on life are chilling to contemplate. Another ominous augury was brought to light in the spring of 1988 when Canadian scientists revealed that a similar hole was apparently developing over the north polar region.

Apparently, the freon used in refrigeration units is also a culprit in the reduction of the ozone shield. There is certain to be a plethora of studies on this matter, and doubtless there will be industrial efforts to slow the shift to other refrigerants because of costs, if a scientific cause-and-effect finger is firmly pointed at freon.

In 1988, the DuPont Corporation announced a major policy change demonstrating corporate responsibility and environmental sensibility. The company will stop manufacturing fluorocarbons and other materials known to deleteriously affect the ozone layer. Such efforts can only be applauded by the environmentally sensitive.

The shielding of the ozone layer at higher elevations is important to protect us from ultraviolet radiation. On the other hand, too much ozone in many large cities is adding to a noxious mix of gaseous pollutants.

Overt Attempts to Modify the Weather

Attempts at cloud-seeding either to dissipate potentially dangerous storms or to bring rain to a parched area have met with varying levels of success. Unsuccessful attempts have been made when atmospheric or cloud conditions were not right. In addition, there has been a certain inhibitory factor of possible lawsuits brought on by the allegation that such a cloud-seeding effort might bring unwanted precipitation to certain areas or precipitation in too large amounts. It does seem that under the right conditions, cloud-seeding, using silver iodide or other hygroscopic nuclei generators, can facilitate the onset of precipitation and enhance the amount received. The short-term effects of such interventions in dry areas would appear to be positive from a human standpoint, but can we be sure? Further, a farmer or rancher in desperate need of rain might be at odds with an entrepreneur promoting outdoor activities best done in the sunshine. Even if there are common needs and desires for precipitation enhancement, rainmakers can do little to increase amounts unless nature sends clouds already pregnant with moisture.

Many climatologists express the fear that widespread cutting of forests in tropical areas could have the effect of modifying global wind circulation patterns and, as a result, change the global weather patterns. Since it appears that such forest destruction will continue, a few decades will reveal whether expressed fears are well founded. Unfortunately, if the doomsayers are correct, the problem is likely to become irreversible rather than self-limiting. In fact, people have reason to be perturbed about the fact that many human activities may create irreversible effects on various facets of the physical environment.

Whether climatological modifications effected by man prove to have global impact, or whether they are restricted to the creation of microclimatological modifications is unknown. There has been and is a human impact on climate. Time will determine the future dimensions and directions of that impact. We have reason to pause and consider carefully the effects of expanding human intervention.

Changes in climate will obviously affect the biota over time. More direct assaults on our biological world produce immediate results. A consideration of some people-induced biological changes follows.

People's Modification of the Biological Environment

Among the most obvious of people's environmental impacts has been the modification of other life forms on this planet. On the one hand, it's clear that people have caused the extinction of many species. The destruction of the passenger pigeon, which darkened the skies with untold millions when Europeans first came to North America, demonstrates starkly that numbers alone are no protection for any species. On the other hand, people as the conservators have protected some species, and, in a few instances, captive breeding has rescued certain species from the brink of extinction. In spite of people's efforts, it is well to recognize that natural extinctions do and will occur.

Domestication

People have drastically altered the character of certain species through careful selection. A Rhode Island Red or a White Leghorn chicken bear only superficial resemblance to their jungle fowl progenitors. A Black Angus or a Jersey differ markedly from the wild cattle from which they descended.

Modern corn (maize) differs significantly from the teosinte from which it apparently derived. Varieties of wheat are quite unlike many of the wild strains of grasses from which they came.

Scores of examples of human modification of "wild" plants and animals to domesticated varieties could be cited. Such modifications have been persistent and pervasive. Certainly, human beings have speeded, interrupted, or changed the direction of evolutionary movement among many species of plants and animals. Most of these changes have apparently been desirable from the human perspective. On the other hand, where we have reduced species diversity, we may have created a kind of vulnerability in lowered plant or animal resistance to disease or parasites.

The modification has occurred through the removal of natural vegetation and the substitution of agricultural crops or other vegetation, through the absolute eradication of entire species of plants or animals, and through genetic manipulation to produce modified races of existing species.

The possibility for creating new life-forms by molecular biologists and geneticists is real. Several examples of such developments have already occurred in the laboratory. If these forms are released in the natural environment, the ultimate impact on other forms is still uncertain.

Stress is placed on the possible mischief of such introductions, but fairness requires consideration of the possibilities for good, at least from the perspective of human beings.

Eradication

The eradication of animal forms began early in human history as the demands for food or shelter caused certain animals to be killed beyond their ability to reproduce themselves—*extinction*. The wheels of modern civilization have speeded that process, and numerous other species of plants and animals will become extinct in the years ahead. The rate of extinction in the world's tropical rainforest is high and accelerating.

The pattern, which has existed in the United States, illustrates the point. As settlers moved across this land, millions of acres of forest were removed both to develop agricultural or grazing opportunities, and to use wood for fuel and construction. A literal vast sea of forests was reduced to islands of forest land, which was in the roughest or least desirable areas. Unwise agricultural practices in some areas has caused some cleared lands to revert to brush and woodland.

In New Zealand, almost entirely covered by forests when Europeans arrived, well over half of the forest land has been lost as lumbering became the preface for agriculture or grazing, and on South Island, erosion has become a major problem because of the removal of the trees. New Zealand is a sparsely peopled country. In a densely settled area like China, most of the original vegetation has been removed or altered except in the most remote and inaccessible places.

In many areas, the reduction of animal populations was a natural concomitant of elimination of habitat. In others, wanton killing left a legacy of death and destruction. When the last passenger pigeon in the United States died in captivity in this country after years of mindless killing, our psyche was injured, not because of what we had done, but because of what we are capable of doing. The California condor literally appears to be on its last wings. It is doubtful that the species can be saved. The hatching of a single condor chick in captivity in late spring of 1988 was cause for considerable rejoicing. Another chick, hatched in the spring of 1989, added one additional tenuous hope to the possibility that the species may be saved. Realistically, however, there is real doubt about whether a breeding program will ever be successful enough to allow a reintroduction of the species to the wilds. Conversely, the pathetically few whooping cranes appear to be making a comeback with human help. We can be cautiously optimistic that the species will survive. We cannot, however, relax our vigilance for a moment, or they, too, will only be found as stuffed specimens in a museum.

The decline of other bird populations, including wild ducks, in several areas of the country is seen as an example of the poor state of environmental health by a

Figure 19.19
Selva areas of the world are rapidly being depleted by the advance of civilization.

Figure 19.20
Accidental fires become more numerous as the number of humans increase. The effects on the biota are pervasive.

number of ecologists. Most believe that pollution of the atmosphere and water, and the reduction in the area of suitable habitat have contributed to the decline in bird counts in many areas.

Biological Introductions

In addition, we have introduced thousands of species of wild plants and animals accidentally or by design to new environments. Some of these introductions have been salutary from a human standpoint, whereas others have had less happy outcomes. A few examples of such introductions as they have influenced the United States, or have the potential to exert such influences, will be described subsequently.

Perhaps the most common bird in the United States is the English sparrow, which was introduced by our forebears early in colonial history. The introduction occurred so long ago and the bird is so ubiquitous that few would recognize it as an exotic species in this country.

In recent times, the cattle egret, which was introduced into Latin America, has made its way to the United States to become a most successful new arrival. These birds apparently do no real harm and may actually be of some value in terms of the insects they consume. As a matter of fact, without birds in their large numbers to act as controls, we would soon be inundated by insect pests of all kinds.

Asian water hyacinths, introduced as an ornamental plant, are another story. They choke many southern waterways and cause great effort and expense in eradication programs. The Mediterranean fruit fly, a new introduction, has caused millions of dollars of damage and costs in eradication programs in the citrus growing states. The tiger mosquito, which apparently came to these shores in 1986, is adding irritation and the threat of disease to several southern states. The march of the killer bees from

Africa to Brazil to Mexico has been chronicled in scientific literature and the tabloids. In early 1988, entomologists predicted that the bees would arrive as colonizers late in 1988 in Texas. Data are still inconclusive as to whether these predictions were borne out by actual occurrences. Isolated colonies have already appeared in California, although it appears that they have been successfully eradicated.

The ordinary honeybee is also under assault from the parasitic attacks of the bee mite, which was inadvertently introduced from infected hives. Florida beekeepers are fighting infestations. If the mite is not controlled, agricultural plants requiring pollination are threatened.

Hundreds of examples of benign and malignant introductions could be chronicled, and an array of dedicated agricultural inspectors at our borders cannot eliminate the introduction of new species either by accident or design. Again, the astute observer can enumerate and describe exotics of his or her own area. Each of us can provide some protection to the environment by being careful not to introduce exotic species into the landscape.

Tinkering with the biota is simply another example of human influence on the environment. As we have manufactured products and moved ourselves from place to place, we have created toxic wastes that have biological impact.

Acid Rain

Acid rain is apparently having a deleterious effect on vegetation. The extent, dimensions, and types of degradation are by no means nailed down, however. There are near hysterical charges and countercharges about the effect of acid rain. Some things are known about it; others aren't. What to do to ameliorate the situation isn't always known, and costs attach to remediation in every instance.

Figure 19.21
Sulphur dioxide emissions contribute to acid rain.
From 1983 Yearbook of Agriculture, *U.S. Government Printing*
Office, Washington, D.C., 1983.

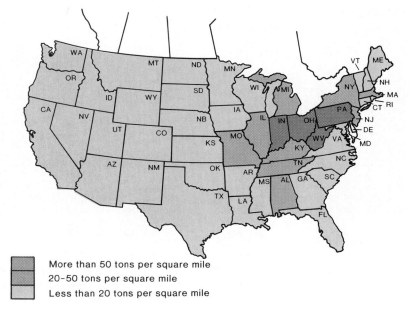

More than 50 tons per square mile
20–50 tons per square mile
Less than 20 tons per square mile

Relationships between the United States and Canada have been strained because American industry is apparently contributing a large share of the smokestack effluent responsible for acid rain in Canada. Because of the economic costs of reducing undesirable emissions, American businesspeople and politicians have dragged their feet in implementing appropriate remedial measures. It's time that we approach this problem head-on both for the Canadians and ourselves. Obviously, the entire world needs to become involved in this aspect of environmental protection.

In virtually every biological niche, there is a steady degradation of habitat and species, which seems destined to make the world a poorer place for our children and grandchildren—poorer for the loss of elements of natural beauty; poorer because of a weakening of the fabric of environmental interconnections; poorer because of the attitudes of our species, which such actions reveal. The great gift of opportunity given with the gift of life, the thrilling fact of a cognitive brain, the awesome responsibility of making wise choices may all be cast on the waste heap of greed, selfishness, and insensitivity. Those who read this book and their fellow billions do, indeed, hold a habitable earth in the palms of their hands.

In a kind of ultimate irony, the improvements in sanitation and medical science have been responsible for the world's population growth. That growth and it's concomitant of environmental pressure threaten the environments that support us.

Correct actions afford enormous opportunities for the years ahead. Continuation of present attitudes and actions portends a gloomy future. As an optimist, I dream of a better earth peopled by wiser citizens in the future. The realist within me, however, warns that we had better become better stewards of our environmental legacy soon, or disaster may overtake those who are to follow us.

Although all the actions necessary to curb a deleterious modification of the environment are unknown, it seems clear that we must do something to slow and ultimately halt the growth of population. Continued growth adds inescapable environmental burdens. Indeed, the expansion of the world's population and all that portends is probably the greatest challenge facing mankind.

It is equally certain that the volume of hydrocarbons burned must be reduced, and those that are used must be burned in cleaner ways. The negative effects of the "greenhouse effect" have already been described.

Figure 19.24
Sunrise or sunset for Homo sapiens?

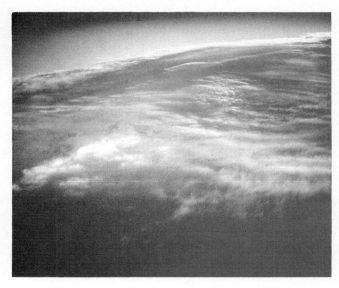

Finally, and not least, we need to make a significant shift to renewable energy resources rather than to continue our preponderant dependence on the fossil fuels. Our systems of production, consumption, and distribution must become renewable systems insofar as that is possible.

Collectively, we must attempt to reduce our impact on the lithosphere, hydrosphere, atmosphere, and biosphere. Living with nature, the human species can occupy this good blue-green earth for millennia. If we continue our assault on the spheres that support us, our great-great-grandchildren may be eking out their survival in a cheerless and sterile place.

The biblical admonition, "Where there is no vision, the people perish," applies now as it always has.

Study Questions

1. Explain, in a general way, the differences between people's creation of a cultural environment and modification of the physical world.
2. How have people caused geomorphological modifications of the environment?
3. What are some examples of geomorphological features or processes, modified as a result of human intervention?
4. Where has the effects of mining on the geomorphological landscape been particularly prevalent in the United States?
5. What effects have human intervention, by accident or design, have on hydrology?
6. What have been the most notable and pervasive of human influences on the soil?
7. What is a paleosol?
8. What is a heat island?
9. What are some major human modifications of climate?
10. Is average cloudiness in industrialized societies greater now than it was 100 years ago? If so, why?
11. What evidence do we have of a thinning of the ozone layer? What is the probable cause? What measures may be necessary to mitigate the problem?
12. What impact have people had on the biological environment?
13. What are some particularly pernicious examples of introductions of plant and animal species into new environments? From the human standpoint, what are some positive examples of such biological introductions?
14. What do you see as the long-term prospects of *Homo sapiens* on "spaceship earth"?

Selected References

Brown, L. R. et al. *State of the world 1988.* New York: W. W. Norton and Company.

Brown, L. R. et al. *State of the world 1986.* New York: W. W. Norton and Company.

DeBlij, H. J., and Muller, P. O. *Geography: Regions and concepts.* 1988. New York: John Wiley and Sons.

Detwyler, T. R. 1971. *Man's impact on environment.* New York: McGraw-Hill Book Company.

Kahan, A. M. 1986. *Acid rain: Reign of controversy.* Golden: Fulcrum.

McKnight, T. L. 1984. *Physical geography: A landscape appreciation.* Englewood Cliffs: Prentice-Hall.

McNeill, W. H. 1976. *Plagues and peoples.* Garden City: Anchor Press/Doubleday.

Nir, Dov. 1983. *Man, a geomorphological agent: An introduction to anthropic geomorphology.* Jerusalem: Keter Publishing House; and Dordrect, Holland: D. Reidel.

————. 1986. *Research for tomorrow: Yearbook of agriculture.* Washington: United States Department of Agriculture.

Appendix

*The plains area of northeastern Oklahoma has been modified by this collapse sink in the zinc-lead mining area of Kansas, Missouri, and Oklahoma.
Kenneth S. Johnson, Oklahoma Geological Survey.*

A. *Descriptive Landform Classifications*

The discussion of landforms in the body of the text has been primarily genetic (i.e., how landforms have developed and evolved in terms of essential lithology, structure, and erosional processes). There is also some value in discussing landforms in terms of their relief characteristics. The late Vernor Finch and Glenn Trewartha, formerly of the Department of Geography at the University of Wisconsin, developed some relief and positional classifications of major landform types in several iterations. They would have been the first to agree that choices of relief divisions were arbitrary and subject to the same shortcomings attending any classification system.

The following brief description is derived from those characterizations. Plains are usually below 2,000 feet (about 600 meters) in elevation with a local relief of less than 500 feet (about 150 meters). Finch and Trewartha described such plains as (1) *flat,* with a local relief of less than 50 feet (about 15 meters); (2) *undulating,* with a local relief of 50 to 150 feet (about 15 to 45 meters); (3) *rolling,* with a local relief of 150 to 300 feet (about 45 to 90 meters); and (4) *rough,* with a local relief of 300 to 500 feet (about 90 to 150 meters).

Plateaus, on the other hand, are broad uplands, usually above 2,000 feet in elevation (about 600 meters), which rise rather abruptly on at least one side from adjacent lower areas. The local relief is typically greater than for plains, although much of the surface is at relatively accordant levels. Plateaus are sometimes described in terms of their position: (1) *intermontane,* or between mountain ranges; (2) *piedmont,* or situated between mountains and bordering plains, or the sea; and (3) *continental,* or upland surfaces that occupy all or a significant part of a continent.

Hills are landforms that usually have a great percentage of their land in slopes and include elevations in excess of 500 feet (about 150 meters). Normally, they have local relief of less than 2,000 feet (about 600 meters). At that upper limit, mountains begin.

Mountains are similar to hills, but they usually may be said to have elevations above 3,000 feet (about 900 meters) and local relief in excess of 2,000 feet (about 600 meters). It has been suggested that mountains might be classified according to local relief as *low,* with a local relief of 2,000 to 3,500 feet (about 600 to 1,060 meters); *rough,* with a local relief of 3,500 to 6,000 feet (about 1,060 to about 1,800 meters); *rugged,* with a local relief of 6,000 to 10,000 feet (about 1,800 to more than 3,000 meters); and *sierran,* with local relief in excess of 10,000 feet (3,000 meters).

Of course, any such classification according to relief and slope is arbitrary, and place name descriptions clearly do not conform to those at suggested relief limits. Many features described as hills in a place name would, according to local relief delineations, be plains. Similarly, sites described as mountains are, in reality, hills, and so on. The Arbuckle Mountains of central Oklahoma, landforms of considerable geological and geomorphological interest, are plains, from a relief standpoint. On the other hand, the Black Hills of South Dakota do possess the relief characteristics of mountains in several sections.

Classifications and descriptions using elevation and relief are probably less useful than genetic discussions because they give no clue as to origins and evolution. Nevertheless, they have some use in descriptive terms, and the informed reader should be aware of their existence. Clearly, it would be possible to establish a plethora of classification using any number of arbitrary relief and slope classifications.

B. Land Survey Systems

Americans in the colonies along the Eastern seaboard were the inheritors of the land survey system of England. This system, *metes and bounds,* is still in use in many parts of the world, although its shortcomings, which will be described subsequently, are numerous and varied. This system employs natural features in the landscape, such as trees, rocks, hills, streams, and similar features, along with distances and directions from these described reference points to establish property boundaries or limits. Because features tend to be transitory (i.e., trees die or are cut, rocks are moved, stream courses change, and the magnetic field changes over time), these boundaries are highly unsatisfactory over long periods and frequently lead to litigation, due to disagreements over precise boundary locations. Further, land divisions are usually irregular, and parcels of property may be oddly shaped, which often militates against mechanized agriculture and makes the selection and construction of transportation routes circuitous and difficult. Obviously, property must be appropriately defined to assure the rights of the owner and to facilitate transfer of such lands to new owners and inheritors.

Fortunately, the founding fathers of the country initiated a new and more rational system for the new territories outside the original colonies. With the Ordinance of 1787, which opened the new lands north and west of the Ohio River, a new, orderly system was developed. The plan, as written into law, provided that "the surveyor . . . shall proceed to divide said territory into townships of six miles square, by lines running north and south, and others crossing these at right angles, as near as may be." The evolution of this system has resulted in a series of townships numbered east and west of a specific principal meridian and north and south of a given base line. Each township, which is six miles on a side, contains thirty-six square miles, or sections. Each section contains 640 acres. Each section may be further subdivided for the purposes of describing locations and property lines. Because meridians converge toward the poles, it is necessary to compensate for this convergence by surveying a new standard parallel every fourth tier of townships north of the base lines. Range lines are reoriented with respect to these standard parallels, so as to maintain the thirty-six square mile township as nearly as possible. These correction line offsets are principally responsible for the jogs, or deviations, in north–south section line roads.

The tiers of townships are numbered consecutively to the north and south of a base line. All townships in the tier immediately to the north of the base line are labeled as "Township 1 North," and all immediately to the south of the base are designated as "Township 1 South." The ranges are numbered consecutively to the east and west of a principal meridian, and they are similarly designated as ranges east and west (e.g., "Range 1 East" or "Range 1 West").

The individual sections of the township are designated by beginning in the northeast corner with 1 and proceeding west through 2, 3, 4, 5, and 6. The section immediately to the south is numbered 7, and 8, 9, 10, 11, and 12 proceed to the east of 7. Section 13 is immediately south of 12 and numbering proceeds westward to 18. This shifting back and forth, alternatively, in the numbering system causes 36 to be located in the southeast corner of the township. In describing the location of a section, it is mentioned first, and the township designation follows. For example: "Sec. 10, T. 2 N., R. 4 W." A part of a section might be described as follows: "The S.E. ¼ of the N.E. ¼ of Sec. 2, T. 3 N., R. 4 E."

The advantages of the township and range system are numerous. Land descriptions are greatly simplified, the use of transitory landmarks is eliminated, the boundaries of sections are usually marked by roads which produces the characteristic grid pattern that exists over much of the country. The Ordinance of 1787 for a variety of reasons, including the development of the land survey system and the concept of the land grant college, was almost as important as that other great document written in 1787 (i.e., the American Constitution). Of course, the territories of the original thirteen colonies are not included in the Ordinance of 1787, along with some subsequent acquisitions of territory, such as Texas, which brought with it the land grant heritage from Mexico.

Appendix 1

Map showing (A) *principal meridians and base lines in the United States and* (B) *relationships of township to base line, and principal meridian, section within a township, and parts of a section.*

Morris M. Thompson, Maps for America, *United States Geological Survey, U.S. Government Printing Office, Washington, D.C., 1979.*

United States Public Land Office Survey
Base Lines and Principal Meridians

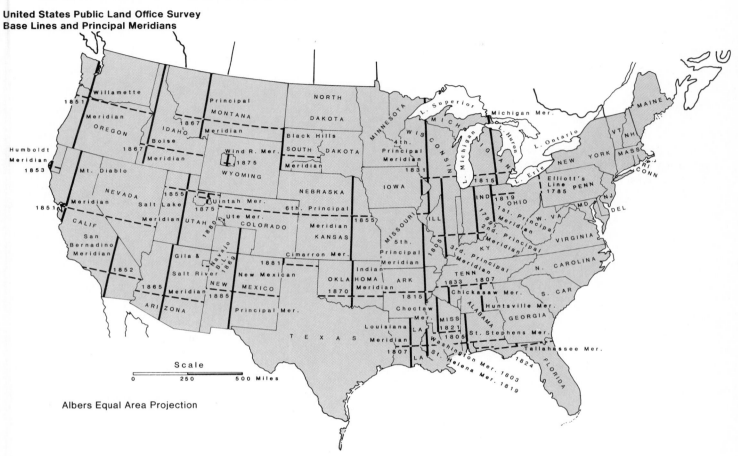

Scale

0   250   500 Miles

Albers Equal Area Projection

A

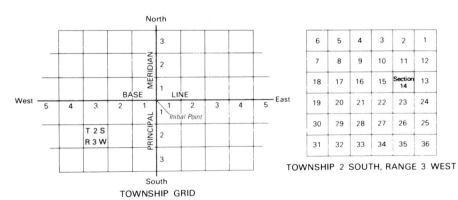

North

TOWNSHIP GRID

West   East
South

TOWNSHIP 2 SOUTH, RANGE 3 WEST

SECTION 14

B

Glossary

Aa a rough, broken type of lava.

ablation reduction of ice or snow through melting, evaporation, and sublimation.

abrasion erosion of surfaces by particles of rock carried by running water, ice, waves, or wind. Surface materials are worn away by particles carried by agents of erosion.

absolute humidity the weight of water vapor in a given volume of air, often expressed in grains/foot³ or grams/meter³.

abyssal deep deep ocean floor typically covered with marine sediments.

acidic (rocks) rocks that have a high percentage of the light minerals, notably quartz.

acid rain rain with a pH less than 5.6 average (average pH for normal rain), believed to be caused largely by dissolved impurities from automobile exhausts and pollutants from the burning of hydrocarbon fuels.

adiabatic heating and cooling heating resulting from compression of a gas or cooling resulting from expansion of a gas, as air descends or rises respectively.

adiabatic lapse rate the cooling or warming of air in a rising or descending parcel at a rate of approximately 5.5°F/1,000 feet or 1°C/100 meters.

advection horizontal heat transfer within the atmosphere.

advection fog a fog developed when a warm, moist air mass moves over a cool surface.

aeolian wind erosion or deposition.

aggradation deposition of materials carried by agents of erosion.

agonic line the line of 0° magnetic declination.

A-horizon the upper horizon in a soil profile; the topsoil.

aimless stream pattern distorted or unrecognizable stream pattern in areas of very low relief or in areas with a heavy sediment burden. Also called *deranged stream pattern*.

air drainage winds *See* katabatic wind.

air mass a large body of air with fairly homogeneous characteristics of temperature and humidity, which may move as a distinct entity for significant distances.

air mass thunderstorm a thunderstorm that develops as the result of convectional updrafts in warm, humid air.

albedo proportion of solar radiation reflected back into space.

alfisols soils with a yellowish-brown topsoil colored by aluminum and iron compounds.

alluvial apron a series of coalescing alluvial fans.

alluvial fan a fan-shaped deposit typically found near the base of mountains in arid regions, resulting from deposits of streams flowing out of mountains into surrounding areas of lower relief.

alluvium stream deposits.

alpine glacier glacier in mountainous region largely contained between valley walls as it moves from higher to lower elevations.

altimeter an aneroid barometer designed to measure altitude.

altitude elevation above mean sea level.

altocumulus middle level clouds arranged in an almost geometric pattern and sometimes referred to as a mackeral sky.

altostratus sheetlike clouds at intermediate elevations often associated with light rain or snow.

analemma a graphic device that depicts sun's declination and variations from mean solar time throughout the year.

andesite igneous rock that is intermediate between the most acidic and basic types.

anemometer a device using a series of cups that rotate at varying speeds measuring changes in wind velocity.

aneroid barometer a device using an evacuated metal diaphragm to measure atmospheric pressure.

angle of repose natural surface inclination of unconsolidated rock particles.

angular unconformity a disjunct in the rock section where steeply dipping rocks are in contact with younger rocks, which are horizontal or gently dipping.

anhydrite a mineral composed of calcium sulfate ($CaSO_4$).

annular drainage pattern a stream pattern that may develop around an eroded dome.

Antarctic Circle 66½° S. latitude.

antecedent stream a stream that preceded geological structures it now cuts across.

anthracite hard coal.

anticline an upfold in rock usually caused by lateral compression.

anticyclone an area of high barometric pressure; a high.

aphelion maximum distance between earth and the sun in the annual orbit.

apparent solar time time determined by the position of the sun.

aquifer a porous and permeable stratum of rock, which holds and allows movement of water within it.

Arctic Circle 66½° N. latitude.

arcuate delta a delta form that is shaped like a fan in plan view.

arête a sharp ridge in glaciated mountains, which separates two cirques or glaciated valleys.

aridisols soils of desert regions, which are high in soluble minerals and low in organic compounds.

arroyo a steep-sided rather flat-bottomed valley in desert areas occupied by ephemeral streams.

artesian spring or well water that flows to the surface under natural pressure.

asthenosphere plastic layer in the earth's mantle, which moves in response to convection-like currents and generates movements in the earth's plates.

atmosphere blanket of air, composed of a mixture of gases, which surrounds the earth.

atoll a ring-shaped coral island.

aurora australis luminous streamers in the Southern Hemisphere night sky resulting from disruption of ionized solar particles by the earth's magnetic field.

aurora borealis the northern lights, comparable to the *aurora australis* but occurring in the Northern Hemisphere.

autumnal equinox time when the noon rays of the sun are vertical at the equator; in the Northern Hemisphere, approximately September 21–22.

axis an imaginary line extending through the poles around which the earth rotates.

azimuth angular distance from north measured clockwise from 0°–360°.

azimuthal projection a map projection in which directions are correct from a given point.

azonal soils immature soils, which exhibit minimal development of horizons.

backshore area of a beach landward of the high water line.

back slope in cuestas or hogback ridges, the slope opposite to the steep escarpment.

back swamp low-lying area between natural levee and valley wall in a river floor plain.

backwasting the meltback of a front of glacial ice.

badlands an area heavily eroded into bizarre forms, usually by running water.

baguio the term applied to a tropical cyclone in the Philippines.

bajada series of interconnecting alluvial fans along the foot of a mountain range.

banding characteristic of certain metamorphic rows in which the minerals are arranged in bands or plate-like stratas.

bar a deposit of alluvial material offshore or in a stream course.

barchan a crescent-shaped sand dune with the points of the crescent pointing in the direction of wind flow.

barograph a clockwork-driven aneroid barometer, which traces atmospheric pressure over a period of time.

barometer a device used to measure atmospheric pressure.

barometric pressure pressure exerted by the atmosphere.

barranca *See* arroyo.

barrier island a wave-built island on coastlines of low relief, which is separated from the mainland by a lagoon.

basalt a fine-grained extrusive igneous rock composed of ferromagnesium minerals and plagioclase feldspar; the extrusive equivalent of gabbro.

base exchange basic ionic exchange from soil to plant.

base level elevation below which a river or stream cannot erode.

base line east-west line from which townships are measured north and south in the township and range land survey system.

basic (rocks) rocks that are composed of a high percentage of ferromagnesium compounds.

batholith largest of the igneous intrusive features.

bayhead bars low mounds of debris deposited at the head of a bay by waves and currents.

baymouth bars low mounds of debris deposited by waves and currents at the mouths of bays.

beach unconsolidated materials at the edge of the sea between the low-tide line and upper limits of wave action.

beach face that portion of the beach between mean lower low water line and the crest of the berm.

bed layer or stratum of sedimentary rock.

beheaded stream a stream that has lost a portion of a tributary as the result of stream piracy.

belted coastal plain an area of cuestas and intervening low lands in tilting alternating weak and resistant rock strata in an area adjacent to the coast.

bench mark a marker with specific elevation above the datum plane inscribed.

bergschrund large crevasse at the head of a glacier between the valley rock wall and the main body of the glacier.

B-horizon the soil horizon beneath the A-horizon; the subsoil.

biogeography the study of ecological developments and biological distributions.

biomes aggregations of discrete plant and animal associations.

biosphere living organisms of the earth.

bird's foot delta a delta form in which deposits of alluvium along major distributaries are separated by arms of the sea.

bituminous coal soft coal.

Bjerknes Norwegian meteorologist who promulgated the concept of fronts during World War I.

block diagram perspective drawing that depicts surface landforms and underlying geological structures.

bolson a desert basin of interior drainage surrounded by mountains.

bora regional name for cold, air drainage wind in Yugoslavia.

boreal forest coniferous forest of high latitudes.

bosses large, irregular igneous intrusions.

bottomset bed the bed of coarse alluvium deposited at the bottom of a delta.

braided stream stream with a series of interconnecting channels, which are periodically blocked and redirected by a heavy load of coarse sediment.

breccia a rock composed of angular fragments cemented together in a fine matrix.

butte small, flat-topped, steep-sided erosional remnant in semiarid or arid regions.

Buys Ballot's law a law which indicates that if one stands with his or her back to the wind in the Northern Hemisphere, the lower pressure is at the left. The reverse is true in the Southern Hemisphere.

calcification soil-forming process of subhumid to arid environments where mineral bases, especially calcium carbonate, accumulate.

calcite common rock-forming mineral composed of $CaCO_3$.

caldera large crater caused by volcanic collapse, apparently because of the withdrawal of large quantities of magma from beneath the surface during the volcanic eruption.

caliche a layer of calcium carbonate deposited at the surface of soil in a dry region because of the evaporation of capillary water.

capillary water soil water, which adheres to soil particles and moves in the direction of greatest capillary tension (i.e., from areas of surplus water to those with a deficit).

cap rock resistant rock, which forms the tops of mesas and buttes and the ridges in cuestas.

carbonation chemical reaction of carbonic acid with earth materials.

carnivore a meat-eating animal.

cartographer a scientist who constructs maps and charts.

cartography the art and science of mapmaking.

casts fecal matter of earthworms.

Celsius scale (sometimes referred to as centigrade) a temperature scale of measurement where 0° is the freezing temperature of water and 100° is the boiling point at standard atmospheric pressure.

centigrade scale *See* Celsius scale.

centrifugal force the force that pulls an object away from the center of rotation.

centripetal drainage pattern a drainage pattern where streams flow into a basin.

change of state the change of a substance from gas to liquid or solid, or the reverse change from solid or liquid to a gas.

chaparral term applied to the low-growing, drought-resistant plants of the Mediterranean climatic regime in the United States.

chart a map used for navigation.

chemical weathering the reduction of rock materials as the result of chemical combinations, producing new, weaker compounds.

chert a siliceous rock often occurring as a lens or bed in limestone.

chinook hot, dry wind descending down the lee-side of a mountain barrier in the western U.S.

C-horizon the broken rock material beneath the B-horizon from which the soil develops.

chronometer a very accurate timepiece.

cinder cone a volcanic cone where cinders, ash, and other solid debris have been expelled from a central vent.

circle of illumination line around the earth dividing daylight from darkness.

circular projection a projection that has the perimeter bounded by a circle.

cirque amphitheater-like basin formed at the head of an alpine valley by the action of ice scour.

cirque glacier a glacier occupying a cirque.

cirrocumulus high, ice-crystal clouds, looking like snowflakes arranged in a geometric pattern.

cirrostratus a relatively thick layer of ice-crystal clouds, which cause a halo effect if sunlight or moonlight shines through.

cirrus high, thin, wispy, ice-crystal clouds.

clastic rock sedimentary rock in which rock fragments are cemented together by compaction or material in solution.

clay inorganic particles usually smaller than .002 millimeters.

climate average conditions of weather over a period of time within a particular region.

climatology the scientific study and analysis of extant climatic patterns.

climax vegetation a vegetation association that ultimately develops in a particular climatic and edaphic region.

climograph a chart that depicts temperature and rainfall conditions.

clouds forms composed of visible water droplets or tiny ice crystals at an elevation of more than fifty feet.

clouds with vertical development cumuliform clouds, which may extend through more than one level of low, middle, and high cloud groups.

coal a mineral fuel composed largely of carbon derived from lithified plant remains.

col a cut made through a mountain ridge by the headward erosion of glacial ice on both sides of the ridge.

cold front contact zone between advancing cooler, denser air mass and a warmer, less dense air mass.

colloids gelatin-like particles of mineral and organic matter in soils.

columnar (structure) cohesion of soil particles into structures resembling miniature columns.

columns calcareous cave deposits in the shape or form of a column.

compass rose a graphic depiction of directions on a map, usually expressed in degrees from north with north being 0° or 360°, east being 90°, south being 180°, and west being 270°.

composite volcano a symmetrical volcanic cone created by alternate eruptions of lava and ash.

compound delta a delta formed by the combined distributaries of two proximate master streams.

compound shoreline a shoreline exhibiting characteristics of both emergence and submergence.

Comprehensive Soil Classification System (CSCS) a soil classification system, sometimes called the Seventh Approximation System, developed by the Soil Conservation Service, which is based on composition and horizonation in soils.

concordant feature an igneous intrusion that is essentially parallel to existing rock materials.

condensation the change of moisture from a gaseous to a liquid state.

condensation nuclei *See* hygroscopic nuclei.

conduction transfer of heat by contact from one molecule to adjacent molecules.

conformal projection a map projection that depicts shapes correctly.

conglomerate a sedimentary rock in which pebbles or cobbles are cemented together in a finer matrix.

conic projection projection of a portion of the earth's surface onto a tangent cone.

coniferous trees bearing cones; they are usually needle-leafed.

consequent stream a stream that develops in response to regional slope of the land.

continental glacier thick ice that covers a very large area and moves out in all directions from several zones of accumulation.

continental plateau a plateau of almost continental dimensions (e.g., Africa).

continental shelf gently sloping surface extending from the shore to the edge of the continental slope. Water over the continental shelf is generally less than 100 fathoms (600 feet or about 180 meters) deep.

continental slope the zone that extends from the margin of the continental shelf to the ocean floor.

contour interval vertical distance between successive contour lines.

contour line a line connecting points of equal elevation.

convection circulation within a fluid or gas where the warmer fluid or gas rises and is replaced by cooler fluid or gas, which sinks.

convectional precipitation precipitation that results as warmed air rises and is chilled to the dew point.

coquina a type of limestone where shells of marine organisms have been cemented together.

coral reef calcareous subsurface ridge formed in tropical waters from the skeletons of marine organisms called coral polyps.

core the hot, dense inner portion of the earth.

Coriolis force the force due to earth rotation, which causes horizontally moving bodies to be deflected to the right in the Northern Hemisphere and to the left in the Southern Hemisphere.

corrasion mechanical erosion by particles being carried by water, ice, or wind.

creep the movement of soil and regolith downslope in response to the pull of gravity.

crevasse stress crack in a glacier occurring most commonly along the margins or near the terminal end of the glacier.

crevasse fill a small ridge of glaciofluvial origin where debris has been deposited in a crevasse by a meltwater stream.

crust the outer layer of the earth.

crystalline rock igneous or metamorphic rocks composed of visible crystal particles. Rocks that cooled slowly have large well-developed crystals, whereas those that have cooled rapidly have very small crystals.

cuesta asymmetrical ridge formed by differential erosion in alternating resistant and weak sedimentary beds.

cumulomammatus pendulous, bulbous, drooping cumulus cloud forms indicating atmospheric downdrafts and severe turbulence.

cumulonimbus anvil-topped clouds of great vertical development, which may be characterized by copious precipitation, hail, strong winds, and lightning.

cumulus clouds of vertical development with flat bases and rounded, cauliflower-like tops.

cumulus castellatus a towering cumulus cloud usually extending to considerable heights.

cumulus congestus a rapidly developing and turbulent towering cumulus cloud.

cuspate delta a delta form in which an arcuate delta has been modified by longshore currents, which have produced cusps.

cyclone a low pressure cell.

cyclonic precipitation precipitation that develops as a result of lifting along a front.

cyclopean stairs irregular step-like features left in a glaciated valley by irregular erosion of the glacial ice.

cylindrical projection a map projection where globe grid is projected onto a tangent cylinder.

deciduous trees that drop their leaves in response to cold or dry season.

deflation the removal of fine surface materials by the wind.

degradation wearing away of land surfaces by the processes of weathering and erosion.

delta a landform resulting from a stream deposit in a sea or lake.

dendritic drainage a stream pattern where tributary streams enter main streams at an acute angle in a branching tree-like pattern.

denuded the weathering and erosion of landform surfaces.

depression contour a contour line with small hachures inside, designed to show depressions, pits, or declivities in the landscape.

deranged drainage a rather helter skelter drainage pattern where streams have had their courses altered by deposition of material often of glacial origin.

desalinization removal of dissolved salts from water to make it potable.

desert pavement a rocky surface left after wind has removed most small particles of material in an arid environment.

desert varnish manganese and iron oxide coatings, which sometimes form on rock surfaces in desert regions.

dew droplets of water on surfaces left when diurnal cooling has reduced air temperature below the temperature at which condensation occurs.

dew point the temperature at which saturation occurs in a given air mass.

D-horizon the unweathered bedrock at the base of the soil profile.

diastrophism tectonic forces (folding, faulting, warping) which distort the earth's crust.

dike an igneous intrusive feature where magma intrudes into an almost vertical crack.

dip the angle made between rock strata and a horizontal plane.

dip angle the angle between the bedding plane of rocks and the horizontal.

dip direction a direction at right angles to the strike in a rock structure.

dip-slip faults movements along fault lines up and down dip in rocks.

disconformity a disjunct where several strata have been removed by erosion, but existing contacts between rock strata are essentially horizontal.

discordant (feature) an igneous intrusion that cuts across layers of existing rock materials.

disintegration physical weathering that reduces materials in particulate size without changing their chemical composition.

distributary a stream that breaks away from a major stream, usually in a delta region.

diurnal daily changes, as, for example, the difference between maximum and minimum temperatures in twenty-four hours.

doldrums area of low pressure adjacent to the equator.

doline a small sinkhole in a karst region.

dolomite a sedimentary rock composed of calcium magnesium carbonate.

dome circular or elliptical uplift in rock.

downwasting the thinning of glacial ice as the result of melting.

drainage basin the area drained by a stream and its tributaries.

drainage density length of all streams in a given region divided by the area.

drainage divide the line or zone separating adjacent drainage basins.

drift all material deposited by glacial ice or meltwater.

drumlin an elongate, egg-shaped hill deposited by glacial ice. The long axis parallels the direction of ice movement, and the steep end tends to face the direction from which the ice came.

dry adiabatic lapse rate the rate of cooling in a column of rising air where condensation has not occurred, resulting from the expansion of air, or, the rate of heating in a column of descending, dry air. The rate is about 5.5°F/1,000 feet or 1°C/100 meters.

dry wash *See* arroyo.

dune a mound of deposited wind-blown sand.

dust devil a small whirling convectional updraft characteristic of areas in the Great Plains during the warm season.

dynamic equilibrium condition within a system where the energy input equals the energy output.

earthflow the movement of water-saturated earth downslope.

earthquake vibratory movements in the earth's crust caused by movements of blocks along a fault line.

easterly wave an easterly bend, or trough, in subtropical latitudes, which may be a precursor to a hurricane.

ecological succession the sequential development of natural vegetation, which leads to the ultimate successor, the climax vegetation.

ecology a science that investigates the relationships between living organisms and their environments.

edaphic refers to soil.

elastic strain deformation of rock, which allows it to return to its original form or shape after the force or load causing the deformation has been removed.

elevation vertical distance above mean sea level.

elevation angle the angle made between the horizon and the line of sight to a celestial body.

elliptical projection a projection, usually equal area, where the perimeter is bounded by an ellipse.

El Niño periodic replacement of upwelling cold water by warm surface water in the eastern Pacific. The phenomenon often generates significant climatic aberrations and biological dislocations.

eluviation the removal of fine particles of soil in suspension from the A-horizon by percolating water.

emergents trees in the selva that protrude well above the general canopy level.

entisols soils with poor horizon development.

environment the combination of physical and cultural features, which serves as a habitat for organisms.

eolian *See* aeolian.

epicenter point on the earth's surface above the focal point of an earthquake.

epipedon the surface layers of soil in the comprehensive soil classification system.

epiphytes plants that attach themselves to other plants, but do not draw sustenance from them.

equal area projection a map projection where each section has an area comparable to the global section of the same scale.

equation of time difference between apparent solar time and mean solar time.

equator an imaginary line extending around the earth midway between the poles; 0° latitude.

equatorial low a zone of low pressure straddling the equator.

equidistant projection a map projection that depicts distances correctly.

equilibrium theory a theory that says the agencies and effects of erosion are always in equilibrium with the landform environment.

equinox date when day and night are of equal length all over the earth (March 21 and September 22 are approximate equinox dates).

equivalent projection map projection where each section has an area comparable to the global section of the same scale.

erg a sandy desert.

erosion the removal and transportation of weathered materials by running water, ice, waves, or winds.

erratic a boulder transported by glacial ice and deposited on a surface with different bedrock, or surface material.

esker sinuous ridge composed of glaciofluvial deposits formed by a stream flowing beneath the ice.

estuarine delta a delta deposit in an estuary, or drowned river mouth.

estuary an embayment where fresh water and salt water mix.

eutrophication excessive growth of plant organisms in a body of water usually facilitated by an excessive input of nitrates or phosphates into the body of water.

evaporation the change of state from liquid to gas.

evaporite a soluble salt deposited in dry areas when waters containing the material are evaporated.

evapotranspiration combined water loss through evaporation from surfaces and transpiration from plants.

exfoliation a weathering process where rock peels off in layers or concentric rings from exposed rock surfaces.

exotic stream a water course that originates in a humid region and flows through a dry one.

extinction the termination of an entire species by natural or man-induced causes.

extrusive rocks molten rocks deposited on the earth's surface.

extrusive vulcanism ejection of molten material or igneous rock at the earth's surface.

eye wall an area of violent updrafts around the margin of the calm eye at the center of a hurricane.

Fahrenheit scale   a temperature scale where 32° represents the freezing point and 212° represents the boiling point of water at standard atmospheric pressure.

fault-block mountains   upthrown blocks of terrain along fault lines resulting in mountainous topography.

faulting   the movement of earth blocks along cracks or joints in the earth's crust.

fault line   an intersection of a fault plane with the earth's surface.

fault planes   zones of weakness in the earth's crust along which rock movements may occur.

fault scarp   the exposed face of rock where one block of rock has been displaced vertically relative to another along a fault line.

fault zones   *See* fault planes.

feldspar   a mineral composed principally of aluminum silicates, which is a common constituent of igneous rock.

felsic rocks   igneous rocks that are high in silica.

fetch   the expanse of open water or relatively smooth land surfaces over which winds blow without interruption by obstructions such as hills or mountain barriers.

firn   compact, granular snow formed from melting and refreezing induced by weight of overlying snow.

first-order stream   the smallest tributary in a watershed.

fjord   a steep-sided, over-deepened glacial valley that was invaded by the sea when the ice melted. Also *fiord.*

flat plain   a plains region with local relief of less than 50 feet (about 15 meters).

floodplain   plain area adjacent to stream composed of stream deposits.

fluvial   processes associated with the work of running water.

foehn   wind caused by adiabatic heating on the lee side of a mountain barrier after condensation occurs on the windward slope; term applied to such a downslope wind in Europe; analogous to chinook wind in the United States.

fog   suspended water droplets in the air adjacent to the surface.

folding   the bending of rock materials resulting principally from compressional forces.

foreset bed   the material deposited at the seaward margin of a delta.

foreshore   the zone that is regularly covered and uncovered by water with the rise and fall of the tides.

fractional scale   type of map scale; an expression of the relationship between distance on a map and the same distance on earth.

fracturing   the breaking of rock when it has been subjected to forces beyond its plastic limits.

front   contact zone between two dissimilar air masses.

frontal precipitation   precipitation that develops as the result of lifting along a front.

frontal thunderstorm   a thunderstorm that develops along a cold, warm, or occluded front as the result of uplifting of air along a contact zone between cold and warm air masses.

frost   deposits of ice crystals on surfaces when the temperature of condensation is less than 32° F or 0° C.

frost wedging   a weathering process where rocks are broken apart by the repetitive freezing and expanding of water in rock cracks or interstices.

fumarole   vent in the earth's surface that emits steam and hot gases.

gabbro   a mafic igneous rock occurring in a pluton; intrusive analogue of basalt.

galactic rotation   the movement of the constituent elements of a galaxy around a fixed point in space.

galaxy   a discrete aggregation of stars.

galeria   *See* gallery forests.

gallery forests   a line of trees along stream banks extending into adjacent grass or scrub land.

geoanticline   a broad regional upfold in rock.

geography   a study of the interrelationships between people and their environment.

geology   the science that analyzes the orogeny, composition, form, and evolution of earth materials.

geomorphic cycle   a concept initiated by William Morris Davis, which holds that landscapes affected by erosion proceed through a sequence from youth, to maturity, to old age.

geomorphology   the study of the orogeny and evolution of landforms.

geostrophic winds   upper level winds where the wind flow parallels the isobars.

geosyncline   a broad regional downfold in rock.

geysers   ejected hot water and steam through a vent in a volcanic region.

glacial flour   very finely divided rock particles resulting from the grinding effect of glaciers on rock.

glacial grooves   deep gouges cut in bedrock as the result of glacial scour.

glacial till   unassorted debris deposited by a glacier.

glaciofluvial deposits   sorted materials deposited by glacial meltwater.

glaze   a coating of ice that develops when rain strikes a surface that has a temperature below the freezing point.

gneiss   metamorphic rock usually characterized by wavy bands of segregated minerals.

gnomonic projection   a map projection where all straight lines represent arcs of great circles.

graben   a down-dropped block between two essentially parallel faults.

granite   a common igneous rock composed of feldspar, quartz, and mica.

granular (structure)   a type of soil structure where particles tend to cohere in irregular clumps or grains.

gravitational water   water that moves downward through the pore spaces of soil after a rain.

gravity transfer   the movement of weathered rock material downslope in response to the force of gravity.

great circle   a circle on the surface of a globe that divides the globe into two equal hemispheres.

great soil groups   a subdivision in the Russian-American soil classification system. There are approximately forty great soil groups. In the CSCS system there are 185 great soil groups.

greenhouse effect   warming of the earth's surface because an increased burden of carbon dioxide in the atmosphere permits the free passage of short-wave radiation to the surface, but inhibits the movement of long-wave radiation from the earth's surface back to space.

Greenwich Mean Time (GMT)   time at 0° longitude used as a reference time for the earth's time zones; also termed Universal Time or Zulu Time.

ground moraine   glacial till deposited beneath the surface of a glacier.

groundwater   subsurface water.

Gulf Stream   warm ocean current flowing northeastward along the east coast of the United States.

gullies   small water channels developed in the early stages of stream development.

guyot   flat-topped, submarine seamount.

gypsum   a hydrated calcium sulfate ($CaSO_4 \cdot 2H_2O$).

gyre   a large scale circulation pattern of major ocean currents, generally clockwise in the Northern Hemisphere and counterclockwise in the Southern Hemisphere.

hachures   short lines parallel to slopes, which are used to depict relief.

Hadley cell   a circulation generated by rising air at the equator flowing poleward at elevation and sinking to the earth's surface in the zone of the subtropical highs.

hail   balls composed of concentric rings of ice, which fall from cumulonimbus clouds.

halite rock salt.

hanging valley a glacial trough formed by a tributary glacier at an elevation above a valley floor cut by a principal valley glacier.

hardpan a compact layer of clay minerals formed in the B-horizon of a soil as the result of illuviation.

harmattan a hot, dry wind of North Africa often bearing clouds of sand or dust.

haystack hills rounded hills characteristic of old age karst.

heat island area where local temperatures are higher than surroundings—especially characteristic of large cities.

hemisphere half of a sphere.

herbivore an animal that subsists by consuming plant material.

high a zone of high barometric pressure; an anticyclone.

high clouds clouds found at elevations above 20,000 feet (6,000 meters).

histosols soils high in organic matter, which developed in waterlogged environments.

hogback ridge narrow, sharp crested ridge formed in resistant rock strata tilted at a very steep angle.

homocline inclined layer of sedimentary rock, often the limb of a fold.

homolosine projection a projection that uses some of the characteristics of a sinusoidal projection and some of the characteristics of a Mollweide projection.

hook a bar with a curved end created by longshore currents.

horizons zones in soils having distinctive physical and chemical characteristics.

horn a pointed ice scour peak resulting from the erosive effect of three or more intersecting mountain glaciers.

horse latitudes *See* subtropical highs.

horseshoe lake a lake in the form of a horseshoe, which has resulted when a meander of a river has been cut off by a change in stream course.

horst an uplifted block between two essentially parallel faults.

humidity the amount of water vapor present in air.

hums *See* haystack hills.

humus semidecayed organic material usually found near the surface of a soil.

hurricane an intense tropical storm where the wind velocity is seventy-four miles per hour or more. Regional names for the same type of storm are baguio, willy-willy, tropical cyclone, and typhoon.

hurricane eye relatively clear area of descending air currents at the center of a hurricane.

hydration attachment, without chemical change, of water molecules to other molecules.

hydrologic cycle the system of water circulation at and above the earth's surface where water evaporates from oceans and lakes, falls as precipitation, and runs off as surface water back to oceans and lakes.

hydrology the study of water origins, patterns, and distribution at or near the earth's surface.

hydrolysis the union of water molecules with other molecules to form new compounds.

hydrosphere the earth's waters.

hydrostatic the equilibrium of fluids.

hygric vegetation of wet environments.

hygrograph a device using a clockwork mechanism and tracing pen to depict changes in relative humidity over a period of time.

hygrometer a device, often using human hair, to measure the relative humidity of air.

hygrophytic vegetation water-loving plants.

hygroscopic nuclei particles around which water droplets condense.

hygroscopic water water that is tightly bonded in a microscopically thin layer around soil particles.

hygrothermograph a device using a clockwork mechanism and tracing pens to depict changes in relative humidity and temperature over a period of time.

hypsometric map a map using color tints between contours to depict elevations.

ice sheet glacial ice covering very large areas, frequently of continental proportions.

ice scour plains areas that have had much of the soil and regolith removed as the result of ice erosion.

igneous rock rock solidified from molten materials.

illuviation the deposition of fine soil particles in the B-horizon by percolating water.

Imperata cylindrica scientific name for cogon grass.

inceptisols immature soils with little or no profile development.

influent stream a stream that loses water by seepage through the channel floor to the water table.

initial stage a newly exposed landscape just beginning to be affected by agents of erosion.

insequent stream a stream that shows little obvious relationship in pattern to regional structure.

insolation energy received from the sun.

instability an atmospheric condition where the lapse rate of ambient air is greater than that within a rising parcel of air. As a result, the parcel continues to rise and may ultimately be cooled to and below the dew point, resulting in clouds and precipitation.

interfluves the upland areas between adjacent streams.

interlobate moraine a moraine developing between lobes or tongues of ice in continental glaciation.

intermediate rocks igneous rocks that are essentially midway between mafic rocks and felsic rocks in chemical composition.

intermittent stream a water course that carries water only part of the time, usually for only a short period after precipitation has fallen.

intermontane between adjacent mountain ranges.

international date line approximately 180° longitude, where each day begins and ends. As one crosses the line going west, the calendar is advanced one day; going east, the calendar is set back one day.

Intertropical Convergence Zone (ITC), (ITZ) an area of low pressure near the equator where trade wind air rises; also known as the intertropical front (ITF).

Intertropical Front (ITF) *See* Intertropical Convergence Zone.

Intrazonal soils soils in the Russian-American classification scheme with well-developed profiles, which have developed in response to some local environmental factor, such as slope, drainage, or parent material.

intrusive rocks molten rock material deposited beneath the surface of the earth.

intrusive vulcanism the intrusion of magmatic material beneath the earth's surface.

inversion a condition where warm air overlies cold air.

ionosphere the term formally applied to the combined mesosphere, mesopause, and thermosphere zones. An area in which molecules are heavily ionized as the result of solar bombardment.

isarithm a line connecting points of equal value; also known as an isopleth.

isobar a line connecting points of equal barometric pressure.

isogonic line a line connecting points of equal magnetic declination.

isohyet a line connecting points of equal rainfall.

isotherm a line connecting points of equal temperature.

jet stream high velocity upper airstreams, which tend generally from west to east. Oscillations, or bends, in the stream north and south are often responsible for invasions of cold or warm air masses.

joints cracks in bedrock.

jungle a thicket of low-growing bushes, lianas, and bamboo in a tropical wet or tropical wet and dry environment.

kame a conical-shaped hillock of assorted glaciofluvial debris, which apparently formed near ice margins from glacial meltwater.

kame and kettle terrain (topography) surface with kames and interspersed kettle holes.

kame terrace glaciofluvial deposits in valleys near ice margins in an area of generally hilly terrain.

kaolin a clay composed of hydrous aluminum oxide.

karst solution topography.

katabatic wind a cold wind flowing down from mountainous or hilly terrain, or from an ice cap to adjacent lower country or the sea.

kettle a depression, often filled with water, which resulted from a buried ice block melting near the ice margin.

knickpunkte an area where resistant rock has created a zone of rapids as the result of stream erosion.

laccolith a toadstool-shaped igneous intrusion.

lagoon a shallow body of water between the coast and an offshore bar or coral reef.

Lambert conformal conic a type of map projection derived mathematically as if a cone cut the earth along two parallels. The resultant projection is conformal, and a straight line drawn on the surface approximates a great circle.

laminar flow a nonturbulent flow in a liquid. It characterizes movement in glaciers.

land breeze wind blowing from land to sea, resulting from more rapid radiational cooling of land areas at night than over adjacent seas.

landslide rapid movement of soil and rock debris downslope.

langley unit of solar radiation equal to one gram calorie per square centimeter.

lapse rate the diminution of air temperature with increased elevation.

latent heat of condensation the release of heat as water changes state from gas to liquid.

latent heat of vaporization the heat that is taken up when water changes from liquid to gas.

lateral moraine a moraine formed at the side of a valley glacier.

laterite an iron- and aluminum-rich layer, which is the end product of weathering in the humid tropics.

lateritic soils soils that are deep in the A-horizon layer and are rich in iron and aluminum.

laterization a soil-forming process in the humid tropics, which results in the leaching of most soluble minerals from the topsoil and the eluviation of most fine particles. Soils resulting are high in aluminum and iron compounds, low in organic material, and generally low in inherent fertility.

latitude angular distance north and south of equator.

lava molten rock at the surface, issuing from craters or vents.

leaching the removal of soluble materials from the upper soil horizons by percolating water.

lee the side of a feature away from the direction of ice movement, or, in the case of wind, the direction away from the wind source.

leeward the direction away from an approaching wind.

legume a plant that is able to fix atmospheric nitrogen with the assistance of bacteria attached to its roots.

lens-like (structure) agglomeratation of soil particles in a structure resembling interlocking convex lenses.

lenticular cumulus lens-like cumulus clouds, which may form along the margins of mountain barriers.

levee deposits of alluvium along a stream bank. An artificial levee may be constructed by man to contain a river within its banks.

liana a tropical vine.

lightning visible electrical discharge within clouds, between clouds, or between clouds and the ground.

lignite brown coal.

limbs (fold) the tilted beds extending downwards from a ridge crest or upwards from a valley bottom.

limestone sedimentary rock composed of calcium carbonate.

lithification a process whereby minerals are cemented together or compacted to produce rocks.

lithology the study of the mineral constituents and stratigraphic relationships of rocks.

lithosols thin, rocky soils.

lithosphere the earth's rocky crust.

living organisms self-replicating, respiring organized organic molecules.

loam a soil texture with approximately equal parts of sand, silt, and clay.

loess material composed of silt or clay particles deposited by wind.

longitude angular distance east and west of the prime meridian.

longitudinal dune a symmetrical ridge of sand parallel to the prevailing wind direction.

longshore current a current parallel to the shore induced by waves striking the shore at an oblique angle.

lopolith an igneous intrusion similar in shape to a laccolith, except that the lopolith is essentially flat topped.

low an area of low barometric pressure; a cyclone.

low clouds clouds found at elevations below 7,000 feet (about 2,000 meters).

low mountain a mountainous region with a local relief between 2,000 and 3,500 feet (about 600 to 1,060 meters).

loxodrome *See* rhumb line.

macchia Italian term for maquis.

macro-organisms living organisms large enough to be seen by the naked eye.

mafic rocks igneous rocks that contain limited amounts of silica.

magma molten rock beneath the earth's surface.

magnetic declination angle between magnetic north and true north.

mallee Australian term for maquis.

mantle that portion of the earth above the core and below the crust.

map a two-dimensional representation of all or a part of the earth's surface.

map projection a systematic arrangement of grid lines on a flat surface.

maquis French term for low-order shrub-form vegetation found in Mediterranean climates.

marble a metamorphic rock composed of crystalline limestone or dolomite.

mass wasting downslope movement of topsoil or regolith due to the force of gravity.

maturity a stage in the erosion cycle when the agents of erosion have had a significant impact in reducing a landscape to base level.

maximum thermometer a thermometer that measures maximum temperatures in a specific time frame.

meander a sinuous loop or bend in a stream course.

meander scar a former oxbow lake, which has been filled subsequently by inorganic and organic debris.

mean solar time solar time corrected to take into account variations in position of the sun at different places in earth's orbit.

medial moraine moraine in the central portion of a large valley glacier, which develops as lateral moraines from two tributary glaciers coalesce.

Mercator projection a modified cylindrical projection, which is conformal and characterized by straight lines crossing meridians at constant angles.

mercurial barometer a device that records rise and fall of pressure in a column of mercury, which equals external pressure.

mercurial thermometer a type of thermometer that uses mercury as the expandable liquid that measures changes in temperature.

meridians north-south lines extending to the poles, which measure longitude east and west of the prime meridian.

mesa a large isolated tabular area of resistant rock that is more extensive than a butte, but less extensive than a plateau.

mesic a vegetative environment intermediate between wet and dry.

mesopause boundary between mesosphere and thermosphere where temperature no longer diminishes with increasing altitude.

mesophytes plants that develop in areas that are neither excessively wet nor excessively dry.

mesosphere a zone in the atmosphere between the thermosphere and the stratosphere in which temperatures diminish with increasing altitude.

mesquite resinous, shrub-form vegetation found in some dry areas of the American west.

metamorphic rock a rock developed from igneous or sedimentary rock, which has been altered by heat and pressure.

meteorology the study of weather.

microclimate climatic conditions in a very restricted area.

micro-organisms living organisms too small to be seen by the naked eye.

midbay bars low mounds of debris deposited by waves and currents along the middle margin of a bay.

middle clouds clouds found at elevations between 7,000 and 20,000 feet (between 2,000 and 6,000 meters).

millibar unit of measure for pressure. One millibar equals a force of 1,000 dynes/centimeter². Standard sea level pressure is 1,013.2 millibars.

mineral a naturally occurring inorganic substance with fairly definite physical and chemical characteristics.

minimum thermometer a thermometer that measures minimum temperatures in a specific time period.

misfit stream a stream that is significantly smaller than the valley it occupies.

mist fogs fogs that develop over water bodies in night and early morning hours as the result of radiational cooling.

mistral the term applied to a katabatic wind in southern France.

mogotes *See* haystack hills.

Mohorovicic discontinuity (Moho) contact zone between earth's crust and mantle.

mollisols soils with dark, organic rich topsoil and high base content.

Mollweide projection an equal area projection in which the equator is twice the length of the central meridian.

monadnock an isolated hilly erosional remnant on the surface of a peneplain in the old age stage of the Davis cycle of erosion.

monocline a step-like fold in rock.

monsoon seasonal winds, which blow toward land in the high-sun season and blow away from land in the low-sun season.

moraine rock debris deposited directly by glaciers.

moulin kames kames formed at the base of a meltwater hole in glacial ice.

mountain breeze air drainage wind, which moves downslope at night in a mountainous region.

mudflows the movement of water-saturated soil along existing channels.

mudstones very soft sedimentary rocks composed of consolidated mud.

muskeg quaking bog.

nappe an overturned, or recumbent fold, which has developed a fault line near the base and which has been moved along the fault zone.

natural bridge a bridge or arch formed by erosion of stream meanders.

natural levee ridge of material adjacent to a stream formed by deposition during flood stage.

natural vegetation vegetation that develops normally in given climatic and soil conditions; the end product of ecological succession.

neap tides tides with minimal range between high and low tides, occurring when earth-sun-moon are in quadrature (i.e., at right angles to each other).

neck igneous rock that has solidified within the vent of a volcano.

needle an erosional remnant in a dry region with almost vertical slopes and negligible surface area.

neutral shorelines shorelines that exhibit no evidence of either emergence or submergence.

névé old granular snow in a zone of glacial ice accumulation.

nickpoint *See* knickpunkte.

nimbostratus stratus clouds of low or middle elevation from which precipitation is falling.

nimbus the cloud designation that refers to rain.

nonconformity a contact point between igneous or metamorphic rocks and overlying sedimentary rocks.

normal fault a fault where the fault plane dips toward the down-thrown block.

normal lapse rate (standard lapse rate) the decrease in temperature with increase in elevation in still air. The average decrease in temperature is about 3.5° F/1,000 feet or 6.5° C/kilometer.

North Atlantic Drift the northeast and easterly extension of the Gulf Stream, which ameliorates the climate of northwestern Europe.

northeast trades winds that blow from the subtropical high to the equatorial low in the Northern Hemisphere.

northern lights (aurora borealis) luminous bands of light of ghostly hue in northern latitudes emanating in the ionosphere apparently resulting from the influence of the earth's magnetic field.

nunataks rocky peaks projecting above the surface of an ice field.

oblate spheroid term applied to the shape of the earth, which is generally spherical, but is slightly flattened at the poles.

oblique-slip faults fault movements partially of dip-slip and partially of strike-slip character.

obsequent stream a tributary stream that flows in a direction generally opposite to a consequent stream.

obsidian volcanic glass.

occluded front the last stage in a normal middle latitude cyclone. A cold front has overtaken and overridden a warm front.

ocean current the horizontal movement of ocean waters in discrete streams or drifts.

ocean floor the bottom of the ocean or sea.

oceanic trench an area of deep ocean water, usually adjacent to island arcs in a zone of subduction.

offshore bar deposit parallel to the beach at the area of the breakers.

old age a stage in the erosion cycle when the landscape has been reduced almost to base level.

olivine a mineral composed of iron magnesium silicates that makes up a significant part of the mantle.

omnivore an animal that feeds on both plants and animals.

orbit path of a celestial body around a focus of revolution.

orogeny time in earth history associated with a major period of landform development, mountain building, or tectonic activity.

orographic precipitation precipitation that occurs on the windward side of a topographic barrier if air rising up and over the barrier is cooled to and below the dew point.

orographic thunderstorm thunderstorm developing along mountain barriers as the result of turbulent uplift along the windward side of such a barrier.

outcrop bedrock exposed at the earth's surface.

outwash materials deposited beyond the limit of glacial advance by glacial meltwater.

outwash plain flat or undulating glaciofluvially deposited surface beyond the limits of ice advance.

overturned fold an asymmetrical fold generated by a compressional force from one direction being of greater magnitude than the force from the opposite direction.

oxbow lake   semicircular lake formed when a river meander has been cut off.

oxidation   chemical union of oxygen with another element.

oxisols   thoroughly leached soils of humid tropics.

ozone O_3; forms a layer in the upper atmosphere, which screens organisms at the surface from harmful doses of ultraviolet radiation.

Pacific ring of fire   the area around the periphery of the Pacific Ocean marked by numerous volcanoes and earthquakes.

pahoehoe   a thick, viscous, taffy-like lava.

paleosols   old soil horizons buried by subsequent deposition.

parabolic dune   a curved dune in a region of abundant sand. The horns, often anchored by vegetation, point upwind.

parallel   a line parallel to the equator, which connects points of equal latitude.

parallel drainage pattern   a pattern of drainage developing on newly exposed or homogeneous slopes where principal streams are essentially parallel to each other.

parallelism of the axis   tendency of the earth's axis to be parallel to itself at every point within earth's orbit.

parent material   debris from which soil is derived.

park savannas   tropical grasslands interspersed with scattered trees, frequently acacias.

paternoster lakes   chain of lakes in a glaciated valley.

peat   compressed and partially carbonized plant remains.

pedalfers   aluminum- and iron-accumulating soils.

pediment   a sloping bedrock surface usually masked with alluvium, located near the base of a stream-eroded mountain range in a dry region.

pedocal   a calcium-accumulating soil.

pedology   soil science.

Peléean eruption   a volcanic eruption where incandescent dust and gas travels rapidly down the volcanic slope; such an eruption takes its name from the prototypical eruption that occurred on Mt. Pelée in Martinique in the early twentieth century.

Peleé's spine   a rocky projection pushed above the crest of Mt. Peleé after the early twentieth century eruption of that volcano.

peneplain   an erosional surface near base level, which theoretically develops in the final stage of the erosion cycle; it literally means almost a plain.

pepinos   irregular, elongate, knobby, residual hills in an old age karst region.

perched water table   water-saturated layer separated from normal water table level by impervious layer of rock.

percolation   movement of water downward through soil or permeable rock.

perennials   plants that live longer than a single season.

periglacial   landform phenomena formed beyond the limit of glacial coverage, but influenced by glacial meltwater or by glacially induced climatic modifications.

perihelion   minimum distance between earth and the sun in the annual orbit.

permafrost   permanently frozen layer of soil in polar and subpolar environments.

permeability   a characteristic of soil or rock, which determines the rate with which water moves through the material.

pH   refers to the acidity or alkalinity of a substance. Scale extends from 0 (extremely acid) to 14 (strongly basic) with 7 representing neutrality.

photogrammetry   a scientific technique in which aerial photographs are used to produce maps.

photoperiod   the length of daylight and darkness.

photosynthesis   the process whereby chlorophyll of plants manufactures carbohydrates from water and carbon dioxide in the presence of light.

phyllite   a type of metamorphic rock that is characterized by a kind of silky sheen.

physiographic diagram   a diagram illustrating landforms by perspective drawing.

physiological aridity   dry conditions created by porous and permeable soils or high evaporation rates rather than lack of precipitation.

piedmont   an area at the foot of a mountain range.

piedmont glaciers   tongues of glacial ice extending from mountains into the terrain at the foot or base of mountains.

pillow lava   a pillow-like form of lava, which develops as lava is extruded beneath the sea.

pingo   a circular mound or hill with an ice core found in permafrost areas.

pitted outwash plain   outwash plain surface with a number of depressions or kettles.

planar projection   a map projection on a plane surface.

plane of the ecliptic   plane of the earth's orbit around the sun.

plankton   small organisms in the sea that are at the base of the aquatic food chain.

planosols   soils with an almost impervious layer of clay at relatively shallow depths.

plant succession   changing plant associations over time, leading to climax vegetation.

plastic strain   deformation of rock beyond its ability to return to its original shape.

plateau   a landform of limited local relief at high elevation.

plate tectonics   theory that a series of discrete plates in the earth's crust move over the surface as the result of convection-like circulation within the earth's mantle.

platy (structure)   a type of soil structure in which individual soil particles agglomerate in plate-like features.

playa   a dry lake bed in a basin of interior drainage.

Pleistocene   an epoch of the Quaternary period characterized by several advances and retreats of continental glaciers.

plowpan   an almost impervious layer of clay created in certain soils by repeated plowing to a constant depth.

plucking   action of glacial ice, which tends to quarry out pieces of rock as the ice moves over it.

plug   *See* neck.

plugdome   a type of volcano subject to explosive eruption and characterized by rapidly congealing acidic lava in the central vent.

plutons   family of igneous intrusive features.

pluvial   a rainy period.

podsolization   a soil-forming process in humid environments with severe winters. Soils produced are characterized by acidity, leaching of bases, aluminum and iron compounds, and eluviation of insoluble particles.

polar easterlies   winds that blow from polar highs to subpolar lows.

Polaris   the north star.

polygonal ground   irregular patterns in the surface of the tundra caused by freezing and thawing in an area underlain by permafrost.

porosity   the space between soil or rock particles, which determines its water-holding capacity.

prairie grassland   tall grasslands of the middle latitudes.

precession   the slow movement of the rotation axis of a spinning body.

precipitation   liquid or solid moisture that falls from the atmosphere to the surface.

pressure   force exerted by interaction of molecules in the air.

pressure gradient   the difference in atmospheric pressure between two adjacent pressure cells or the difference in pressure from the center to the margins of a pressure cell.

prevailing westerlies   *See* westerlies.

prevailing wind   predominant wind direction at a given location.

primary rocks   igneous rocks.

prime meridian   the 0° meridian, which is the reference line for measurement of longitude. This meridian passes through the Royal Observatory at Greenwich near London, England.

principal meridian   north-south line from which townships are measured east and west.

prismatic (structure)   agglomeration of soil particles into structures resembling miniature prisms or crystals.

psychrometer   a device using two identical thermometers, except that one bulb is encased in a wet sock or sleeve, used to measure relative humidity.

pumice   glassy, light-colored, low-density scorria.

pyroclastic materials   solid materials thrown from a volcanic vent during an eruption.

quarrying   *See* plucking.

quartzite   a metamorphic rock composed of quartz.

radar   a device using radio waves to detect physical or climatological features. The term derives from *radio direction and ranging*.

radial drainage pattern   a drainage pattern where streams radiate out from the center of a high elevation.

radiation   emission of waves, which transmit energy through space.

radiation fog   fog produced as the result of the radiation of heat from near the surface during nighttime hours.

rain   liquid precipitation.

rain gauge   a cylindrical collector used to measure the amount of precipitation that has fallen.

rain shadow   the area lying on the leeward side of a topographic barrier.

recessional moraine   end moraine deposited at a location back of the terminal moraine.

rectangular drainage pattern   drainage pattern developed in response to joint patterns in rock where the joints intersect each other at a high angle.

refraction   the bending of a ray of light, heat, or sound as it moves from one medium to another.

reg   a rocky desert resulting from deflation.

regolith   mantle of weathered rock material above the bedrock.

rejuvenation   in Davis's concept of cycle of erosion, the return of an older area to youth as the result of uplift or the introduction of a new energy element into a system.

relative humidity   the ratio, expressed as a percentage, between the amount of water vapor present in the air and the amount in saturated air at that temperature.

relief model   a three-dimensional model designed to depict changes in landscape relief.

remote sensing   the collection of information about an environment from a distance, using photography, radar, infrared, or other bands within the electromagnetic spectrum.

Rendzina   an intrazonal limestone soil characterized by its red A-horizon.

representative fraction   type of map scale; an expression of the relationship between distance on a map and the same distance on earth.

resequent stream   a stream developed on exhumed rock strata, which flows in generally the same direction as a consequent stream.

reverse fault   a fault in which the hanging wall block moves upward relative to the footwall block.

revolution   the path of the earth in its orbit around the sun, which takes approximately 365¼ days.

rhumb line   a line drawn on the surface of a map, which cuts all meridians it crosses at a constant angle.

rhyolite   an igneous rock, which is the extrusive equivalent of granite.

ria shoreline   an irregular shoreline resulting from the drowning of river mouths during a period of submergence.

Richter scale   a scale used to measure the intensity of earth tremors. As the number on the Richter scale increases by one, the tremor is ten times as great.

rift valley   a graben or lowland trench.

rill   a small erosion channel.

rip current (riptide)   a narrow band of seaward-moving water, which results from the piling up of water by incoming waves.

rocdrumlins   drumlins deposited over a solid, rocky core.

rôche moutonnée   an asymmetrically scoured glacial boulder with a steep side facing away from the direction of ice movement.

rock cycle   the modification of igneous rock to sedimentary and metamorphic rock, which may be remelted along subduction zones to produce more igneous rock.

rock flour   finely ground rock debris within and beneath a glacier.

rolling plain   a plain with local relief between 150 and 300 feet (about 45 to 90 meters).

Rossby waves   north-south oscillations or undulations of prevailing westerlies, which shift storm tracks within the zone of the westerlies.

rotation   the turning of the earth on its axis. One rotation requires approximately twenty-four hours.

rough mountain   a mountainous area with a local relief between 3,500 and 6,000 feet (about 1,060 to 1,800 meters).

rough plain   a plain with a local relief between 300 and 500 feet (about 90 to 150 meters).

rugged mountain   a mountainous area of local relief between 6,000 and 10,000 feet (about 1,800 to more than 3,000 meters).

runoff   flow of water off a land surface.

rusting   the combination of iron with oxygen.

salinization   soil-forming process in desert regions, resulting in the accumulation of soluble salts.

saltation   the bouncing of rock particles along a surface by water or wind.

salt wedging   splitting of rock particles by expansion or growth of salt crystals.

sandstone   common sedimentary rock in which quartz grains are cemented together.

Santa Ana   dry foehn wind of the desert southwest of the United States.

satellite imagery   photographs or images of the earth's surface taken from orbiting manned or unmanned spacecraft.

saturation   temperature at which moisture capacity of air is reached.

savanna   tall, coarse grass vegetation of the wet-dry tropics.

scale   ratio between map distance and comparable distance on the earth.

schist   foliated metamorphic rock resulting from the effects of heat and pressure.

schlerophyll   type of tough-leafed forest characteristic of Mediterranean climatic regimes.

scirocco   desiccating desert wind of Africa and the Middle East, which blows into surrounding humid areas.

scoria   volcanic ejecta with numerous pore spaces created by escaping gas.

sea arches   arches cut in rocky headlands by the erosion of waves.

sea breeze   wind blowing from sea to land during the day because differential heating results in slightly warmer temperatures over land. The rising air is replaced by somewhat cooler and denser air from the sea.

seafloor spreading   movement of oceanic crust in opposite directions away from mid-ocean ridges. One of the mechanisms that move crustal plates.

secondary rocks   sedimentary rocks.

second-order stream   a stream formed by the joining of two first-order streams.

sedimentary rocks compacted fragments, organic remains, or chemical precipitates that produce rocks beneath the surface of the sea.

seismograph instrument used to detect earth tremors.

selva the tropical rainforest.

sensible temperature human sensations relative to cold or warmth; temperature, wind, and humidity may combine to cause temperatures to feel colder or warmer than they actually are.

sextant device used to measure elevation of celestial bodies to determine locations on the earth.

shale a sedimentary rock composed of clay particles cemented together.

shield an area composed primarily of complex igneous and metamorphic rocks, which has remained above sea level for protracted periods.

shield volcano volcanic cone with relatively gentle slopes created by repetitive lava flows from a central vent.

shoreline of emergence a shoreline that is rising relative to the adjacent water.

shoreline of submergence a shoreline that is declining in elevation relative to the adjacent water level.

sialic rocks rocks of the upper crust, which are high in aluminum and silicon.

sidereal time time determined by reference to distant stars.

sierran mountain a mountainous area with a local relief in excess of 10,000 feet (about 3,000 meters).

sill igneous intrusion in the form of a sheet between rock strata.

silt particles from 2 to 50 micrometers in diameter.

siltstones soft sedimentary rocks composed of lithified silt.

simatic rocks rocks of the lower earth's crust composed primarily of silicon and magnesium compounds.

sinkhole surface depression resulting from the work of solution in areas of soluble rock.

sinusoidal projection an equal area projection where meridians are derived from sine curves.

slate a fine textured metamorphic rock with distinct cleavage.

sleet pellets of ice formed when rain falls through a zone of the atmosphere where the temperature is below freezing.

slope vertical rise or fall of land in a specific horizontal distance.

small circle any circle on the earth's surface cut by a plane that does not pass through the earth's center.

smog combination of chemical pollutants and particulate matter, especially over urbanized areas.

snow precipitation in the form of hexagonal ice crystals.

soil creep slow downslope movement of soil in response to the pull of gravity.

soil families a subdivision of the great soil groups in the Russian-American classification system or a subdivision of subgroups in the CSCS system.

soil horizon a zone within the soil with distinctive physical and chemical characteristics.

soil orders the most general level of soil classification. The three principal soil orders are the zonal, intrazonal, and azonal.

soil phases subdivisions of soil types.

soil profile vertical cross section of soil exposing the various soil horizons.

soil series a subdivision of soil families in both the Russian-American and the CSCS system.

soil structure characteristic agglomerations of individual soil particles in a particular soil type.

soil subgroups divisions of the great soil groups in the CSCS soil classification system.

soil suborders in the CSCS soil classification system there are approximately forty-seven suborders.

soil texture characteristics of individual particles of a soil.

soil types a subdivision of soil series within the Russian-American classification system.

solar constant the rate at which insolation is received at the outer margins of the atmosphere.

solifluction the slow movement of water-saturated materials above the bedrock downslope due to gravity.

solstice dates at which the sun's vertical rays at noon reach their maximum extent north and south of the equator (i.e., at 23½° N. on about June 22 and at 23½° S. on about December 22).

solum the soil proper (i.e., the A- and B-horizons).

solution dissolved materials carried by running water, or, the dissolving of soluble rock materials by water.

solution cavities cavities or channels that develop in soluble rock.

sonar a technique in which reflected sound waves are used to determine the nature of submarine topography.

southeast trades winds blowing from the subtropical high to the equatorial low in the Southern Hemisphere.

specific humidity the weight of water vapor in a given weight of air.

speleothems calcareous cave deposits.

spit a wave depositional feature extending from mainland and projecting into an adjacent body of water.

spodosols soils with a heavily leached topsoil overlying subsoils that are more compact and are stained by aluminum and iron compounds.

spring surface flow of water where water table is exposed.

spring tide particularly high tide occurring when sun, earth, and moon are in line (i.e., at the time of the new and full moon).

squall line a line of storm clouds that often precede a rapidly advancing cold front.

stability a condition in which the adiabatic lapse rate is greater than the standard lapse rate. Such air resists rising.

stack an offshore rock isolated by wave erosion.

stalactite a deposit of calcium carbonate extending downward from the ceiling of a cave.

stalagmite a deposit of calcium carbonate extending upward from the floor of a cave.

standard atmosphere normal atmospheric pressure at sea level, about 14.7 lbs./in.^{2} or 1,013.2 millibars.

stationary front discontinuity surface between dissimilar air masses where there is essentially no forward movement.

steppe short, bunch grass characteristic of semiarid areas.

stereographic projection a projection developed as if a light source were projecting from a point on the earth's surface opposite to the point of tangency of the surface being projected upon.

stereoplotter a technical device using stereopair aerial photographs to produce contours directly.

stock an irregularly shaped igneous intrusion.

stomata microscopic openings in plant leaves through which carbon dioxide is absorbed and moisture is lost through evapotranspiration.

storm a disturbance of the atmosphere characterized by strong winds and frequently by precipitation and often by thunder and lightening.

storm surge rise in sea level during a hurricane created by low atmospheric pressure and strong winds.

stoss the side of a glacial scour feature facing the direction from which the ice came.

strata layers of sedimentary rock.

stratocumulus low clouds, which represent the merger of numerous cumuliform clouds.

stratopause upper limit of stratosphere.

stratosphere layer of atmosphere above the troposphere characterized by fairly constant temperature and a concentration of ozone.

stratus low, gray, sheet-like clouds.

stream capacity a maximum load that can be carried by a stream with a specific discharge.

stream discharge the amount of water flowing past a given point in a specific time period.

stream frequency the number of streams in a particular landscape.

stream gradient the slope of a stream valley that affects the velocity of a stream. Steep gradients contribute to rapid velocity and gentle gradients to slow velocity.

stream piracy the capture of a tributary stream in one system by a more aggressive eroding stream in another.

striae scratches in bedrock created by rocks frozen in the basal portion of a glacier moving over the surface.

strike compass direction of rock outcrop or fault line. Strike is perpendicular to rock dip.

strike-slip faults movements along fault lines parallel to the strike in rocks.

subduction the downward movement of an oceanic plate in contact with a continental plate.

sublimation change of state from gas to solid or from solid to gas.

submarine canyon a steep-sided canyon beneath the sea, which has been cut in the continental shelf.

subpolar lows belts of low pressure lying between the westerlies and polar easterlies.

subsequent stream a stream that develops along a zone of crustal weakness, usually as a tributary to a consequent stream.

subtropical highs belts of high pressure at approximately 30° N. and 30° S.

succulent a xerophytic plant characterized by thick, spongy, water-holding leaves or stems.

superimposed stream a stream flowing on a surface quite different from the one on which it originated.

superposition in an undisturbed series of sedimentary rocks, the youngest are at the top of an exposure and the oldest are at the bottom.

supersaturated (air) air that has a relative humidity of more than 100 percent, and condensation has not occurred.

suspension materials carried in suspension by moving water.

synclinal mountains elongate ridge resulting from differential erosion in a syncline.

syncline a downfold in rock.

synoptic chart a map showing weather conditions over a broad area at a given time.

syzygy planetary configuration when moon, sun, and earth are in line. There are especially high tides at such times.

taiga the subarctic coniferous forest.

taku an Alaskan katabatic wind.

talus rock debris broken from a rock slope by weathering, which has accumulated at the foot of the slope.

talus aprons coalescing alluvial cones at the base of a steep topographic slope.

tarn lake a cirque lake.

tectonic forces those forces in the earth's crust that modify the surface.

temperature a measure of the thermal energy present in a solid, liquid, or gas.

temperature gradient rate of change of temperature in any horizontal direction from a site.

temperature inversion an anomalous condition where temperature increases with increasing elevation.

temperature lag the deferral of maximum seasonal heating or cooling beyond the solstice caused by the time needed to heat or cool the earth's atmosphere.

tephra pyroclastic material explosively ejected from a volcanic vent.

terminal moraine debris deposited at the maximum point of ice movement.

terraces areas of relatively level land above the existing flood plain of a river marking levels of former erosion or deposition.

terra rosa a variety of red soil that develops over limestone, especially in areas with a Mediterranean climate.

tertiary rocks metamorphic rocks.

thermocline a zone within the ocean marking a sharp decline in temperature.

thermograph a device using a metallic strip connected through gear linkages to a pen and a clockwork mechanism, which traces variations of temperature on a graph over a period of time.

thermometer a device, usually consisting of a reservoir of alcohol or mercury within a graduated glass tube, which measures temperature by expanding when heated and contracting when cooled.

thermosphere uppermost of four layers of the atmosphere where temperatures increase with increasing elevation.

thrust fault a fault where one block overrides another along an inclined fault plane.

thunder sound resulting from rapidly expanding and heated air along a lightning strike.

thunderstorm severe storm accompanied by strong winds, copious precipitation, lightning, thunder, and sometimes, hail.

tidal range difference in water level at a particular site between high and low tides.

tide periodic rise and fall of sea level in response to the gravitational attraction of the sun and moon.

tierra caliente term applied to elevations in mountainous tropical elevations in Latin America latitudes where elevation is generally below about 3,500 feet (about 1,000 meters). It means hot land.

tierra fria mountainous zones in tropical latitudes of Latin America with elevations generally between 8,000 and 12,000 feet (about 2,500 meters to more that 3,600 meters). It means cold land.

tierra helada mountainous areas in tropical Latin America above 12,000 feet (more than 3,600 meters). It means ice land.

tierra templade mountainous zones in tropical latitudes of Latin America with elevations above about 3,500 feet to approximately 8,000 feet (more than 1,000 meters to about 3,500 meters). It means temperate land in Spanish.

till unassorted debris deposited by glacial ice.

till plains plains covered by deposits of glacial debris.

tilth workability of soil.

tombolo an island connected to the mainland by a sandbar.

topset bed bed of fine alluvium deposited at the top of a delta.

tornado a violent, localized storm with extremely low pressures and very high winds.

tornado warning an alert issued by the weather service that a tornado has been sighted in a particular vicinity.

tornado watch an alert issued by the weather service that conditions are right for possible tornado formation.

township an area of thirty-six square miles in the township and range land survey system.

traction the slow movement of larger pieces of material along a stream bottom as the result of the hydraulic force exerted by the water.

tradewinds winds blowing from the subtropical high pressure belt toward the equatorial low, or intertropical convergence zone.

transcurrent fault a horizontal or strike-slip fault.

transform faults the portion of a fracture zone between adjacent offset segments of a mid-oceanic ridge.

transpiration the loss of moisture through plant stomata to the surrounding atmosphere.

transverse dune a relatively straight, elongate dune form oriented perpendicular to the prevailing wind.

travertine calcium carbonate deposits, often in sheet-like form, in caves.

tree line the upper limit of tree growth in a mountainous region, or the maximum poleward extent of tree growth in high latitudes.

trellis drainage pattern stream pattern where tributaries enter master streams almost at right angles in an area of symmetrically folded rock.

tropical cyclone term applied to violent subtropical storm in south Asia, analogous to hurricane or typhoon.

Tropic of Cancer 23½° N. latitude.

Tropic of Capricorn 23½° S. latitude.

tropopause boundary between the troposphere and the stratosphere.

troposphere area of atmosphere closest to earth's surface, which generates weather.

trough a zone of lower pressure in the atmosphere.

tsunami tidal wave created by earthquake or volcanic eruptions.

tuff layers of volcanic ash that have been consolidated into porous rock.

tundra vegetation characteristic of high latitudes poleward of the taiga, and consisting of shrubs, mosses, lichens, and low-growing flowering plants.

turbidity condition of water laden with quantities of sediment.

typhoon violent subtropical storm in the western Pacific; analogous to a hurricane.

ultisols soils leached of bases that develop in hot humid environments.

unconformity a disjunct in the geologic record where the rock above a line or zone is significantly younger than that below it.

underfit stream a stream that is unusally small compared to the valley it occupies.

undulating plain a plain with a local relief between 50 and 150 feet (about 15–45 meters).

ungraded stream a stream that has not yet reached base level.

uniformitarianism theory that the processes now operating to modify the landscape have operated in the same way in the past.

unloading the removal of a load of ice or regolith from a segment of the earth's crust.

unstable (air) air that tends to rise because the normal lapse rate is greater than the adiabatic lapse rate.

upslope fogs fogs developing along lower slopes of mountain regions as the result of upslope movement of air.

upwelling upward movement of cold water to replace warmer surface waters, which have been moved offshore by winds.

urstromtäler ancient stream valleys cut by marginal streams flowing along the glacial ice front.

U-shaped valley the profile of a valley eroded by a glacier.

uvalas a large sinkhole in a karst region.

valley breeze movement of air upslope during daylight hours.

valley glaciers glaciers generally restricted to valleys in mountainous regions.

valley train outwash plain of a mountain or valley glacier.

Van der Grinten projection a type of map projection bounded by a circle in which the entire world is shown on a single sheet.

vapor pressure the percentage of atmospheric pressure contributed by ambient water vapor present in a given parcel of air.

varved clays alternating bands of dark- and light-colored clays of glaciolacustrine origin.

veins small cracks filled with magma.

ventifact a faceted rock created by the abrasive actions of wind.

vernal equinox the day when the sun's rays are vertical at noon at the equator, on or about March 21 in the Northern Hemisphere.

vertisols clay soils with little horizon development because of alternating periods of saturation and drying.

vesicles small cavities in igneous rocks created as gas escapes while the lava is molten.

vulcanization igneous activity along zones of crustal weakness, or along the contact zone between moving tectonic plates.

wadi *See* arroyo.

warm front a discontinuity surface between cold and warm air where the warm air is displacing the cold air.

warping general uplift or settling of the earth's crust over a broad area.

water gap transverse gorge cut through folded mountains by running water.

watershed the area supplying runoff water to a master stream and its tributaries.

waterspout a funnel cloud with a vortex of air and quantities of water; occurs over open water.

water table the upper limit of the subsurface area, which is saturated with water.

water vapor the gaseous form of water.

wave-built terrace a more or less level area seaward of the wave-cut terrace built up of materials eroded from the wave-cut terrace.

wave-cut bench gently sloping surface cut at the base of a sea cliff by the erosive action of waves.

wave-cut cliff an escarpment eroded at the sea's edge by surf and waves.

wave-cut terrace a bench cut at the seaward margin by surf and waves.

wave height the vertical distance between a wave crest and a wave trough.

wave length the horizontal distance between successive wave crests or wave troughs.

wave refraction the bending of waves, which approach the shore as a result of drag along the bottom.

weather atmospheric conditions at a given time.

weathering the breakdown of rock materials in place.

Weisenboden a half bog soil.

westerlies the wind belt between the subtropical highs and the subpolar lows.

wet adiabatic lapse rate the rate of cooling within a rising column of air when condensation is occurring.

willy-willy the term applied to a tropical cyclone in the South Pacific.

wind the horizontal movement of air from areas of higher pressure to areas of lower pressure.

wind gap notch in a ridge cut by a stream, but no longer occupied by a stream.

wind vane device that indicates wind direction.

windward side from which the wind comes.

xeric vegetation of dry environments.

xerophyte vegetation-type characteristic of arid regions, which is able to resist moisture loss through evapotranspiration.

yazoo tributary a tributary stream that parallels the course of a main stream for a substantial distance before entering it.

youthful stage a time in the cycle of erosion when processes of erosion have begun to wear down the surface of the land.

zenith a point directly overhead.

zenithal projection a type of map projection in which all directions from a given point are true.

zenith angle angle between the line of sight to a celestial body and a point directly overhead.

zonal soils soils in the Russian-American system, which have developed in primary response to climate and natural vegetation.

Moraines, 222
Moreau River, 227
Moscow Principal Meridian, 15
Mosses, 282
Moulin kame, 226
Mount Adams, 175, 176
Mountain breeze, 90
Mountains, 355
Mount Baker, 175
Mount Fuji, 170
Mount Haleakala, 176
Mount Hood, 175, 176
Mount Hualalai, 176
Mount Jefferson, 175
Mount Katmai, 175
Mount Kilauea, 175
Mount Lassen, 176
Mount Mayon, 170
Mount Mazama, 172
Mount Olympus, 109
Mount Pelée, 175
Mount Rainier, 175
Mount Saint Helens, 169, 172, 176, 204
Mount Shasta, 176
Mount Tambora, 174
Mudflows, 188
Mudslides, 188
Mudstones, 144
Mukden, China, 111
Musk ox, 330

Namib-Kalahari Desert, 125, 126
Nappe, 161
Nashville Basin, 209
Nashville, Tennessee, 106
Natural gas, 322
Natural levee, 208
Nautical Almanac, 17, 19
Neck, 179
Needles, 253
Neon, 44
Neptune, 12
Neutral buoyancy, 56
Neutral shorelines, 277
Nevada del Ruiz, 175
Névé, 219
New Madrid, Missouri, 152
Newton, Sir Isaac, 22
Ngorongoro Crater, 162
Niagara Falls, 209
Nickpoint, 201, 203
Nile Delta, 213
Nile River, 210, 213, 279, 340
Nile Valley, 210
Nimbostratus clouds, 54
Nitrogen, 44, 283
Nitrogen-deficient soils, 7
Nitrous oxides, 7
Nomads, 3
Nonconformity, 165
Normal fault, 162
Normal lapse rate, 45
Norman, Oklahoma, 75
North American Deserts, 126
North Atlantic Drift, 52, 86
Northeast trades, 62
Northern lights (aurora borealis), 45
Nunatak, 233
Nyasa Lake, 162

Oak, 320
Oblate spheroid, 13
Oblique photographs, 39
Oblique-slip faults, 162
Obsequent stream, 197
Obsidian, 144
Occluded front, 68
Ocean currents, 52, 85–86
Ocean floor, 267
Ocotillo, 328

Oder River, 227
Offshore bar, 274
Ogallala formation, 239, 344
Ohio River, 227
Oimekon, USSR, 112
Okhotsk, USSR, 113
Old age (erosion cycle), 202
Old Faithful, 241
Olive, 322
Olivine, 140
Olympic Penninsula, 109
Omega form, 45
Organic compounds, 283–84
Orographic precipitation, 58
Orographic thunderstorms, 58
Orthoclase feldspar, 141
Orthographic projection, 25
Otter, 322
Outwash plains, 225–26
Overturned fold, 161
Oxbow lake, 207
Oxidation, 184–85, 291
Oxisol, 297
Oxygen, 44
Ozone, 348–49
Ozone layer, 45

Pacific Gas and Electric Company, 241
Pacific Ring of Fire, 176
Pahoehoe, 167
Paleoclimates, 148
Paleosol, 282, 346
Pampas, 325
Panama Canal, 340
Pangea, 217
Parabolic dune, 260
Parallel drainge pattern, 200
Parallelism of the axis, 14
Parallels, 16
Paraná Plateau, 167
Parent material, 282
Paricutín, 170, 175
Paris Basin, 209
Park savannas, 324
Passenger pigeon, 350
Patagonian Desert, 126
Paternoster lakes, 230
Peat, 144
Pedalfer, 293
Pedocal, 293
Pedology, 4, 282
Peléean eruption, 175
Pelée's spine, 175
Peneplain, 202, 203
Penguins, 330
Pensacola, Florida, 106, 236
Pepinos, 245
Perched water table, 237–38, 295
Percolation, 237, 285
Perigee, 271
Periglacial phenomena, 227–28, 255
Perihelion, 14
Permafrost, 115, 295, 330
Permeability, 235, 236
Permian glacial deposits, 215
Permian glaciation, 216, 217
Peruvian-Atacama desert, 126
Petroleum, 322
pH, 288
Pheasant, 325
Phosphorous, 283
Photogrammetry, 39
Photoperiod, 314
Photosynthesis, 267
Phyllite, 147
Physical weathering, 182–84
Physiographic diagram, 35
Physiological aridity, 166, 329
Picher, Oklahoma, 341
Piedmont glaciers, 229
Piedmont plateau, 355

Pillow lava, 167
Pines, 322
Pingoes, 331
Pipe, 179
Pitted outwash plain, 226
Plagioclase feldspar, 141
Plains, 355
Planar projection, 25
Planosols, 294
Plastic strain, 161
Plateaus, 355
Plate tectonics, 140, 147–52, 162
Platy soil structure, 286
Playa, 255, 261
Pleistocene glaciation, 216, 217–18, 221, 255, 323
Plowpan, 345
Plucking, 220
Plug, 179
Plugdome volcano, 170
Pluto, 12
Plutons, 177
Podsol, 294
Podsolization, 292
Polar air masses, 67
Polar bear, 330
Polar (E) climates, 115–16
Polar continental air masses, 67
Polar easterlies, 65
Polar high, 65
Polar ice cap (EF) climates, 116
Polaris, 16, 19–20
Polar maritime air masses, 67
Pollution, 348
Polyconic projection, 28
Polygonal ground, 330
Polyp (coral), 277
Pompeii, 175
Poplar, 322
Porosity, 236
Possibilism, 7
Potash, 266
Potassium, 7, 283
Prairie chicken, 325
Prairie dog, 325
Prairie grasslands, 325
Prairie soils, 295–96
Precession, 14
Precipitation, 55
Predator, 325
Pressure, 61
Pressure gradient, 61, 62
Prevailing westerlies, 63
Prickly pear, 328
Primary activities, 3
Primary rocks, 141
Primates, 319
Prime meridian, 15
Prismatic soil structure, 286
Profile, 34
Pronghorn, 325
Psychrometer, 51
Pumice, 142
Pyroclastic materials, 169

Quadrangle, 34
Quail, 325
Quarries, 341
Quarrying, 220
Quartzite, 147
Quaternary activities, 3
Quinine, 316

Radar, 39
Radial drainage pattern, 200
Radiation, 48
Radiation fog, 52
Radioactive materials, 266
Rain, 59
Rainforest, 315, 318
Rain gauge, 59

Rance Estuary, 271
Recessional moraine, 222
Rectangular drainage pattern, 200
Red and yellow podsolic soil, 294
Red Sea, 152, 162
Redwood, 321
Reef, 277
Reelfoot Lake, 152
Reg, 258
Regolith, 252
Rehabilitative legislation, 341
Rejuvenation (stream), 203
Rekyavik, Iceland, 241
Relative humidity, 51
Relict vegetation, 218, 323
Relief, 33–37
Relief map, 33
Relief model, 34
Remote sensing, 39–40
Rendzina soil, 296
Reproductive industries, 343
Reptiles, 325
Resequent stream, 197
Reverse fault, 162
Revolution, 14
Rhine River, 227
Rhine River Valley, 262
Rhinoceros, 324
Rhizomes, 8
Rhone River, 213
Rhone Valley, 91, 262
Rhumb lines, 26
Ria coast (ria shoreline), 276
Richter, C. F., 154
Richter scale, 154
Rift valleys, 162
Rill, 193
Rocdrumlin, 222
Rôche moutonées, 221
Rock cycle, 147
Rock slides, 188
Rocky Mountains, 67, 88, 237, 252
Rodents, 330
Rolling plain, 355
Roosevelt, Franklin D., 271
Rosewood, 316
Rossby waves, 63
Rotation, 13–14
Rotifers, 284
Rough mountain, 355
Rough plain, 355
Rubber, 316
Rudolf Lake, 162
Rugged mountains, 355
Russian-American Soil Classification, 293
Rusting, 185

Sage, 326
Saguaro cactus, 328
Sahara desert, 126
Sahel, 126, 303
Salt and Gila River Valleys, 289
Saltation, 194, 258
Salt dome, 158
Salt flats, 255
Salt marsh, 274
Salt plug, 158
Salt wedging, 182
San Andreas Rift, 152
Sand, 222
Sand seas, 259
Sandstone, 144
Sandy Hook, 274
San Francisco Peak, 177
San Francisco quake, 152
Santorini, 172–74
Satellite, 40, 347
Satellite imagery, 40
Saturated air, 51
Saturn, 12
Savanna, 8, 324
Scale, 24
Schist, 147